Hydrocarbon Contaminated Soils and Groundwater

Analysis

Fate

Environmental and Public Health Effects

Remediation

Volume 1

Paul T. Kostecki & Edward J. Calabrese
Editors

Charles E. Bell
Technical Editor

Library of Congress Cataloging-in-Publication Data

Hydrocarbon contaminated soils and groundwater:
Analysis, fate, environmental and public health effects, and
remediation/Paul T. Kostecki and Edward J. Calabrese, editors.

p. cm.
Bibliography: p.
Includes index.
I. Kostecki, Paul T. II. Calabrese, Edward J.

ISBN 0-87371-383-4 91-1991

LEWIS PUBLISHERS, INC.
121 South Main Street, Chelsea, Michigan 48118

PRINTED IN THE UNITED STATES OF AMERICA

Preface

Hydrocarbon contamination of soils and groundwater is not a new or emerging issue. Groundwater contamination has received major regulatory and research attention since the early 1970s, especially as it relates to drinking water and the public health. While soil contamination has received relatively less attention historically than groundwater contamination, this has changed since the late 1970s with the national awareness of the Love Canal and Times Beach incidents.

It is only more recently that soil contamination from hydrocarbon products and *its relationship* to groundwater quality has begun to be an important issue. This focus resulted in the mid-1980s primarily from the regulatory concern about leaking underground storage tanks, but has escalated rapidly on many fronts. One such case has been at the national level with the creation of the Association for the Environmental Health of Soils (AEHS). AEHS was established as a point of crystallization for the multidisciplinary parties interested in soil contamination issues including: chemists, geologists, hydrogeologists, engineers, modelers, toxicologists, regulatory scientists, and lawyers.

This book, the proceedings from the first annual West Coast Conference on Hydrocarbon Contaminated Soils and Groundwater (February 1990, Newport Beach, CA) is designed to serve as a vehicle of information dissemination and exchange among those concerned about effectively dealing with the challenges of contaminated soils and groundwater. As the first in a series of future proceedings, the book represents the beginning of a continuing and reliable source for state-of-the-art information. The volume assesses the field with respect to identification of various current perspectives including state (Chapters 1 and 5), county (Chapter 2), and municipal (Chapters 3 and 4), as well as emerging issues such as sharing responsibilities for cleanups (Chapter 6) and the remediation oil field wastes (Chapter 7). Technical assessment is also provided in the areas of analysis (Chapters 8 and 9) and site assessment (Chapter 10), environmental fate (Chapter 11) and modeling (Chapter 12), remedial technologies (Chapters 13–16), and risk assessment and management (Chapters 17–21).

The book is predicated on the recognition that the eventual accepted solutions to the challenge of contaminated soil can only be found through the proper integration of sound, scientific information and rational debate within the context of good societal and regulatory judgment. Thus, this book will serve as a useful source of technical information as well as a guidance document for muncipal, county, state, and federal regulators; scientific researchers; industry executives, engineers, and staff with environmental responsibilities; environmental attorneys; and environmental consultants.

In addition, the book will provide students with a broad perspective that will allow for a balanced appreciation of the problem while providing information which will enhance their professional development.

Donald Rothenbaum
Kleinfelder, Inc.

Paul Kostecki
University of Massachusetts at Amherst

Acknowledgments

We wish to thank all the agencies, organizations, and companies that sponsored the conference. Without their generosity and assistance, the conference and this book would not have been possible.

Sponsors

American Environmental Management Corporation
Applied Geosciences
Atlantic Richfield Company
California Department of Health Services
Chevron Corporation
EA Engineering, Science & Technology
Edison Electric Institute
EPA/Office of Underground Storage Tanks
Hanby Analytical Laboratories
ICF Technology
International Technologies Corporation
Kleinfelder Incorporated
Lawrence Livermore National Laboratory
Lockheed Aeronautics
Los Angeles Water & Power
McLaren
Port of Los Angeles
Roy F. Weston
Shell Oil Company
Southern California Edison
Texaco USA
Toxic Treatments (USA)
Western States Petroleum Association
Woodward-Clyde Consultants

In addition, we express our deepest appreciation to the members of the Scientific Advisory Board who volunteered their valuable time to provide guidance and encouragement.

Scientific Advisory Board

We also wish to thank Linda Rosen, Ann Novosel and the entire conference staff for their tireless efforts in making the conference a success and Vivian Collier for her patience and assistance in the production of the proceedings. Last, but not least, we want to thank the 373 attendees.

Paul T. Kostecki, Associate Director, Northeast Regional Environmental Public Health Center, School of Public Health, University of Massachusetts at Amherst, received his PhD from the School of Natural Resources at the University of Michigan in 1980. He has been involved with risk assessment and risk management research for contaminated soils for the last five years, and is coauthor of *Remedial Technologies for Leaking Underground Storage Tanks* and coeditor of *Soils Contaminated by Petroleum Products* and *Petroleum Contaminated Soils,* Vols. 1, 2, and 3. Dr. Kostecki's yearly conferences on hydrocarbon contaminated soils draw hundreds of researchers and regulatory scientists to present and discuss state-of-the-art solutions to the multidisciplinary problems surrounding this issue. Dr. Kostecki also serves as Managing Director for the International Society of Regulatory Toxicology and Pharmacology's Council for Health and Environmental Safety of Soils (CHESS) and Executive Director of the newly formed Association for the Environmental Health of Soils (AEHS).

Edward J. Calabrese is a board certified toxicologist who is professor of toxicology at the University of Massachusetts School of Public Health, Amherst. Dr. Calabrese has researched extensively in the area of host factors affecting susceptibility to pollutants, and is the author of more than 270 papers in scholarly journals, as well as 12 books, including *Principles of Animal Extrapolation, Nutrition and Environmental Health,* Vols. I and II, *Ecogenetics, Safe Drinking Water Act: Amendments, Regulations and Standards, Petroleum Contaminated Soils,* Vols. 1, 2, and 3, and *Ozone Risk Communication and Management.* His most recent books include *Multiple Chemical Interactions* and *Air Toxics and Risk Assessment.* He has been a member of the U.S. National Academy of Sciences and NATO Countries Safe Drinking Water committees, and most recently has been appointed to the Board of Scientific Counselors for the Agency for Toxic Substances and Disease Registry (ATSDR). Dr. Calabrese also serves as Chairman of the International Society of Regulatory Toxicology and Pharmacology's Council for Health and Environmental Safety of Soils (CHESS) and Director of the Northeast Regional Environmental Public Health Center at the University of Massachusetts.

Contents

PART I
PERSPECTIVES ON HYDROCARBON CONTAMINATION

1. Looking Past Soil Cleanup Numbers, *Jeffrey J. Wong, G. Michael Schum, Edward G. Butler, and Richard A. Becker* 1

2. Regulatory Approaches to Hydrocarbon Contamination from Underground Storage Tanks, *Seth J. Daugherty* 23

3. Private Sector Perspectives on Hydrocarbon Contamination, *John J. Hills* 65

4. Unique Problems of Hydrocarbon Contamination for Ports, *Donald W. Rice* 71

5. Review of State Cleanup Levels for Hydrocarbon Contaminated Soils, *Charles E. Bell, Paul T. Kostecki, and Edward J. Calabrese* 77

PART II
EMERGING HYDROCARBON CONTAMINATION ISSUES

6. Sharing Responsibilities: A New Approach to Groundwater Remediation Involving Multiple Potentially Responsible Parties, *Jean B. Kulla and Angelo J. Bellomo* 91

7. Challenges Encountered in Hydrocarbon Contaminated Soil Cleanup, *Andrew C. Lazzaretto* 99

PART III
ANALYTICAL METHODOLOGIES AND SITE ASSESSMENT FOR HYDROCARBON CONTAMINATED SOILS AND GROUNDWATER

8. Sampling and Analysis of Soils for Gasoline Range Organics, *Jerry L. Parr, Gary Walters, and Mike Hoffman* 105

9. A New Method for the Detection and Measurement of Aromatic Compounds in Water, *John D. Hanby* 133

10. A Critical Review of Site Assessment Methodologies, *Douglas A. Selby* 149

PART IV
ENVIRONMENTAL FATE AND MODELING

11. Where Do Organic Chemicals Found in Soil Systems Come From?, *James Dragun, Sharon Mason, and John Barkach* .. 161

12. Fate of Hydrocarbons in Soils: Review of Modeling Practices, *Marc Bonazountas* 167

PART V
REMEDIAL TECHNOLOGIES FOR HYDROCARBON CONTAMINATED SOILS AND GROUNDWATER

13. Soil Vapor Extraction Research Developments, *George E. Hoag, Michael C. Marley, Bruce L. Cliff, and Peter Nangeroni* 187

14. Bioremediation of Hydrocarbon Contaminated Surface Water, Groundwater, and Soils: The Microbial Ecology Approach, *Michael R. Piotrowski* 203

15. Recycling of Petroleum Contaminated Soils in Cold Mix Asphalt Paving Materials, *Terry C. Sciarrotta* 239

16. Soil Venting at a California Site: Field Data Reconciled with Theory, *P. C. Johnson, C. C. Stanley, D. L. Byers, D. A. Benson, and M. A. Acton* 253

PART VI
RISK ASSESSMENT AND RISK MANAGEMENT

17. Human Health Risks Associated with Contaminated Sites: Critical Factors in the Exposure Assessment, *Jayne M. Michaud, Alan H. Parsons, Stephen R. Ripple, and Dennis J. Paustenbach* 283

18. Health Risks Associated with the Remediation of Contaminated Soils, *Gayle Edmisten-Watkin, Edward J. Calabrese, and Robert H. Harris* 293

19. A Preliminary Decision Framework for Deriving Soil Ingestion Rates, *Edward J. Calabrese, Edward S. Stanek, and Charles E. Gilbert* 301

20. Development and Application of a Decision Tool for Managing Petroleum Contaminated Soils, *Katherine K. Connor, Jennie S. Rice, and Justin L. Welsh* 313

21. A Progress Report on the Council for Health and Environmental Safety of Soils—CHESS, *Paul T. Kostecki and Edward J. Calabrese* 331

Glossary of Acronyms 339

List of Contributors 341

Index 345

Hydrocarbon Contaminated Soils and Groundwater

CHAPTER 1

Looking Past Soil Cleanup Numbers

Jeffrey J. Wong, G. Michael Schum, Edward G. Butler, and Richard A. Becker, Toxic Substance Control Program, California Department of Health Services, Sacramento, California*

INTRODUCTION

The primary goal in the remediation of hazardous waste sites is the protection of the public health and the environment. The public demands and deserves adequate protection through representation by their federal, state, and local regulatory agencies. The public's trust in any institutional risk management policy, e.g., the extent and magnitude of remedial efforts required at hazardous waste sites, must be maintained by having policy goals that are: (1) protective, (2) attainable, (3) enforced equally, and (4) reflective of the greater public need. The intent here is to: (a) describe the need for the development of a structured and informed analytic process in the setting of risk management policy and the making of risk management decisions, and (b) describe how the use of generic soil cleanup numbers that are not site-specific may forestall the development and maturation of such a process. Effective risk management in the area of hazardous waste site cleanup must be coupled to a process in which the nature and extent of the hazardous waste contamination is defined prior to the proposal of a final remedy. Informed public health and environmental risk management policy must be founded

*The opinions expressed in this paper are those of the authors and are not necessarily representative of the California Department of Health Services.

on a stable, knowledgeable, and experienced infrastructure of appropriate scientific and engineering staff and resources.

HISTORICAL FRAMEWORK FOR RISK ASSESSMENT AND RISK MANAGEMENT

At the request of the United States Congress, the National Academy of Sciences (NAS)[1] examined the regulatory role of the risk assessment process in the formulation of risk management policy within the federal government. The NAS recommended that regulatory agencies ". . . maintain a clear conceptual distinction between assessment of risks and consideration of risk management alternatives. . . ."

The NAS provided further recommendation that uniform guidelines for risk assessment be developed to ". . . promote clarity, completeness, and consistency . . . and enable regulated parties to anticipate governmental decisions." These recommendations are applicable to not only the federal government, but also state and local government.

In the area of risk assessment, significant progress has been made toward establishing the recommended guidance. In 1986, the United States Environmental Protection Agency (U.S. EPA)[2] in accepting the recommendations of the NAS, published five guidelines for assessing the health risks of environmental pollutants. Earlier in 1985, the California Department of Health Services (CA DHS) published risk assessment guidelines for chemical carcinogens.[3] While the approaches described in these documents will evolve as the result of continued scientific debate, many of the concepts have become embedded in the risk assessment approaches supporting hazardous waste site cleanup.[4-6] These processes will continue to mature in reflection of the increasing understanding of the various underlying scientific principles.

DOES RISK ASSESSMENT BRING ONLY BAD NEWS?

To establish the need for risk management decisions, the analytic process of risk assessment is used to integrate information on the nature and extent of contamination, toxicological data on the contaminants present, dose-response relationship between potential exposure levels and toxic effects, and site-specific exposure conditions to assess the existence and nature of potential adverse health or environmental impacts (see Figure 1.1). In concept, this process can not only forward calculate the potential level of unacceptable health impact, e.g., cancer risk, associated with an initial reservoir of chemical contamination, but also provides a vehicle to back calculate from a level of acceptable risk to the associated residual reservoir of chemical. The difference between the reservoirs of exposure

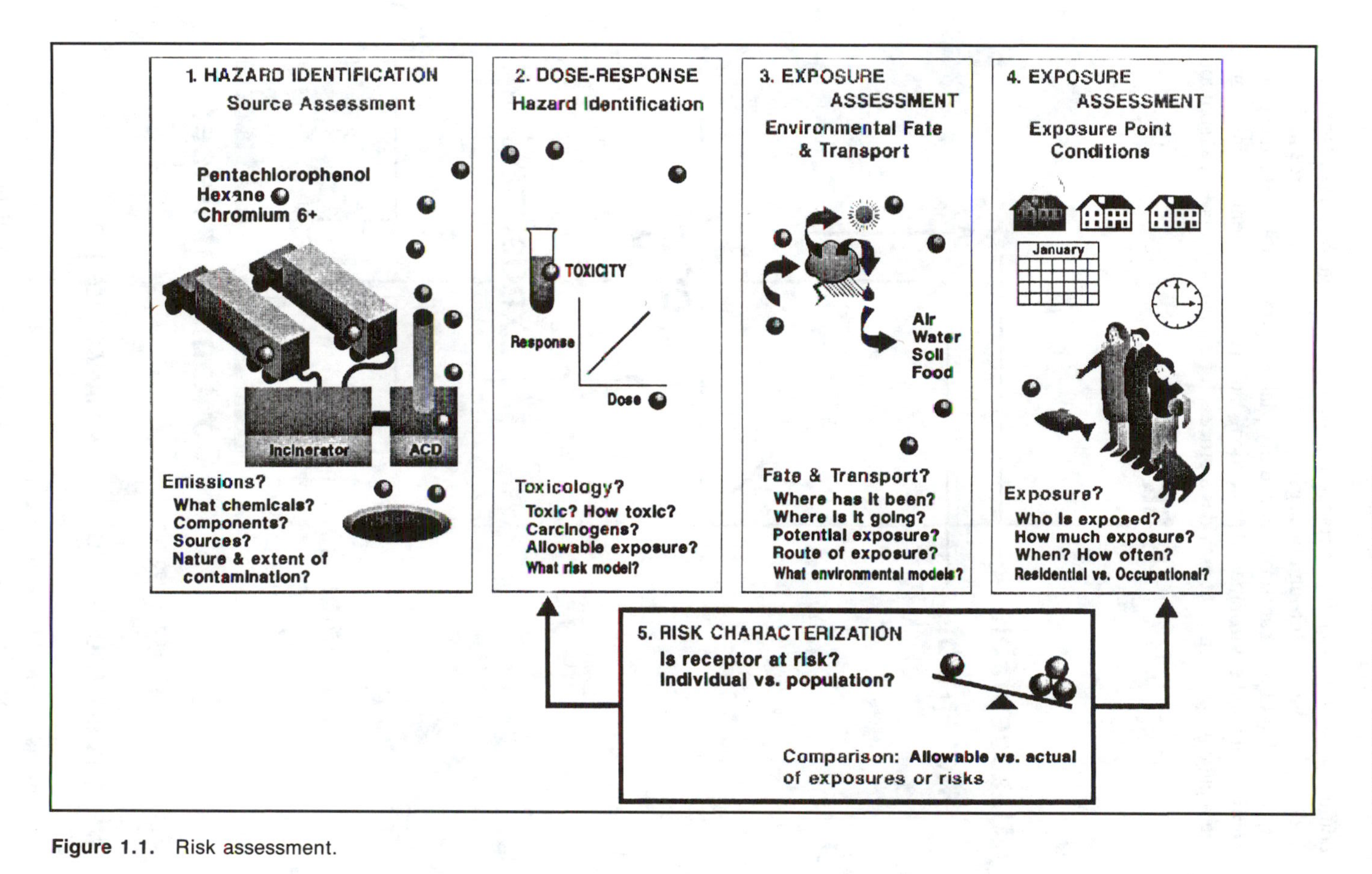

Figure 1.1. Risk assessment.

could represent the risk-based component of a risk management goal (see Figure 1.2). Risk reduction through management of the contaminant source and/or the exposure conditions can serve as equally viable options to achieve the goal of public health and environmental protection. Thus risk assessment not only frames the regulatory problem, but provides a process for choosing the potential solutions.

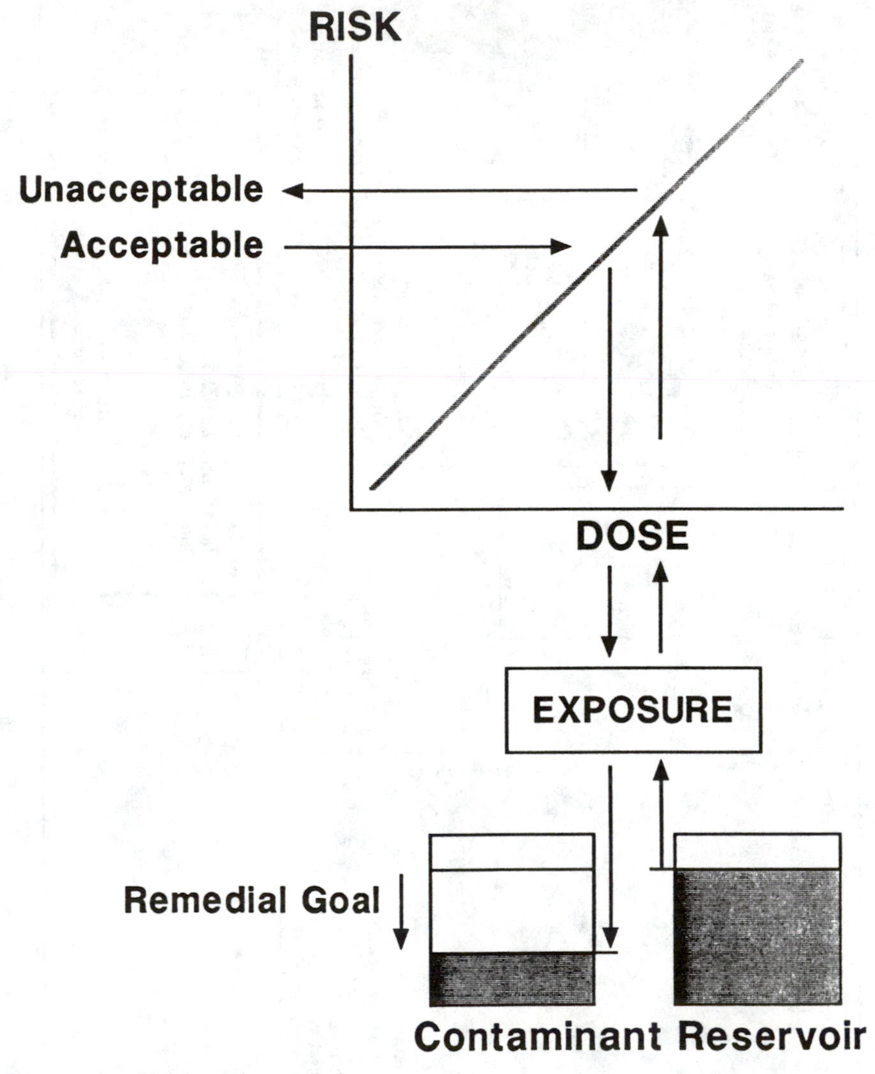

Figure 1.2. Risk-based remediation.

WHAT IS SO TOUGH ABOUT RISK MANAGEMENT?

There has been a growing recognition of the various flaws in current risk management policies. The lack of a regulatory definition of de minimis and significant risk has been identified as one of the many restraining factors in the further maturation of risk management as a process.[7-9] Lave and Males[10] further warn that the current risk management framework is too simplistic to adequately address the actual complexities of scientific and sociopolitical considerations. They go on to state that the current system ignores fundamental problems of structure, goals, and priorities, and they emphasize that little legislative guidance for balancing protection and cost-effectiveness in making cleanup decisions is provided to the federal Superfund program. This would appear to make the system ''inefficient'' as it tries to read the mind of the Congress. Recent criticism of hazardous waste site cleanup programs[11,12] should be seen as harbingers of problems within current risk management practices and the need for a soul-searching analysis of the process.

State programs share with their federal counterparts similar problems. Multiple state agencies are responsible for the protection of public health and the environment. Empowered by different legislative mandates, each operates with a program or agency-specific bias. The process of making risk management decisions is usually vague (often forcing the regulated community to attempt to read the regulatory agencies' minds), often differs between agencies, and may provide a conflicting regulatory maze for the regulated community. Risk management at the state level would benefit greatly from a statewide risk assessment process and risk management policy.

A regulatory institution must balance the public demands for absolute protection or ''zero risk'' against the real limits of public and private resources. The process of choosing the potential solution or risk management option is intended to not only consider issues uncovered during the risk assessment phase, but also to integrate considerations, such as engineering feasibility, financial resources, community needs, and real or potential benefit of the proposed risk reduction solution (see Figure 1.3). The demanding policy areas of such a process, by virtue of their vulnerability to public and political controversy, encompass (1) the assignment of appropriate weight to each of these considerations, and (2) the integration of these considerations into a risk management decision.

Discord in risk management policy begins with the demand for an absolute finding from the risk assessment process. The use of a particular mathematical model for carcinogenic dose-response extrapolation and risk estimation represents a critical juncture in determining the need for risk management decisions (see Figure 1.4). Toxicologists or other scientists specializing in risk assessment are aware, for example, that cancer risks at low exposure levels cannot be measured directly either by animal bioassay or by human epidemiologic studies, and therefore various mathematical models are employed to extrapolate from high to low dose.[2] Further, since no single mathematical model is recognized as the most appropriate, linearized multistage procedures, described by U.S. EPA, are usually employed to provide plausible estimates of the upper limit for risk. The true value cannot be determined and may even be zero. But the application of conservative assumptions assures that the approach is not likely to lead to an underestimate of actual risk. It is here that risk managers often misinterpret the role of

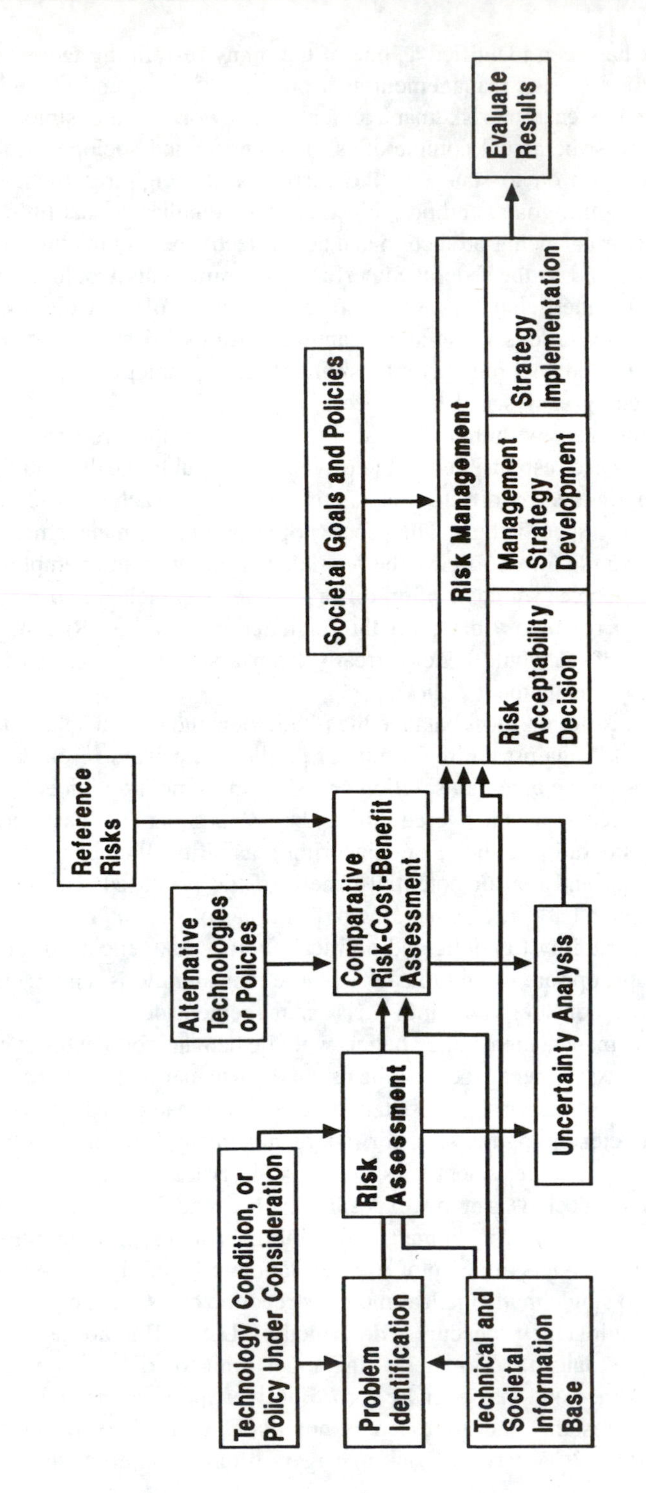

Figure 1.3. Risk assessment and management.[11]

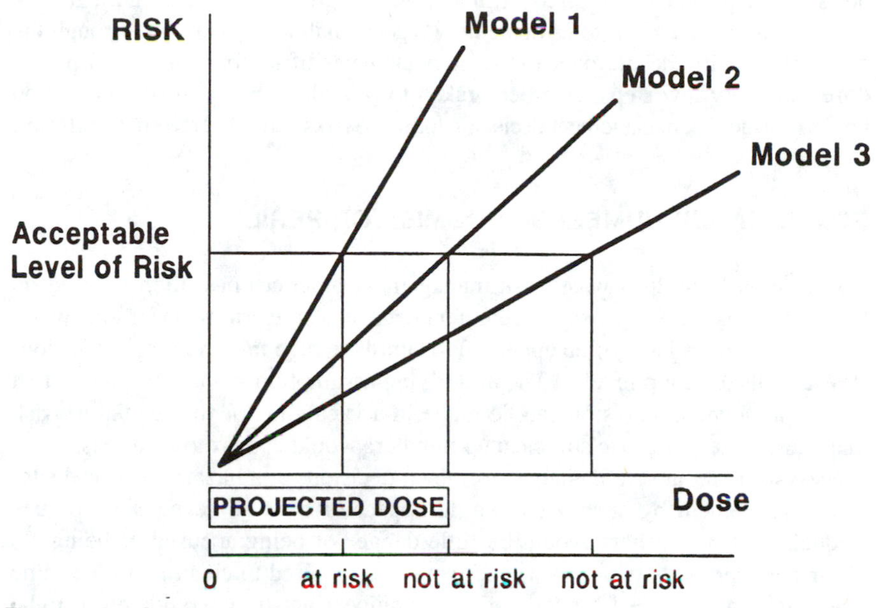

Figure 1.4. Cancer risk extrapolation.

conservative assumptions and approaches, and claim that the "sacred" boundary between risk assessment and risk management has been breached. The obvious allegation is that such embedded assumptions and approaches are disguised risk management considerations and lead to an intentional overestimation of risk. Based upon this premise, risk managers often attempt to engage in debate about the risk assessment process as a way to relieve the risk management burden.

Beyond considerations of risk, factors such as feasibility and cost do not appear to be considered within an equally rigorous and analytic paradigm similar to that observed during the risk assessment phase. Federal risk management policy, under the Comprehensive Environmental Response, Compensation, and Liability Act (CERCLA), with goals to regulate individual excess lifetime cancer risk to within a range of 10^{-4} to 10^{-7} (1 excess cancer case in 10,000 exposed individuals to 1 excess cancer case in 10,000,000 exposed individuals) without stating what decision rules are to be applied to support a choice within this range, could be open to presumably arbitrary and capricious implementation. Without the employment of a structured analytic paradigm of analysis, the risk management system is left vulnerable to the inconsistent or random consideration of the various factors. The cost of such policies will be the misdirection of resources, incomplete protection of public health and the environment, and the loss of institutional credibility, public trust and standing before the trier-of-fact. Therefore risk management must begin and end with a process of problem analysis.

Unfortunately, a sequel to the effort on risk assessment by the National Academy of Sciences for risk management has not occurred, and the path by which such decisions are made remains dimly lit. It appears that great strides in scientific methodologies have been taken to estimate how big or small a risk may be (through the constant scientific debate over the various elements of its foundation and procedure), but very few steps have been taken to provide public policy guidance on how to pursue risk management decisions for those risks that all agree are significant.

SOIL CLEANUP NUMBERS—PROMISE OR PERIL?

Debate on hazardous waste risk management practices has often focused on ''How clean is clean?'' Risk management decisions concerning hazardous waste sites and facilities have often appeared to simply evolve into the direct question, ''What is the cleanup level?'' The underlying assumption in this question is that a risk management decision can be made in a vacuum. For the regulatory risk manager, a list of generic soil cleanup numbers would appear to be a reasonable process for making soil cleanup or removal decisions at a hazardous waste site. Risk management by such a list would appear to treat all responsible parties ''equally.'' If true, there would be little danger of being accused of being arbitrary or capricious, since everyone would be required to clean up to the same concentration. The need for scientific and engineering study would appear to be substantially diminished since there would be little need for technical interpretation. Scheduled removal or treatment could begin right away, since, for the responsible parties and their contractors, the goals of the regulatory agency would be stated with crystal-like clarity. Time would not be wasted waiting for a regulatory decision. Issues would apparently dissolve down to administrative items such as time, schedule, and cost.

But the reduction of risk management policy to predetermined solutions, in the form of generic soil cleanup numbers, chosen for, as yet, not fully investigated problems will potentially follow a path inconsistent with the recommendations of the NAS. Instead of pursuing the development of a structured process to support risk management policy decisionmaking, the development of list(s) of generic soil cleanup numbers would endorse the concept that the manipulation of the risk assessment process to achieve current risk management goals is a viable public policy approach. It will be difficult to set generic soil cleanup goals in the mere light of the potential combinations of chemicals that may be present as a hazardous waste site mixture. This is readily apparent with petroleum products. Efforts to manage potential health risks associated with hazardous waste sites through the fixed ''gunsight'' of generic soil cleanup numbers may unknowingly be aiming at targets out of range of resources and inconsistent with legal requirements.

The best application for generic soil cleanup numbers would appear to be in the case of those sites involving limited numbers of chemicals, area of contamination, and exposure pathways. Unfortunately, since the number of sites fitting this prerequisite will be limited, so will the overall utility of the numbers. But once used, the way is paved for their indiscriminate use and the danger of becoming embedded in institutional dogma. First, it can be envisioned that schedules will lead to demands that the most expedient methods for decisions be used, e.g., generic soil cleanup numbers. Second, the continued use of generic soil cleanup numbers will potentially eliminate any incentive or opportunity for the understanding of the role, and thus the appropriate implementation, of risk assessment. Third, the growth of a work force lacking sufficient knowledge about the role of the risk assessment will likely lead to the impairment in the ability to make informed, and therefore flexible, risk management decisions. This will lead to the erosion of: (1) the concept that protection of public health and the environment is the primary goal and not time and schedule; (2) the role of risk assessment as a process of public health problem definition, and therefore a potential vehicle for identifying appropriate solutions; and (3) the incentive for the development and maturation of both a responsive and informed risk management and public health policy.

The indiscriminate use of criteria is not without precedence. In regard to U.S. EPA policy on dioxins in soil, Paustenbach et al.[13] pointed out that despite the numerous site-specific caveats embedded in the Centers for Disease Control's (CDC) methodology for the assessment of risk, many regulatory agencies have adopted the CDC guidelines of 1.0 ppb as a generic standard for cleanup regardless of the circumstance at the site. Past practices within the State of California have included (1) the use of 1,000 ppm total petroleum hydrocarbons (TPHs) as a generic cleanup goal for petroleum contaminated sites, and (2) the misuse of the total threshold limit concentration (TTLC) and the soluble threshold limit concentration (STLC) hazardous waste classification values prescribed in Title 22 of the California Code of Regulations, as generic cleanup goals. The use of 1,000 ppm TPHs has been applied without regard for adequate hazard identification or the nature and extent of contamination. Risks from petroleum mixtures dominated by xylene and toluene may be different than those posed by benzo(a)pyrene and related polycyclic hydrocarbons. In reference to the consideration of chemical mixtures, and at odds with this practice, the U.S. EPA under National Primary Drinking Water Regulations[14] determined it would not be appropriate to set drinking water criteria for "total volatile organic chemicals." The origin and scientific justification of the 1,000 ppm TPHs value is vague. In the second example, TTLC and STLC values are intended to provide waste classification criteria needed to determine waste disposal requirements. The underlying scientific assumptions used in the derivation of these values and the lack of appropriate exposure considerations and assumptions make TTLC and STLC values irrelevant in the determination of cleanup goals. Regardless of the original legal intent, examples of the misuse of TTLC and STLC values as generic cleanup goals persist.[15] It is not suggested here that either of these past practices represent an intentional breach in the responsibility to protect public health and the environment. Instead, each is speculated to result from a lack of understanding of the relationship between site-specific risk assessment and risk management. It is obvious that improvements can be made.

ORIGINS OF GENERIC SOIL CLEANUP NUMBERS?

The concept of generic soil cleanup numbers appears to be an offspring of past environmental quality criteria. By convention, these criteria are expressed as some permissible concentration of compound in a medium of exposure. For example, both CA DHS and U.S. EPA are involved in the development of drinking water criteria referred to as Maximum Contaminant Levels (MCLs). The process of setting MCLs initially involves review of the toxicological literature, followed by the use of scientifically based regulatory procedures for the calculation of allowable daily dose. Then, an entirely health-based MCL value is derived by dividing the allowable daily dose by a generic allocation for daily exposure (e.g., 2 liters/day). The final MCL value is based not only upon health considerations, but also reflects considerations of feasibility and public and environmental policy (see Figure 1.5).

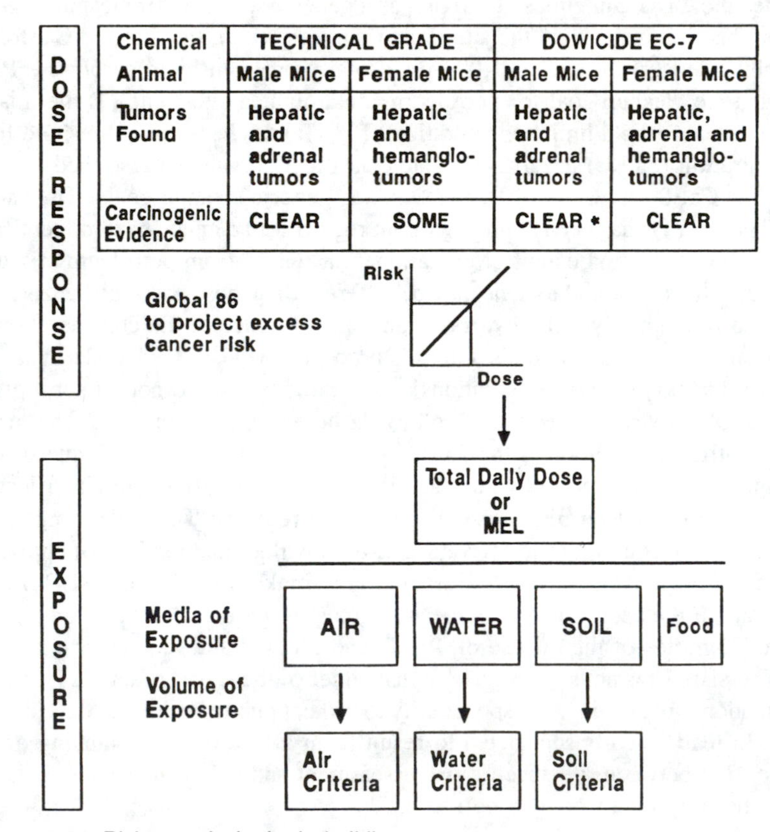

Figure 1.5. Risk appraisal criteria building.

In its hazardous waste management program, CA DHS has a process to develop criteria for not only water, but also for air and soil contact, known as Applied Action Levels, or AALs (see Figure 1.6). But unlike MCLs, the AALs are not intended for broad use as environmental quality criteria, but for the appraisal of risk for those exposures to contaminated media originating from hazardous waste sites and facilities.[4] AALs do not take into account cost, feasibility, or public health or environmental policy. The AALs are analogous to the Reference Dose (RfD) values developed by the U.S. EPA in that they are entirely health-based criteria. Therefore, the MCLs are risk management values and the AALs are for the assessment of risk. AALs for soil contact do not represent cleanup goals. They represent concentrations in soil above which a risk management decision may be needed, strictly limited to exposure scenarios involving *only* direct contact (e.g., ingestion and dermal pathways).

Misconceptions about soil cleanup numbers for hazardous waste sites appear to begin with the faulty premise that soil can be treated as a single medium of exposure, similar to the MCL process. Contrary to the thinking underlying the MCLs, soil itself not only serves as a medium of exposure, but also serves as

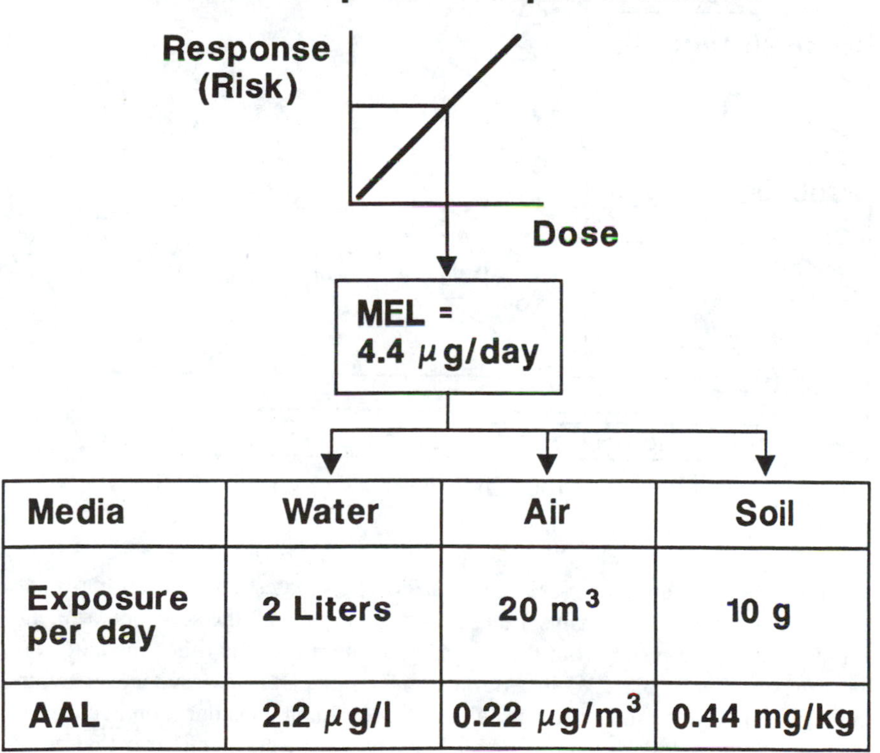

Media	Water	Air	Soil
Exposure per day	2 Liters	20 m^3	10 g
AAL	2.2 μg/l	0.22 $\mu g/m^3$	0.44 mg/kg

Figure 1.6. Applied action levels.

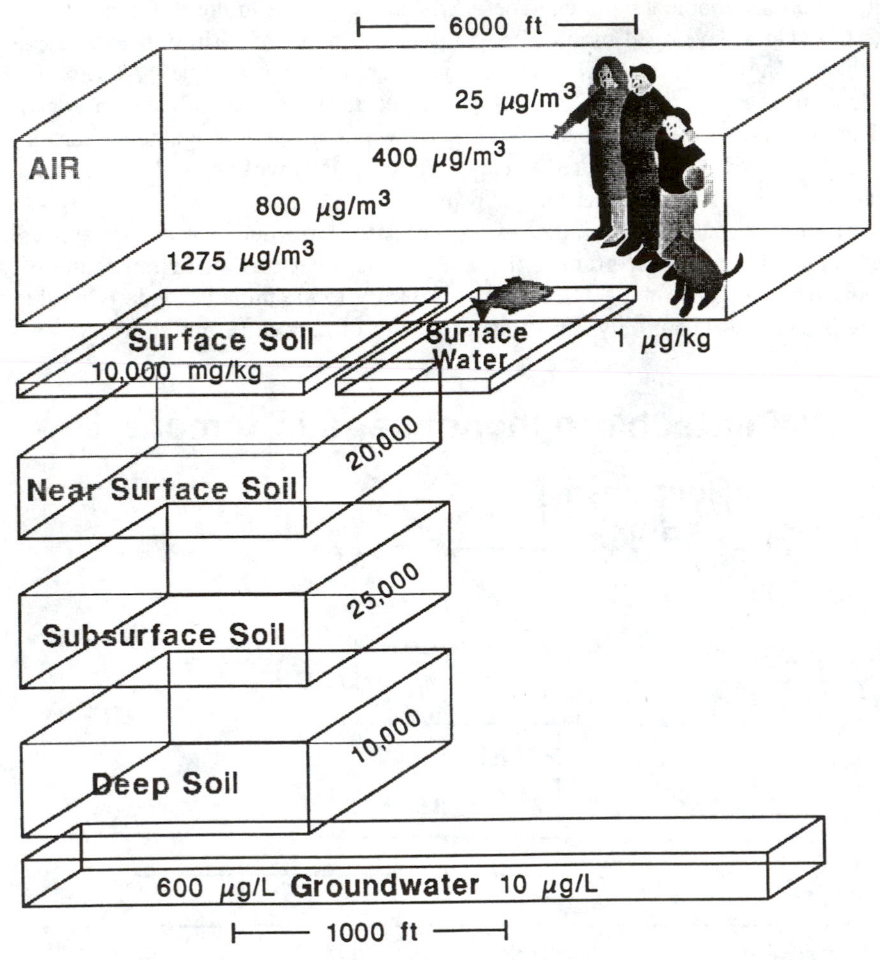

Figure 1.7. Environmental compartments.

a reservoir of contaminants (see Figure 1.7). The greatest concentration of, and therefore mass of contaminants are usually associated with the soil compartment. Contaminants can move from the soil compartment into the other media of exposure (see Figure 1.8). While both air and water can also serve as reservoirs of contamination for soil, this is usually not a dominant consideration in cleanup. Another obvious underlying difference between soil, water, and air is that broad generalizations about the physicochemical nature of water and air matrices are more likely to be valid than those about soil. The concept of setting risk management goals in the context of concentration criteria is most applicable to air and water, since generalizations about exposure pathways can be made. This would even be true for soil if the route of exposure were limited to direct contact and/or ingestion. But this approach is of limited utility for soil by virtue of the further need to consider it not only as a medium of exposure, but also as a reservoir of contamination of other media.

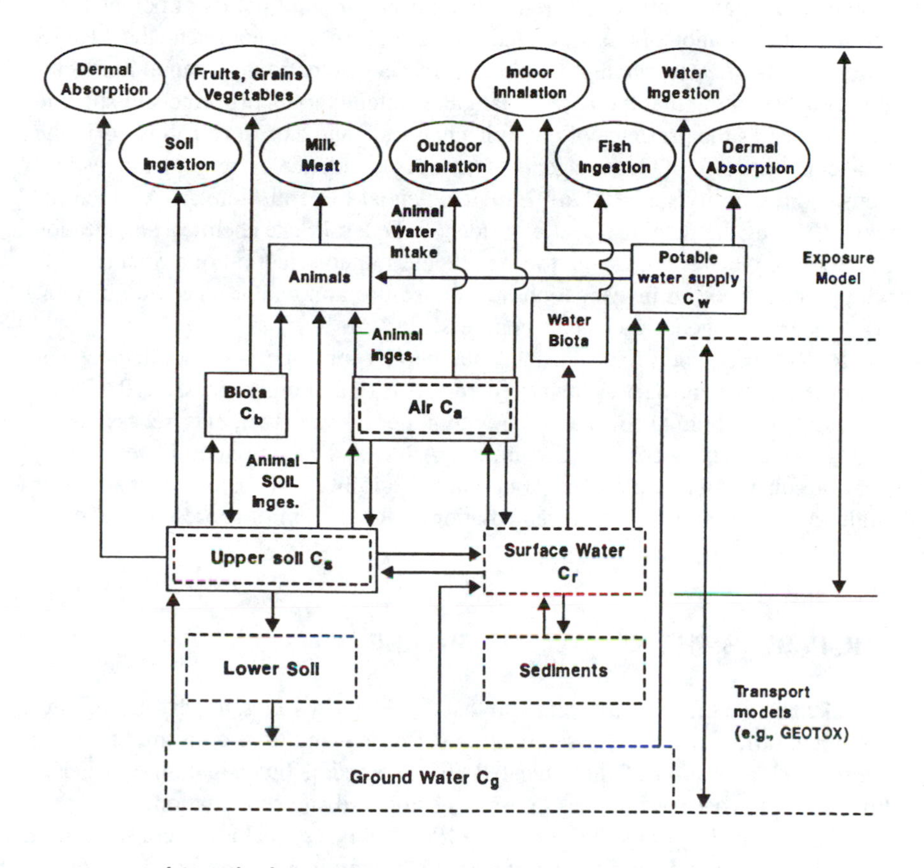

Integration between environmental transport and transformation processes and the assessment of human exposure.

Figure 1.8. Multicompartment exposure from References 17 and 18.

ARARS OR ERRORS

The perceived need for generic soil cleanup levels may be perpetuated by misconceptions about U.S. EPA's "applicable or relevant and appropriate requirements" or ARARs. ARARs are usually environmental quality criteria promulgated under other federal or state programs. U.S. EPA[18] states, "ARARs will define the cleanup goal when they set an acceptable level with respect to site-specific factors'' and ''. . . cleanup goals for some substances may have to be based on non-promulgated criteria and advisories . . . rather than ARARs because ARARs do not exist for those substances or because ARARs alone would not be sufficiently protective in the given circumstances, e.g., where additive effects from several chemicals are involved.'' It is further stated that at the completion of the remedial action, ''ARARs . . . must be attained for hazardous substances, pollutants, or contaminants. . . .'' remaining on site or at all points of potential exposure. Interpretation of the above indicates U.S. EPA does not intend the ARARs requirements to establish an inflexible list of generic national cleanup numbers. It is clearly evident that the ARARs language intends to incorporate site-specific factors such as the presence of multiple chemicals and exposure pathways in the final determination of cleanup goals. Therefore, guidance on remedial investigations and feasibility studies[19] to ''Initiate potential Federal/State ARAR identification'' during the scoping phase and ''Identify Federal/State chemical and location specific ARARs'' is not asking for the predetermined selection of a final remedy without benefit of an investigation into the nature and extent of contamination. ARARs are not generic cleanup numbers.

ARARs, like generic soil cleanup numbers, cannot be a substitute for, nor dominate, a structured risk management decisionmaking process. ARARs do represent an additional consideration within the risk management process, along with factors such as cost and feasibility. ARARs, like engineering feasibility, provide constraint to the process. But while feasibility and cost may set an upper limit on the extent of potential remediation, ARARs do not represent the lower limit of cleanup goals.

THE PAST IS THE KEY TO THE FUTURE

Risk managers frequently complain that the detail involved in performing site-specific multipathway risk assessment is too costly and time-consuming, and is simply an elaboration of the typical RI/FS (Remedial Investigation and Feasibility Study) process which has, at times, hampered site remediation. However, we feel that health protective screening level analyses, including consideration of environmental fate and transport of toxic contaminants, can be made with a modest investment of time and resources. Risk management decisionmaking at waste sites or facilities must not simply be concerned with the concentration of some toxicant per kilogram of soil, but must also consider the total mass or volume of toxicant present at the site, and the size of the interface between the contaminated soil reservoir and other media of exposure. For a given contaminant, the greater the potential extent of exposure, the greater the need to reduce the reservoir of exposure (see Figure 1.9).

Cleanup decisions can be pursued through the back calculation of acceptable soil concentrations from the evaluation of the dose received by a human receptor, in units of mg/kg/day, when the following conditions are met. First, the dose

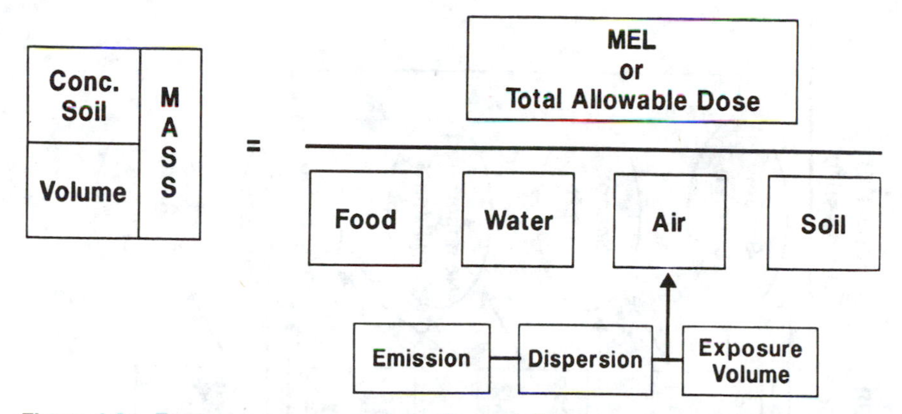

Figure 1.9. Total exposure approach to site remediation.

received from each potential exposure pathway must be estimated in a scientifically defensible manner (see Figure 1.10). This implies that a site-specific exposure scenario has been defined which includes all possible exposure pathways that may exist at a given site. The types of exposure pathways are themselves dependent on the types of receptors which may be present at the site, e.g., factory workers at an industrial facility vs children playing in a residential backyard or a playground. The types of receptors and potential exposure pathways are determined by the nature of the site land use, both pre- and postremediation. Second, suitable environmental fate and transport models must be available, or informed decisions must be made about contaminant dilution in air and groundwater. Third, current, accurate toxicological values must be available to describe an acceptable daily dose of a chemical which is used to back calculate a health protective soil concentration. Last, default and/or recommended values necessary for exposure calculations can be utilized given certain site-specific assumptions. These default parameters can be replaced with actual site data, as it becomes available.

The following categories of input exposure parameters can be used to provide an adequate assessment of dose resulting from soil contamination:

- nature and extent of soil contamination
- current and/or proposed land use
- current and /or proposed receptor population
- potential exposure pathways
- soil, meteorological, geological, and groundwater characteristics

These conditions lead to a fundamental axiom: as the number of possible exposure pathways increases, the calculated health-protective soil concentration will decrease. As a corollary to this, risk management decisions can be made about site remediation to reduce the number and types of exposure pathways to attain health-protective soil cleanup levels which are economically and technologically feasible.

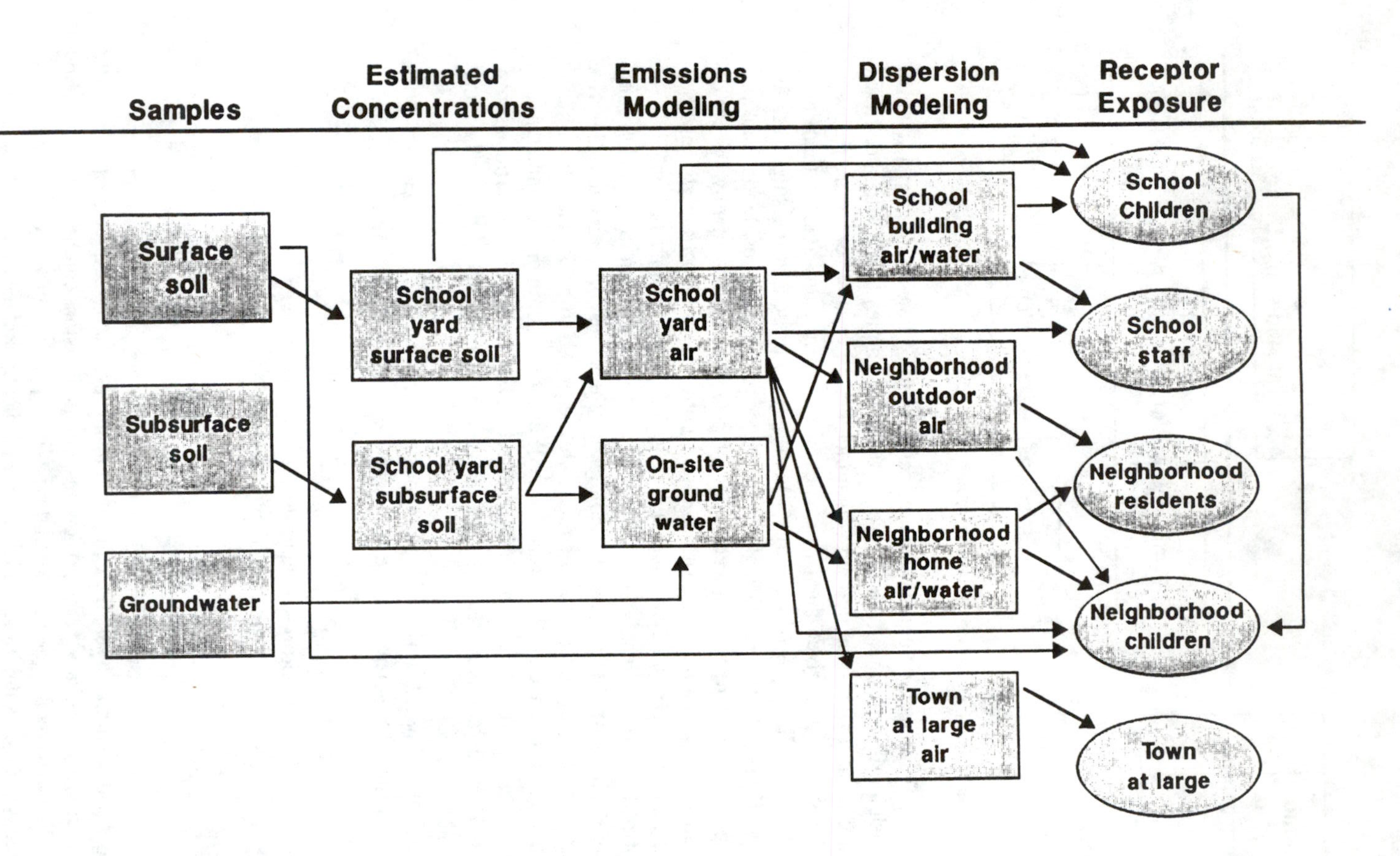

Figure 1.10. Exposure considerations – suburban site.

This approach is not novel. Variations of this process are currently either being used or recommended in risk assessment guidance documents for U.S. EPA,[5,6,20-22] by the Canadian Ministry of Environment,[23] by the U.S. Army,[24] by the Pennsylvania Department of Environmental Resources,[25] and California Department of Health Services' Toxic Substances Control Program.[4,26]

Non-site-specific soil cleanup levels may unintentionally limit risk management flexibility and not be protective of public health. How will cost of remediation be reflected in a list of generic soil cleanup levels? How will single values be set to account for the potential combinations or mixture of chemicals that may be present? Where schedule is at issue, the use of generic soil cleanup levels will likely foster the practice of excavation until the level is achieved. Such practice will invariably consume scarce public and private resources, and may ultimately prevent attention from being focused on more serious problems.

Risk managers should welcome approaches achieving remediation which utilize site-specific cleanup goals for four primary reasons. First, these approaches provide maximum flexibility. Risk management decisions can be made about land use, for example, to reduce the predicted exposure to a level that is considered safe or acceptable (the two are not synonymous). Second, it can provide the greatest cost savings since site-specific data and assumptions will, in most cases, provide values that will be less conservative than those values which must be used to generate any type of generic soil cleanup level. Third, the process rewards data collection, since the use of conservative assumptions in a generic cleanup number may indicate values so low that excavation is the only option, thus a process approach can predict what volume or mass can be left behind. Fourth, the structured paradigm expanded to account for factors such as cost and feasibility will permit the risk manager to test various risk management options.

The demand for generic soil cleanup numbers fails to give priority to the concept that the permanent remediation of hazardous waste contamination (and therefore risk reduction) will require careful consideration of the nature and extent of contamination. Protection of public health and the environment may not always require the remedial alternative of "moving dirt," but it should mean source reduction. Source reduction can be achieved through interdiction of the exposure pathway (e.g., a clay cap). Relocation of the biological receptor may also be a viable option. It is suggested that the risk management mindset change its focus from generic non-site specific soil cleanup numbers as a way of lessening the difficulties in making risk management decisions to (1) the development of structured public policy and guidance in risk management and reduction, and (2) the development of a stable infrastructure of resources to support informed site investigation, risk assessment, and risk management.

SUGGESTIONS FOR THE EARLY 1990s

Solutions to the challenges of future risk management will not be easy. The slow pace of hazardous waste site remediation has resulted from many causes.

A major problem has been the apparent complexity of risk assessment and risk management and the lack of adequately trained personnel. Generic soil cleanup numbers would reduce the burden of the task by the elimination of apparent complexity. However, as described, there is no scientific basis to arrive at such cleanup numbers which could be applied to all sites. Therefore, a more defensible policy would be to work on the tools, policies, and resources needed for risk management.

Risk assessment must remain a viable tool of public health and environmental policy. Informed scientific debate should continue to ensure that conservative assumptions do not inadvertently lead to the perception of the existence of public health problems where none really exist. Focused scientific studies should be conducted to fill in gaps in our knowledge of toxicology and exposure assessment. Most importantly, vigilance must be maintained against intrusions into the risk assessment process by risk management considerations that would make risk management policy myopic to serious public health and environmental problems.

Another tool which holds great promise is an expert-based system for risk management which incorporates scientifically derived multipathway risk assessment processes, technical and engineering considerations for site characterization, current public policy needs, economic and social limitations, and political considerations. This expert-based system for risk management could prove to be a significant advance over current practice. An expert system could be used to query the risk manager for information regarding the site, document the input information, direct the risk manager to obtain additional information, and indicate when a particular specialist should become involved. In addition, the system could compute exposure levels, magnitudes of risk, and suggest remedial options. Most important, the system would illustrate the risk assessment thought process in a way that would be far more beneficial than just reading existing guidance documents. The system would provide structure to, and serve as a record of, risk management decisions. This should result in enhanced consistency of risk management within a dispersed regulatory institution.

To provide clarity to the risk management process, rather than simply focusing on generic soil cleanup numbers for each potential chemical contaminant, a regulatory agency may choose to simply regulate the upper limit of exposure by all plausible routes to all relevant chemical contaminants found at a hazardous waste site or facility to some fixed risk level, e.g., 10^{-6} individual excess lifetime cancer risk. This would define a de minimis or significant risk and provide a clear requirement to the responsible party about the goal desired by the regulatory agency and the public, yet not intentionally hinder the range and flexibility available for risk management options. This will demand that a clear understanding of the nature and size of the source be ascertained. It will maximize efforts to control the largest sources of risk, take into account the presence of mixtures, and hopefully place the goal of residual soil concentrations in perspective.

There are a number of indications that the above would serve as a viable public policy. U.S. EPA under CERCLA is reported to be making steps toward this direction in proposing the narrowing of the risk management range from 10^{-4} to 10^{-7} to 10^{-4} to 10^{-6}.[27] One of the reasons cited was the absence of decisions using the 10^{-7} value. It should be noted that the U.S. EPA under the National Primary Drinking Water Regulations (*Federal Register,* 1987) already employs a "target reference risk range" for carcinogens of 10^{-4} to 10^{-6} for the setting of MCLs. In California, under Title 26, Section 22-12703, of the California Code of Regulations, implementing the ballot initiative Proposition 65, the regulations identify 10^{-5} individual excess cancer risk to represent "no significant risk."

Acknowledging the limited resources of the various regulatory agencies and the potential impediments attendant to such constraints, it is suggested that, for sites of limited complexity, (1) willing responsible parties should be allowed to proceed on their own with site investigation and remedial activities, and (2) self-certify that the hazardous waste site has been remediated to the point that public health and the environment is protected. Such activities will have to conform to guidance supplied by the regulatory agency. Guidance should span subjects from who may author a multimedia risk assessment, to the construction of ground-water monitoring wells. Continuous regulatory oversight is not envisioned to occur in this process, but adherence to such guidance would be rigorously audited. Sites found to be deficient would be subject to enforcement actions. Penalties for noncompliance should take the form of both criminal, civil, and administrative sanctions, and should be so severe as to render their use very rare. Again the goal is to achieve a clear understanding on the part of the responsible party as to the expectations of the regulatory agency. This sharing of responsibilities with the private sector will allow regulatory agencies to focus staff and resources on abandoned hazardous waste sites and the more serious public health and environmental problems.

In terms of resources, a stable infrastructure of knowledgeable and experienced staff and policies for both risk assessment and risk management must be established and maintained by both the regulatory agencies and the private sector. Risk management must be based upon informed scientific, technical, and public health and environmental policies. Conditions should not be propagated which lead to inappropriate allocation of resources such as requesting a geologist to write or review risk assessments or create frustrations so great as to force the use of generic soil cleanup numbers. In concept, methodologies of risk management should be made as uniform as possible across regulatory agencies. Like risk assessment, a structured analytic paradigm of analysis is needed to support risk management decisionmaking. Initiative and momentum are needed in both of these areas. Products of risk management policy should not be a reflection of absolutes, but a compromise between competing issues with the goals of protection of public health and the environment in the forefront.

ACKNOWLEDGMENTS

The authors would like to thank S. DiZio, W. H. Soo Hoo, R. A. Howd, T. E. McKone, S. T. Reynolds, N. Ozaki, and A. Wolfenden for their helpful discussions and critical review.

REFERENCES

1. National Academy of Sciences, "Risk Assessment in the Federal Government: Managing the Process," (Washington, DC: National Academy Press, 1983).
2. United States Environmental Protection Agency, "The Risk Assessment Guidelines of 1986," Offices of Health and Environmental Assessment. EPA/600/8-87/-45. August, 1987a.
3. California Department of Health Services, "Guidelines for Chemical Carcinogen Risk Assessments and Their Scientific Rationale," State of California Health and Welfare Agency, November, 1985.
4. California Department of Health Services, "California Site Mitigation Decision Tree," Alternative Technologies and Policy Development Section, Toxic Substances Control Program, May 1986.
5. United States Environmental Protection Agency, "Superfund Public Health Evaluation Manual," Office of Emergency and Remedial Response, EPA/540/1-86/060. October, 1986a.
6. United States Environmental Protection Agency, "Risk Assessment Guidance for Superfund. Human Health Evaluation Manual—Part A. Interim Final," Office of Solid Waste and Emergency Response. OSWER Directive 9285.7-01a. September, 1989.
7. Byrd, D., and L. B. Lave. "Narrowing the Range: A Framework for Risk Regulators," *Issues in Science and Technology,* Summer, 1987, pp. 92–100.
8. Rodricks, J. V. "Comparison of Risk Management in U.S. Regulatory Agencies," *J. Hazardous Mat.* 21:239–253 (1989).
9. Young, F. E. "Risk Assessment: The Convergence of Science and the Law," *Regulatory Toxicology and Pharmacology* 7:179–184 (1987).
10. Lave, L. B., and E. H. Males. "At Risk: The Framework for Regulating Toxic Substances," *Environ. Science and Technology* 23:386–389 (1989).
11. Lawless, E. W. "Risk Assessment Methodology for Hazardous Waste Management." Draft Final Report. U.S. EPA Contract EQ4C15. Prepared for U.S. EPA Office of Policy, Planning and Evaluation and the President's Council on Environmental Quality. July, 1986.
12. United States Congress, Office of Technology Assessment (U.S. OTA), "Coming Clean." OTA-ITE-433. Government Printing Office. Washington, DC. October, 1989.
13. Paustenbach, D. J., H. P. Shu, and F. J. Murray. "A Critical Examination of Assumptions Used in Risk Assessments of Dioxin Contaminated Soil," *Regulatory Toxicology and Pharmacology* 6:284–307 (1986).
14. United States Environmental Protection Agency. 40 CFR Parts 141 and 142. "National Primary Drinking Water Regulations—Synthetic Organic Chemicals; Monitoring for Unregulated Contaminants; Final Rule." *Fed. Reg.* 52(130):25689–25704. July, 1987b.
15. Fishman, B. E. "Risk Assessment of Pesticides in Former Agricultural Soils," Proceedings of the Fifth Annual Hazardous Material Management Conference/West. November, 1989.
16. McKone, T. E. "Methods for Estimating Multi-Pathway Exposures to Environmental Contaminants," AD UCRL-21064. Lawrence Livermore National Laboratory. University of California. June, 1988.

17. McKone, T. E., J. J. Wong, and S. M. DiZio. "Soil Remediation Levels: A Multimedia Approach for Defining Waste-Site Cleanup Goals." Paper presented at the 1989 Meetings of the Society of Risk Analysis at San Francisco, California. October, 1989.
18. United States Environmental Protection Agency. "CERCLA Compliance with Other Laws Manual: Draft Guidance." Office of Emergency and Remedial Response. EPA/540/G-89/106. August, 1988.
19. United States Environmental Protection Agency. "Guidance for Conducting Remedial Investigations and Feasibility Studies Under CERCLA. Office of Emergency and Remedial Response. EPA/540/G-89/004. October, 1988.
20. United States Environmental Protection Agency. "Development of Advisory Levels for Polychlorinated Biphenyls (PCBs) Cleanup." Office of Health and Environmental Assessment. EPA/600/6-86/002. May, 1986.
21. United States Environmental Protection Agency. "Guidance for Establishing Target Cleanup Levels in Soils at Hazardous Waste Sites." Office of Health and Environmental Assessment. Internal Review Draft. OHEA-E-201. May, 1988.
22. United States Environmental Protection Agency. "Guidance for Establishing Target Cleanup Levels in Soils at Hazardous Waste Sites." Office of Health and Environmental Assessment. Review Draft. OHEA-E-201. March, 1989.
23. Environment Canada. "Contaminated Soil Cleanup in Canada," Volume 2. Interim Report on the 'Demonstration' Version of the AERIS Model (An Aid for Evaluation of the Redevelopment of Industrial Sites). Decommissioning Steering Committee. December, 1988.
24. Small, M. J. "The Preliminary Pollutant Limit Value Approach: Manual for Users." Technical Report 8918. U.S. Army Biomedical Research and Development Laboratory. Fort Detrick, MD. 1988.
25. Pennsylvania Department of Environmental Resources (PADER). "Risk Assessment/Fate and Transport Model." Bureau of Waste Management. Draft Document. October, 1989.
26. California Department of Health Services. "Guidelines for Determination of Soil Remediation Levels." Toxicology and Risk Assessment Unit, Technical Services Section, Toxic Substances Control Program. Draft Document. In preparation, 1990.
27. Environmental Policy Alert. "Superfund Pollutant Risks—EPA Likely to Accept Greater Human Health Dangers," January 24, 1990, pp. 23–25.

CHAPTER 2

Regulatory Approaches to Hydrocarbon Contamination from Underground Storage Tanks

Seth J. Daugherty, Orange County Health Care Agency, Santa Ana, California

INTRODUCTION

Action or lack of action by the appropriate regulatory agency is often the most important factor in determining remedial action or closure requirements for hydrocarbon contaminated sites. The position of the regulatory agency is important because of strong laws and public support for pollution control. Regulatory agencies are also being placed in key positions in some areas due to the desire for agency clearance prior to transfers of real estate. Perhaps the most important reason that regulatory agencies occupy such a key position is that the public health impacts of hydrocarbon contamination often are uncertain at the regional level and unpredictable on a site-specific basis. This situation results in a failure to indicate a clear direction or to achieve a consensus on a proper course of action. Thus, satisfaction of agency requirements becomes the most important factor in addressing contaminated sites.

The diversity of regulatory criteria is well known statewide and well documented nationally.[1,2] In California, the diversity of approaches is due to: (1) that very lack of a clear understanding of the true impact of hydrocarbon contamination; (2) lack of state or federal standards for soil cleanup, and state water quality objectives that are not always achievable; (3) vagueness in the underground storage tank law; and (4) the number and diversity of agencies enforcing the underground storage tank regulations.

Until recently the California Health and Safety Code Section 25298(c)(4) required the responsible party to demonstrate to the local agency that there was no significant soil contamination resulting from a discharge from the tank. There was no further guidance regarding the level of soil contamination considered to be significant, or as to what constituted a proper demonstration. Section 25298(c)(4) has recently been changed. Now the responsible party must demonstrate to the agency having jurisdiction that the site has been investigated to determine if there are any present or were any past releases, and, if so, that appropriate corrective or remedial actions have been taken. These changes shift the emphasis from determining what is significant soil contamination, to implementing appropriate corrective action without specifying environmental media. However, neither appropriate corrective action nor what constitutes a proper investigation is further defined.

There are 100 local agencies enforcing the California underground storage tank law. Most are health agencies (50) and fire departments (35), although some are agriculture departments (5), special hazardous substances control programs (3), and agencies with primary responsibilities for public works (2), emergency services (2), air pollution control (1), planning (1) and solid waste (1). This diversity of agencies, with a variety of primary functions and staff with different backgrounds, has resulted in a diversity of approaches to hydrocarbon contamination. Some agencies may require only mitigation of hydrocarbons to levels that would not cause an explosion. Others may evaluate excess lifetime cancer risk due to exposure to benzene vapor. In general, most agencies attempt to protect groundwater, which is the stated intent of California underground storage tank regulations.

Existing state and federal guidance, including the Environmental Protection Agency (EPA) documents such as the Risk Assessment Guidance for Superfund[3] and the Superfund Exposure Assessment Manual[4], as well as the California Site Mitigation Decision Tree Manual[5] are useful at the conceptual level but less so at the local operational level. Of all the guidance documents, the California Leaking Underground Fuel Tank Field (LUFT) Manual[6] comes closest to addressing hydrocarbon contamination in a state-of-the-art manner useful at the operational level.

Much of the local agency effort has been directed toward determining acceptable cleanup levels for soil or, in some cases, determining the proper point of application of groundwater cleanup standards. On one hand, state and federal guidance documents, as well as the technical literature, make a strong case for science-based state-of-the-art risk assessments. On the other hand, both regulators and the regulated strongly desire numerical cleanup levels for more practical and definite guidance. Science-based risk assessments require numerous judgments to develop, and at our present level of understanding, involve inherent high uncertainty. Moreover, it is apparent that desirable cleanup levels may be impractical to attain in a reasonable period of time.

The primary focus of this chapter is public health considerations relating to small sites contaminated with petroleum hydrocarbons from underground storage tanks. An attempt is made to describe the regulatory perspective of the leaking underground storage tank problem and to develop the best regulatory position, especially for petroleum hydrocarbon contamination. I will identify some basic regulatory concerns, review contaminant fate from a regulatory perspective, describe some regulatory approaches, and then present an approach that considers the present state of the science as applied to the local operational level. This approach basically recognizes public health and safety criteria as desirable goals, but if health-based goals cannot be reasonably met, then a shift to criteria based on a reasonable application of appropriate technology is suggested.

BASIC REGULATORY CONCERNS

Uncertain Exposure and Effects

Multiple Exposure Pathways and Public Safety

The most immediate concern at a leaking tank site is public safety. Hydrocarbons may migrate long distances along subsurface structures such as sewer lines and other utilities trenches. The distinction between immediate and longer-term corrective actions is appropriate for potential public safety hazards. Removal of recoverable free product and highly contaminated soil should begin as soon as possible. Potential public health effects, especially long-term effects, are less obvious. Routes of environmental exposure include inhalation, ingestion, and dermal absorption. Exposure pathways may include direct contact, inhalation of hydrocarbon vapors, and ingestion of contaminated water and food. Direct contact may be prevented by limiting access to contaminated sites or by appropriate land use regulation, and appears to be a readily definable, controllable, and limited concern, at least in urban areas. Hydrocarbon exposure due to unintentional contamination of food also appears to be limited.

Subsoil to Groundwater to Drinking Water Supply. Exposure via contaminated drinking water, especially groundwater, is of great concern, and the stated purpose of much of the hydrocarbon and underground storage tank regulations. The relatively high environmental mobility and propensity of the aromatic hydrocarbons to partition into water, together with the drinking water quality standards often applied as groundwater quality objectives, makes this a well known regulatory concern.

Subsoil or Groundwater to Indoor or Outdoor Air. The potential public health impacts of exposure to vapors due to hydrocarbon contaminated sites is less recognized and understood. The overall effects of short-term vapor exposures due to tank removals, spoil piles, and remedial actions appear to be insignificant.[7]

However, the case for concern regarding incremental additional excess lifetime cancer risk associated with benzene vapor exposures in the air inside of buildings, due to contaminated soil or groundwater beneath a proposed building site, can be made.

Nature of Hydrocarbon Contaminants

Petroleum hydrocarbon fuels may contain hundreds or even thousands of individual constituents of several different chemical classes.[8] Each of these substances occurs in the liquid, dissolved, adsorbed, or vapor phase according to its physical and chemical properties. Chromatographic effects and differential multiphase partitioning, selective solubilization, and other "weathering" effects are characteristic of hydrocarbon contamination.[9,10]

Subsurface Heterogeneity

Irregularities in the subsurface can exist on both the micro and the macro scale. Differences in hydrogeologic characteristics such as permeability can vary over eight orders of magnitude in sedimentary strata and fractured bedrock.[11] Complex hydrocarbon distribution patterns may be expected in the subsurface.[12,13]

Carcinogenicity

Some petroleum hydrocarbon constituents, most notably benzene, are considered to be carcinogens.[14] Benzene is listed as a carcinogen in the California Safe Drinking Water and Toxic Enforcement Act of 1986 (Proposition 65 regulations). The California drinking water Maximum Contaminant Level (MCL) for benzene is 1 μg/L (1/ppb), while the California Department of Health Services health-based Applied Action Level (AAL) in air is 0.66 μg/m^3 (0.2 ppbv) at the 10^{-5} excess lifetime cancer risk level.[15] Comparable levels for air of 0.3 and 0.4 ppbv can be derived from the Proposition 65 regulations and EPA methodology, respectively. As fresh gasoline generally contains from 1% to 5% benzene, source to receptor attenuation factors of as much as 50 million to one for water and 250 million to one for air may be needed to reach levels considered to exert no significant effect. Most of the public health concern and uncertainty associated with petroleum hydrocarbons is due to benzene. A systematic review of the public health implications of petroleum contaminated soils may be found in Calabrese et al.[16]

Cumulative Effects

Most evaluations of contamination consider single site source to receptor or source to resource impacts. However, populated areas have many widespread contaminated sites above groundwater basins where cumulative effects are difficult to evaluate.

Determination of Acceptable Cleanup Levels

This includes the pervasive problem of the conceptually ideal but sometimes difficult to obtain levels derived from quantitative site-specific risk assessments, as opposed to the readily usable but often inappropriate broadly applied numerical cleanup levels. Evaluations of these two regulatory approaches from the industry viewpoint may be found in Bauman.[17,18]

Public Expectations of Protection

Public perception of risk is more related to familiarity, equitability, and controllability than to a calculated risk level.[19,20] Evidence that Californians expect a high degree of protection from discharges which may affect groundwater and from exposure to carcinogens may be inferred from the overwhelming passage of Proposition 65, the Safe Drinking Water and Toxic Enforcement Act of 1986. From this perspective, public agencies, especially health agencies, would prefer not to leave any amount of carcinogen, such as benzene, remaining in the ground or the groundwater in an essentially uncontrolled condition.

Need for Documentation

The need for documentation may be expressed on at least two levels: (1) a thorough delineation of the contamination, and (2) a defensible justification for site completion or intermediate ''sign off'' by the regulatory agency. Documentation may also take at least two basic forms: (1) a demonstration that contamination has been removed; or (2) a conservative science-based predictive evaluation that any remaining contamination either will not spread or be of no significance if it does spread. Ideally, documentation includes a state-of-the-art-and-science justification for closure or other action that is clearly in conformance with established risk management policy.

Large Number of Sites

As of January 1990, the nine California regional boards reported 13,270 underground storage tank (UST) leak cases with 11,134 open. The five top counties alone had 5,718 cases. Individual agency staff are often responsible for overseeing 50 to 200 sites at any time. Regulatory approaches must recognize this reality.

Need for Timely Decisions

Regulators are called upon daily to give direction or agency requirements to responsible parties and environmental consultants. A basic regulatory framework must be in place to give guidance in a consistent and intelligible manner.

HYDROCARBON FATE AND MODELING FROM THE REGULATORY PERSPECTIVE

An attempt is made to extract some of the conclusions from recent scientific research as applicable to the regulation of small hydrocarbon contaminated sites. Reviews of the literature relating to the solubilities and natural attenuation of petroleum hydrocarbons may be found in publications of the American Petroleum Institute.[21,22] The transport of organic contaminants in aquifers was covered by Mackay et al.[12] Dragun[13] detailed comprehensive descriptions of the fate of hydrocarbons, especially for the unsaturated zone. Bonazountas[23] provided a comprehensive treatment of modeling as related to petroleum products in soil.

Table 2.1 gives an overview of environmental fate processes in the subsurface and their regulatory implications. The main points are categorized below.

Multiphase Partitioning and the Importance of Nonaqueous Phase Liquid (NAPL). Hydrocarbons can exist in four separate phases: the liquid phase, sometimes called the bulk, mass or NAPL phase; dissolved in water as aqueous phase either in groundwater or in the pore water surrounding particles in the unsaturated zone; in the gaseous phase as vapor in soil gas; and as the adsorbed phase held onto solid surfaces of particles. It should be pointed out that liquid phase can exist as an emulsion and that there are several possible types of adsorbed phases. Both emulsions and contaminants adsorbed onto colloidal solids could be mobile. However, in this chapter, the term "liquid phase" refers either to hydrocarbons "floating" on the water table, or to the basically immobilized and stationary liquid trapped within subsurface formations. After liquid phase hydrocarbons have drained, or after recoverable "free" product has been removed, a substantial amount of hydrocarbon will remain[24] in the subsoil as liquid phase residual saturation. Experimentally derived gasoline residual saturation may range from 35,000 to 120,000 mg/kg (ppm) or about 10% to 30% of pore volume;[25] although environmental values could be more on the order of 6% of pore space due to weathering.[26]

Liquid phase hydrocarbon residuals, known as globules, droplets, or ganglia, become trapped within the subsoil and are long-term sources of contamination, causing plumes to grow and persist.[27] They cannot be removed by the usual pump and treat remediation, although pumping groundwater will prevent the spread of dissolved phase product. In a mathematical simulation of dissolution kinetics, Hunt et al.[28] predicated that ganglion lifetimes are on the order of decades to centuries for sizes expected from gravity or viscous emplacement. Further, these ganglia cannot be mobilized by groundwater extraction and may produce dilute waste steams of massive volume. Dissolved contaminant levels leaving a site with NAPL were predicted to be near the solubility limit.

Certain hydrocarbons are somewhat soluble in water, notably the aromatic hydrocarbons, especially benzene. The solubility of benzene in water is usually given as 1780 mg/L, although the maximum calculated concentration in an aquifer immediately beneath a gasoline release has been estimated to be about 62 mg/L

(62,000 ppb).[29] In a mathematical modeling analysis, Baehr[30] concluded that aromatic constituents of gasoline dominate the hydrocarbon mass partitioned into unsaturated zone water, and that the significant hydrocarbon flux to the atmosphere does not result in a commensurate decline in groundwater contamination potential due to the enrichment of aromatic compounds relative to other hydrocarbons. He further concluded that diffusive transport in the unsaturated zone is a significant transport mechanism which can cause aqueous and vapor plumes to spread away from the immiscible liquid source, resulting in increasing groundwater contamination potential. Moreover, Falta et al.[31] in a theoretical investigation concluded that density-driven flow may dominate the gas phase transport of some organic vapors, possibly including benzene, resulting in groundwater contamination remote from sources.

The significance of adsorption in natural systems is not clear, although adsorptive potential is generally considered to be related to organic carbon content of the adsorbing substrate. Using generally accepted principles to evaluate adsorption phenomena, Roy and Griffin[32] classified benzene as "highly mobile" in soil; toluene and o-xylene were classified as "medium"; while m- and p-xylene were considered to have "low mobility." A critical review of the standard approach to analysis of adsorption may be found in Dragun.[13] Broader reviews of the significance of liquid phase hydrocarbons in the subsurface, including halogenated organics, and the implications for national groundwater policy may be found in Mackay et al.[12] and Freeze and Cherry.[33]

Transport processes. Subsurface contaminant transport phenomena include advection, dispersion, and diffusion. These processes are quantitatively well described, especially for solutes in ideal porous media. However, subsurface macro and micro level heterogeneities confound analyses in natural substrates.

Transformation Processes. The most significant transformation process for organic chemicals in the subsurface is generally recognized to be bacterial biodegradation. Biotransformations are generally considered to depend on many interacting factors[13,21,22] with the rates of biodegradation being generally unpredictable at present.[9,34] Contact between oxygen and hydrocarbons is generally considered to be the rate limiting step,[35] although anaerobic degradation of petroleum hydrocarbon constituents has been reported in microcosm studies.[36,37] As Kindred and Celia[38] point out, different processes and different microbial consortia may dominate in different areas of the plume.

Field studies indicate that dissolved phase gasoline components can be effectively biodegraded but rates vary. Barker et al.[39] introduced 1,800 L of groundwater spiked with about 2.4 mg/L benzene, 1.7 mg/L toluene, and 3.5 mg/L xylene into a shallow sand aquifer. All of the mass was lost to biodegradation in 434 days; only benzene persisted beyond 270 days. Biodegradation probably followed zero order kinetics as dissolved oxygen availability was rate limiting. Chiang et al.[40] calculated a natural attenuation rate for benzene of 0.95% per day in a contaminated aquifer. In this study, models, laboratory, and field data indicated strong

Table 2.1. Analysis of Environmental Fate of Subsurface Petroleum Hydrocarbon Releases from a Regulatory Perspective

Process	Phenomena	Analysis	Reference
Mass movement of bulk nonaqueous phase liquid hydrocarbon (NAPL).	Saturated flow from contaminant source. Immobilization when trapped at residual saturation where capillary forces balance gravity or hydraulic head.	Hydrocarbons penetrate unsaturated zone and spread in capillary zone (water table). Complex distribution patterns in heterogenous subsurface NAPL may remain as a long-term source of contamination.	9, 13, 25, 27, 28, 30, 44, 46, 48
Multiphase partitioning in unsaturated subsoil.	Selective solubilization, volatilization and adsorption of individual chemical constituents of hydrocarbon mixture to soil pore water, soil gas, or onto organic carbon.	Partition and distribution coefficients suggest aliphatic hydrocarbons move to soil gas and aromatic hydrocarbons move to soil water and soil gas.	30
Movement through unsaturated subsoil pore water.	Aqueous diffusion to ground water or surface.	Slow process but may be significant.	13, 30
Movement through soil air.	Gaseous diffusion to the atmosphere or indoor air, possible density driven vapor flow.	Fairly rapid process; hydrocarbons lost to atmosphere; loss reduced and contamination remains if site is paved; may be public health impacts.	13, 30, 31
Adsorption-desorption.	Retardation, mainly on organic matter. Possible sorption to mineral surfaces.	May not be significant due to low organic matter content of subsoils; aromatic hydrocarbons, especially benzene, have relatively high mobility.	32
Infiltration through vadose zone due to rainfall, irrigation, leaky water lines.	Dissolution and advection-dispersion due to intermittent aqueous flow.	Potential to move hydrocarbons to groundwater, especially aromatics; obviated by effective barriers; concentration may be reduced by dispersion.	

continued

Table 2.1. Continued

Process	Phenomena	Analysis	Reference
Biological Transformation.	Biodegradation of hydrocarbons ultimately to CO_2 and H_2O. Rate and effectiveness depends on many factors.	Undoubtedly occurs to some extent under some, perhaps most, conditions; may help contain hydrocarbons in some aerobic subsurface environments; rates of degradation not readily predictable.	9, 13, 35, 36, 37, 38, 39, 40, 42, 43, 47
Immobilization due to "impervious" tight subsurface formation.	Aqueous flow slower in fine grained formations.	Extent and effectiveness of retardation often not know; most sedimentary formations not really impervious; NAPL hydrocarbons change clay structure to increase permeability; undetectable macro pore flow may occur.	13, 28
Chemical Transformation.	Oxidation, reduction, hydrolysis.	Generally not considered significant for hydrocarbons in the subsurface, although not well understood.	
Groundwater Flow.	Dissolution followed by saturated flow with advection and dispersion.	Generally relatively slow in sedimentary formations. Public health and resource impact if human exposures or aquifer contamination occurs.	

correlations between dissolved oxygen and dissolved hydrocarbon degradation. In his review of the literature, Dragun[13] reported seven different biodegradation rates for benzene with half-lives generally on the order of weeks to months.

Biodegradation may only be effective around plume margins, however. Hunt et al.[41] reported that approximately 12,000 gallons of gasoline still remain 10 years after a leak with an estimated original loss of 9,000 to 17,500 gallons. In a field investigation of a creosoting site contaminated mainly with polycyclic aromatic hydrocarbons, Borden et al.[42] found that horizontal mixing with oxygenated formation water results in biodegradation of trace quantities of hydrocarbon at the plume edges and reduces the area of the plume, but has little effect on high concentrations in the plume center. Exchange with the unsaturated zone appears to supply oxygen to the plume center. At another creosoting site, Wilson et al.[43] also found biodegradation to be rapid at the plume margin and controlled by the oxygen supply. They concluded that once adaptation occurs, biotransformation can be considered instantaneous in slow groundwater flow, and there is no need to characterize biological activity. They also pointed out that there is a need for tools to recognize whether adaptation has occurred and to delineate the time and conditions needed for adaptation. In theoretical simulations, Borden and Bedient[35] concluded that biodegradation will typically result in a zone of reduced hydrocarbon between oxygenated formation water and the plume. Hydrocarbons plumes will thus appear more narrowed in plan view than expected due to transverse mixing with the oxygenated formation water and resulting biodegradation.

In summary, natural biodegradation appears to limit the spread of plumes in some situations if oxygen is available. However, the main source of contamination is not always effectively degraded under natural conditions. Moreover, biodegradation in aquifers can vary substantially with both space and time and is not related in a simple way to geology or hydrology.[34]

Mathematical Modeling. When all the transport and transformation processes are combined to simulate fate of multicomponent contaminants in time and space in porous media, quantitative descriptions become quite complex. A few comprehensive models have been developed.[44–46] Biological processes in the subsurface have been modeled.[35,38,47] These models may, however, result in elegant systems of equations that require unavailable input parameters and prohibitive computational effort to model a specific site.[48] Mathematical descriptions in the scientific literature appear to be directed toward advancing theory or characterizing contaminant behavior under different conditions. Skopp[49] distinguished between two motivations for developing theories: to obtain detailed descriptions for predictive value, and to develop qualitative understanding through identification of key parameters and interrelationships. The latter motivation seems most appropriate to the present state of science.

Available mathematical computer models have not been verified or validated for hydrocarbons in natural porous media. Hillel[50] pointed out many basic sources of uncertainty associated with modeling organics in soil, including layering, variable moisture, and unstable flow. Callahan[51] indicated that predictive exposure assessments can have limited accuracy, can be misleading, are difficult to validate, and rely on data that may not be available. Indeed, subsurface irregularities such as highly variable sorting and layering of particles, and macropore or other channeled flow appear to make it inherently improbable to accurately predict contaminant movement in detail. Moreover, well-known and available models such as SESOIL cannot simulate the transfer of hydrocarbon constituents from the nonaqueous liquid phase to the dissolved aqueous phase in, for instance, infiltrating rainfall.[52] Models are, of course, useful to assess generic effects of contaminants or impacts over larger areas. At this point, however, simplified expressions derived from first principles appear to be as useful as more elegant computer models for the evaluation of contaminant fate and transport at small sites.

Evidence from Case Studies and Groundwater Monitoring. There is no doubt that gasoline leaks can cause serious environmental impacts in certain circumstances. In Orange County, for example, an accidental leak from an aboveground tank released an estimated 37,000 gallons of gasoline over a two-day period. Gasoline constituents penetrated 90 feet of vadose zone and were transported through or around a thin silty clay zone at 100 to 105 feet. Benzene concentrations were as high as 10,000 ppb in onsite monitoring wells, and as high as 100 ppb (generally around 20 ppb) in a municipal production well. Three City of Anaheim wells had to shut down. Other cases of extensive contamination due to gasoline leaks have been reported in North Florida[53] and Long Island, N.Y.[54]

On the other hand, of 4,765 small water systems (less than 200 service connections) sampled in California, only 5 had benzene contamination and 6 had toluene contamination.[55] Benzene was found in only 9 of 2,947 large water system wells in concentrations up to 1.1 μg/L with a median value of 0.2 μg/L.[56] Moreover, of the 11,651 open leaking underground storage contamination cases reported for California in October 1989, only 58 impacted drinking water.[57]

Conclusions. Impacts of hydrocarbon contamination depend on the rate of transport, the processes of transformation, and dilution (dispersion, diffusion). A case for caution can be made from projections based on the physical chemistry of contaminants and mass transport. Biotransformation probably occurs to some extent in most cases, but it is difficult to quantify its effect on contaminant plumes. The conservative position is to assume it's not effective unless demonstrated to be so. There are documented cases of severe pollution due to petroleum hydrocarbon leaks, but the known impacts do not yet appear commensurate with the number of known contaminated sites. In short, the impacts of hydrocarbon contamination appear to be a race to the well screens or human breathing zone between transport and transformation processes.

Some conclusions from the regulatory viewpoint regarding multiphase partitioning and entrapment of liquid phase hydrocarbons include:

1. The most important action for remediation of hydrocarbon contaminated sites is to remove liquid phase hydrocarbons that are trapped within the

subsurface to eliminate a source of contamination that can exist for decades. These trapped hydrocarbons cannot be removed by the usual "free product" extraction methods. Employing secondary recovery methods such as steam injection or bioremediation may result in spreading of contamination. Distribution patterns of NAPL can be irregular and patchy, making contamination difficult to locate. NAPL should be assumed to be present anytime hydrocarbon contamination has spread over the water table. Considerable NAPL will remain after completion of recoverable free product removal activities.

2. If the liquid phase hydrocarbons are not or cannot be removed, then the prospects for achieving drinking water standards are poor, especially if product such as gasoline has reached the capillary or saturated zones. The best that can be expected from the usual pump and treat methods over the short term is containment of dissolved phase plumes.
3. Attempts should be made to retrieve substrate core samples from the saturated zone and carefully analyze them to determine if liquid phase product exists. In general, methods should be developed to better distinguish between phases.

The following points may summarize the present state of understanding of hydrocarbon contamination as applicable to local level regulatory concerns: (1) principles of physical chemistry and mass transport theory indicate a cause for concern; (2) impacts appear to depend upon the processes of transport and transformation, which in turn depend upon rate of contaminant input and process rates; (3) input and process rates are rarely known, and, in addition, heterogeneity and multiphase partitioning of complex hydrocarbon mixtures results in high uncertainty with respect to quantitative analysis of environmental fate; (4) it is unlikely that all hydrocarbon contamination will be located if a large release occurs into a complex subsurface and; (5) it is unlikely that cleanup of all groundwater to drinking water standards will be feasible, or that capillary and unsaturated zone contamination can be cleaned up to levels that will guarantee that MCLs will not be exceeded in groundwater in the future, especially if nonaqueous phase liquid hydrocarbon is present.

REGULATORY APPROACHES

There are few evaluations of regulatory methodologies in the literature. Brown[58] critically reviewed five state and federal methods documents for defining cleanup levels at hazardous waste sites. Kostecki et al.[59] reviewed the same or successor documents plus two others from a health risk assessment viewpoint.

Methodologies in these documents were generally concerned with exposure at the receptor. Soil cleanup levels were generally based on direct contact or ingestion. Transfer among environmental compartments, if considered, was sometimes based on equilibrium distribution coefficients. From the groundwater cleanup perspective, Schiffman[60] critically evaluated basic groundwater cleanup policy. He considered regulatory problems aimed at cleaning up contaminated groundwater to be misguided and considered a more rational approach to be: (1) establish public health and resource protecting priorities, including delineation of well head protection zones and critical recharge areas; and (2) establish a program to control plumes of contaminated groundwater. Freeze and Cherry[33], in a far-reaching editorial, pointed out that attempts at aquifer remediation have failed and that there is little likelihood that new technologies will provide solutions because of geological complexity compounded by difficulties imparted by contaminants in NAPL. They urged a move away from blanket cleanup of subsurface contamination to focus on protecting valuable water supply aquifers and remediating sites where significant health risk reductions can be achieved. Mackay and Cherry[27] also emphasized that pump-and-treat remediation is best considered a management tool to prevent continuation of contaminant migration when NAPL are present, and it is appropriate to view such approaches as remediation in perpetuity.

Rubel et al.[61] pointed out that cleanup of soils on a health risk evaluation basis requires the largest proportion of soils removed relative to visual or RCRA criteria, and stated that regulatory agencies should establish an agreed-upon risk assessment model. Daugherty[62] presented an overview and evaluation of the California Leaking Underground Fuel Tank Field Manual. Henry and Hansen[63] presented a case for the no-action alternative at a fuel leak site in southern California. Bauman[17,18] evaluated approaches to hydrocarbon contaminated soil; pointed out the problems of using numerical cleanup levels; and considered site-specific risk assessment as the better alternative.

The following is a classification of operational level regulatory approaches to hydrocarbon contamination for small sites. All of these approaches have been used to varying extents by California local underground storage tank regulatory agencies. It is important to understand that regulatory approaches may be classified in other ways and that the various approaches are often combined or used in sequence.

Site-Specific Case by Case

To some extent, this approach is nearly universal, and recognizes that every site has a unique setting, special hydrogeologic features, a unique history, and patterns of contamination not precisely duplicated elsewhere. The real questions are how structured it is and who does it. At its best, the case-by-case approach allows a realistic analysis of the significance of contamination. However, due to the absence of a defined process (at least one that is accessible to the public) and inherent scientific uncertainties associated with any site-specific evaluation, the case-by-case approach may depend more on the regulatory philosophy of the agency or its staff, or the values, personalities, and strength of resolve of the individuals involved.

Health Based Risk Assessment[44]

The quantitative site-specific health-based evaluation is, in concept, considered to be the best approach to site impact analysis. For small petroleum hydrocarbon contaminated sites, the assessment could focus on benzene. Acceptable concentrations at the human receptor can be obtained from Department of Health Services (DHS) Applied Action Levels,[15] EPA Carcinogenic Potency Factors,[64] or the California Safe Drinking Water and Toxic Enforcement Act of 1986. Individual excess lifetime risk calculations can be made and relative risk can be estimated. However, for reasons stated earlier, the processes break down at the environmental fate portion of the exposure assessment where uncertainty is high. The problem of estimating contaminant transport and fate from the soil to the groundwater, through the water distribution system, to the consumer tap, or the breathing zone, remains. Although the framework for risk assessments is well developed, the need for numerous professional judgments and inherent uncertainty associated with the toxicological and exposure assessments can result in very different conclusions by equally competent and well-intentioned individuals. Agencies may also be concerned that the process could be used to justify no action where some action is warranted. Of great concern to all agencies are the resources of time and expertise required to generate quantitative risk assessments for thousands of sites. Adequate regulatory oversight is essential to ensure competency and to properly represent the public interest. At present, the risk assessment process appears best suited for special categories of circumstances, such as residential development over contaminated sites, use of contaminated material for roadbeds, or for selected well-characterized sites.

Resource Protection

The California Regional Water Quality Control Boards appear to apply this strategy as a water quality objective or cleanup goal. Large areas of the groundwater basins have been formally designated for the beneficial use of municipal and domestic supply in the basin plans. The desired cleanup level in areas so designated is the MCL or drinking water action level. These levels may be applied at the first groundwater encountered below ground surface even if the groundwater is perched, is located in a formation of low hydraulic conductivity, is highly saline, or otherwise unfit for human consumption without treatment. This position is justified by the possibility of hydraulic connectedness of the uppermost groundwater with the underlying aquifers, and the state nondegradation policy. However, cleanup to drinking water levels is not always possible.

Application of Appropriate Corrective Action

The purpose of site evaluation is to determine if remedial action is needed. Decisions are then required to identify what constitutes a reasonable application of appropriate technology and to determine when maximum possible reduction of contamination has occurred. It should be recognized that, in some cases, it will be impossible to effect reduction of contamination to levels that clearly meet conservative health-based criteria.

Groundwater Monitoring and Empirical Study

Groundwater monitoring has often been used to indicate the degree of stability of the contaminant plume, to determine the effectiveness of remediation efforts, and to determine if contaminated soil will affect groundwater. Groundwater monitoring can, in effect, be an empirical environmental fate study and used, if extended over a period of years, to determine if natural transformation and attenuation mechanisms are effectively containing contaminants. The effectiveness of groundwater pump-and-treat operations may be indicated by determining if contaminant concentrations increase after the system is shut down. Periodic sampling of contaminated soil, to determine whether the contaminant plume is spreading or whether contaminant levels are decreasing, is little used due to complex patterns of contaminant distribution in the vadose zone.

Combination of TPH/BTXE Categorized According to Selected Site Conditions and Generic Exposure Assessment Based on Mathematical Environmental Fate Models (the California LUFT Manual)

The California LUFT Manual[6] was developed following input from a 40-person task force from various agencies. Basically, a "Leaching Potential Analysis" based on bottom of excavation (tank and piping) samples and site physical features is used to develop three categories of total petroleum hydrocarbons (TPH) and benzene, toluene, xylene, and ethylbenzene (BTXE) values. The Leaching Potential Analysis is used for initial site screening. If the site is not screened out at this level, a "General Risk Appraisal" is performed. The General Risk Appraisal is based on a conservative exposure assessment using a generic scenario and the internally interfaced SESOIL and AT123D models of the EPA Graphics Exposure Modeling System (G.E.M.S.). Samples of subsoil through the vadose zone are analyzed for BTXE and compared to tables of acceptable cumulative concentration limits, which are the concentrations of BTXE that may remain in the soil column and not cause drinking water action levels to be exceeded at a point 10 meters down gradient at the top of the aquifer. The cumulative concentrations are based on modeling simulations and the acceptable contaminant concentrations vary with rainfall and depth to groundwater.

The California LUFT Manual has been considered the most comprehensive document of its kind[2], and has been evaluated in the literature.[18,59,62] In practice, the LUFT Manual is rather conservative both with regard to development of the generic scenario and the structure of the environmental fate model. Although apparently well received by the regulated community due to the desire for greater structure in the regulation of fuel tank leaks, strict adherence to the present LUFT Manual methodology will not likely guide all, or perhaps even most, sites to closure.

Semiarbitrary Numerical Standards

Semiarbitrary standards have been used by local and regional agencies as a practical way of dealing with soil contamination. These criteria have been based on concentration due to the difficulty of determining mass, and have been directed primarily toward the protection of groundwater from contaminants in soil. They are generally applied to samples collected from the bottom or sides of an underground tank or piping excavation. The inappropriateness of regulation based solely on soil contaminant concentrations or on any constant concentration multiplier is easily understood by pointing out that the soil cleanup level is the same for a minute spill in a remote area with no groundwater resources whatsoever, as it is for a massive spill directly over a shallow, unconfined drinking water aquifer.

No system of numerical levels is always practical and defensible. There are simply too many conditions and situations to account for. Accounting for the many factors that could influence a cleanup level decision would complicate any system beyond comprehensibility. Strict application of arbitrary criteria under some scenarios would be inappropriate, or even absurd. However, numerical systems could be useful for prioritization or screening levels. A screening level is considered as the level at which no further demonstration of insignificance of the contamination is necessary. However, unless numerical levels are applied at the biological receptor or justified by verified environmental fate analysis, they are based on tenuous rationale. This can easily be pointed out, especially for soil levels, by asking why a value of one, or two, or ten, or twelve and three fifths above or below the "acceptable" value was not selected. Thus, numerical soil cleanup levels broadly applied throughout a jurisdiction or large area, are clearly risk management rather than risk assessment criteria. As Bauman[17] has stated, "Such numerical standards are typically based on limited field data generated by analytical techniques of questionable accuracy, or based on modifications of existing water quality standards that may have no relevance to soil contamination."

The following categories of numerical soil cleanup levels have been used in California and elsewhere over the past several years.

Zero, Non-Detectable or Background, or Drinking Water Standards. Cleanup to these levels is always a safe position. However, this alternative is not realistic in cases where substantial contamination has occurred and dispersed. In addition, high variance in the unsaturated zone, and differentially permeable zones in the saturated zone, may result in very difficult sampling and statistical problems to determine if these criteria were really achieved.

Hazardous Waste Criteria. The California hazardous waste criteria were used as cleanup levels by some local agencies due to their enforceability and the derivation of some heavy metal and pesticide limits[65] by the soil to groundwater attenuation factor of 100 taken from Battelle Memorial Institute and EPA studies. However, the hazardous waste criteria were never intended as hydrocarbon site cleanup levels. A major problem is the inappropriateness of relevant hazardous waste criteria tests such as flashpoint and aquatic toxicity to hydrocarbon contaminated soil. Indeed, hazardous waste levels for volatile substances in soil have not yet been clearly established.

Total Petroleum Hydrocarbon (TPH). Levels of 10, 100, and 1,000 ppm TPH were given as prioritization levels in the San Francisco Regional Board Guidelines for fuel leaks.[66] Board guidance to the effect that soil with over 1,000 ppm (parts per million) TPH should generally be removed were widely misinterpreted to justify leaving any soil with lower TPH levels in place. The successor document to these guidelines, the Tri-Regional Recommendations,[67] provides guidance that indicates the need for further investigation based on a TPH level of 100 ppm and other site features. Although this document clearly states that 100 ppm TPH is not a cleanup level, it in fact becomes one in situations where no further action is required.

There is little justification to regulate on the basis of TPH as the results of this test do not identify specific chemicals, thus precluding rational environmental fate or toxicological analysis. More importantly, the aromatic components benzene, toluene, ethylbenzene, and xylene (BTEX) are generally considered to be the most mobile and toxic hydrocarbon constituents. Moreover, Baehr[30] in a theoretical modeling study, demonstrated that the groundwater pollution potential of soil could be completely defined by its aromatic hydrocarbon content.

Combination of Conditional TPH and BTEX. For a time, Orange County used a soil cleanup standard of both 100 ppm TPH and less than 0.1 ppm BTEX as a soil screening level, providing that there was no evidence of a release and the site was not near areas such as water district recharge basins. The intent was to allow closure of sites subject to overfilling and incidental spillage. The 0.1 BTEX level was a perceived, readily attainable quantitation limit based on reports of analyses by commercial laboratories.

Constant Attenuation/Availability Factor Multiplier. This approach simply multiplies the drinking water standard by another number to obtain a soil cleanup level. The number most commonly used as an attenuation factor is 100, based on studies by Battelle Laboratories and the EPA. However, as Marshack[68] points out, while both studies selected a 100-fold attenuation to conservatively represent average attenuation of waste constituents to groundwater, both stress that the actual degree of attenuation depends on waste and site-specific conditions, while neither details which waste and site conditions are best approximated by the 100-fold factor.

DISCUSSION OF REGULATORY APPROACHES: NUMERICAL STANDARDS, RISK ASSESSMENT, REMEDIATION TECHNOLOGY AND MONITORING

Regulatory approaches should be firmly based on scientific findings and should also be practical. However, science has not yet shown the best approach to the regulation of hydrocarbon contaminated sites. Scientific theory indicates a cause for concern and science-based risk assessment may demonstrate a concern. Most scientifically derived applicable and relevant or appropriate requirements (ARARs) are applied at the point of human exposure, while most contamination is in soil and upper regions of groundwater formations. As pointed out previously, lack of understanding of rates of contaminant transformations and the heterogeneity of natural subsurface media result in high uncertainty in analyses of pollutant fate and transport. Empirical evidence does not, so far, demonstrate broad-scale serious impacts from petroleum hydrocarbons except in cases of large-scale and long-term releases. The seriousness of large-scale releases often depends on human controlled factors such as accidents, improper installation, and long-term undetected leaks. Moreover, hopes to effectively address underground storage tank leaks at tens of thousands of small sites must recognize the constraints of human and material resources.

At present, there appear to be three basic types of criteria for site cleanup: (1) broadly applied simple or single value numerical standards such as the popular 100 ppm TPH together with BTEX concentrations, often applied to samples collected from the bottom of excavations; (2) site-specific health-based risk assessments which may in fact generate numerical cleanup levels as a result of a more sophisticated analysis, including quantitative prediction of contaminant fate and transport, and (3) remediation technology-based criteria which includes the operation of a remediation system to the point of diminishing returns, often followed by groundwater monitoring. Site-specific risk assessments can be further divided into quantitative predictive evaluations and empirical monitoring or sampling studies of groundwater resources or biological receptors. Each type of criteria has a corresponding site mitigation strategy, to be used singly or in combination with others. Another strategy, environmental monitoring (most commonly of groundwater), can be used by itself as an empirical contaminant fate assessment, or as a way to verify the effectiveness of corrective actions or the veracity of predictions.

Numerical levels have been used primarily as risk management tools to trigger further investigations, prioritize sites, or establish a criteria for the no-action alternative. Numbers are easy to use and are definite. However, due to the great diversity among sites, at some point a single-minded application will become inappropriate and unreasonable. Much effort has been expended on the quest to determine the ''correct'' cleanup number for all sites, without achieving a compelling consensus for any single value or combination of values. If broadly applicable numerical criteria are to be used, it is less important what specific numbers are chosen than how they are used. Numerical standards should be conservative, based on science insofar as possible, and considered to be flexible values that can indicate that the no-action alternative is acceptable without further documentation.

Numerical criteria using contaminant concentrations in the subsoil are subject to two fundamental difficulties. First is the chemical analysis itself, both with respect to variability of quality control among individual laboratories, and with regard to variability due to the various extraction procedures and matrix effects of the different subsoils. Secondly, inherently high variance among samples may be expected under the best sampling and quality control protocol. Samples collected within a few feet of each other can yield concentration values differing by several orders of magnitude due to field sampling technique, sample handling, and real patchiness of hydrocarbon distribution. The question arises as to the meaning of an individual subsoil sample. Numerical levels may be better applied to uppermost groundwater, especially if perched or located in low conductivity formations. There is perhaps a better chance of obtaining representative and accurate contaminant concentrations in groundwater than in subsoils.

It is important to distinguish between numerical levels that are broadly applied with little scientific justification and those which may be based on more rigorous and defensible analyses. The former are usually applied to the bottom of excavation samples and are properly used only as risk management screening tools to indicate the need for further investigation. The latter may apply to contamination throughout a soil column provided that the specific site under consideration has features that correspond to the generic case from which the numerical levels were developed. The only guidance document that appears to make this distinction is the California Leaking Underground Fuel Tank Field Manual.[6]

The risk assessment approach is conceptionally ideal. But to be meaningful it must be quantitative because acceptable exposure levels are quantitative. (Qualitative risk assessments usually reduce to selective emphasis of general principles whose applicability to a specific site is unsupported, i.e., the negotiated case-by-case method). However, a quantitative risk assessment is dependent upon a meaningful exposure assessment, which is in turn dependent upon a reliable environmental fate model, which apparently does not exist for the level of detail desired. Furthermore, if benzene is present, risk assessment often leads to the unpopular conclusion that some risk of exposure will remain, or that some amount of resource will be degraded. Finally, the effort required for application of risk assessment to thousands of small sites would probably exceed available human resources, at least in the absence of a simplified standard methodology. Development of a standardized and detailed risk assessment procedure seems unlikely and would inherently pervert the value of a site-specific approach. More fundamental critiques of risk assessment methodology and relative risks from some carcinogens in the environment may be found in Ames[69] and Lehr.[70]

Some considerations seem especially germane to regulatory approaches: (1) that the development of broadly applicable numerical levels or a defined evaluation process is apparently beyond the reach of the present state-of-the-art-and-science of environmental impact analysis, (2) that risk management decisions at the state or federal level specifying numbers or a simple unambiguous process will not be forthcoming soon; and (3) that there are a number of cases where no reasonable effort will remove contamination to conservative health-based standards. A certain amount of contamination is going to remain in place despite our best efforts.

By process of elimination, it appears that a shift in emphasis to technologically based cleanup criteria is indicated, balanced of course by health-based considerations. This criteria could be expressed as reaching the upper asymptote of the cumulative removal curve along with stabilization of relative proportions of the contaminant mixture, for a well-designed remediation system. The stabilization must be at equilibrium with the maximum possible removal of aromatic hydrocarbons and other "lighter end" contaminants. Conservative health-based standards that reflect conservative science-based judgments could be considered as loosely applied goals. But if these goals are not met, then they drop back to criteria based on reasonable application of appropriate technology. This approach also recognizes that we will try to remove contaminants insofar as possible and hence addresses the issues of abuse of the risk assessment process and the cumulative effects consideration. Risk assessment could be used to support a decision to justify no significant effects of contaminants left in place after remediation. If a site clearly met conservative health-based criteria, no remediation would be required. Periodic monitoring over time could also be used in the sense of an empirical real-time risk analysis. Certain caveats, such as strict requirements for conservative health-based risk assessments for residential developments over contaminated sites, focusing on the vapor exposure pathway, should be included. Finally, health-based criteria for soil and upper groundwater could be strictly applied if they could be clearly defined and quantified at a specific site.

A problem with a practical remediation technology-based approach is the question of what to do if clearly acceptable cleanup levels cannot be reached. The only obvious answer is to adopt the strategy of cutting the losses and preventing the further spread of contamination by continuing extraction and treatment of groundwater or subsoil gas. Such corrective action could result in a long-term, commitment. The question of what constitutes a "reasonable" application of "appropriate" corrective measures also remains. However, the regulatory role here would be to require a professional state-of-the-practice demonstration that remedial actions that may spread contamination (e.g., bioremediation, steam injection) are effectively controlled and that environmental sampling or monitoring programs are required to establish effectiveness of corrective action. The regulatory agency would not need to specify the type of remediation other than to insist upon proven technologies, or controlled and properly permitted demonstrations of innovative technologies.

Best available technology (BAT) and best available control technology (BACT) have been long used for surface water discharges and air emissions. Such approaches appear to be recognized by well-known scientists[27,33,60] and the U.S. Environmental Protection Agency. In describing EPA's approach, Haley et al.[71]

included a scenario for treatment in which ". . . groundwater extraction continues until contaminant levels in the extracted water approach a constant value or asymptote (e.g., contaminant mass is no longer being removed at significant levels), at which point portions of the plume that remain above the cleanup levels are managed through containment and the use of institutional controls." However, a criteria driven by best practical corrective action would undoubtedly require a demonstration of completion that is much closer to the state-of-the-science than is common at present, with respect to environmental and chemical engineering, as well as hydrogeology. Keely[72] has distinguished between conventional, state-of-the-art and state-of-the-science site remediations.

A SIMPLIFIED METHOD FOR CHRONIC EXPOSURE ASSESSMENT IN INDOOR AIR

Changes of land use from service stations to residential and other commercial developments has led to concerns with long-term low level human exposure to benzene vapors. The following section attempts to address chronic exposure to vapors in indoor air in a manner accessible and useful to local agencies. Analysis of the chronic vapor exposure pathway is appropriate for sites where reasonable efforts have been made to remove contamination, but some residual benzene remains trapped in soil or dissolved in groundwater (or in pore water in the unsaturated zone), and occupied buildings are proposed. The analysis is restricted to benzene due to its carcinogenicity and low acceptable exposure levels. Only the single source incremental increase of benzene vapors due solely to contamination remaining beneath the site is considered. In other words, other sources such as ambient air background and diet are not considered. The evaluation is based solely on physical-chemical principles that estimate diffusive flux and assume equilibrium conditions. The pollutant transport equations were taken from Bastian and Johnson.[73]

An overview of the procedure is shown in Figure 2.1. The first step is to estimate the concentration of the contaminant in soil air either by direct measurement, from the mole fraction for liquid phase hydrocarbons in soil, or from the Henry's law constant for hydrocarbons dissolved in groundwater. The mole fraction of benzene in gasoline may be conservatively taken as the decimal fraction by assuming that the molecular weight of benzene is the same as that of gasoline. Alternatively, the mole fraction could be taken as the concentration of benzene divided by the total petroleum hydrocarbon concentration. The Henry's law constant could also be used for soil if TPH concentrations are low, by assuming all benzene is dissolved in soil pore water. It is important to understand that the actual concentration does not matter when calculating soil gas concentrations by mole fraction, as the soil gas is assumed to be saturated with benzene vapor in proportion to its mole fraction, and gasoline is assumed to be in the liquid state. Direct measurement is conceptually ideal but confounded by problems relating to the signal to noise ratio as the level of concern may be as low as 0.2 ppbv,

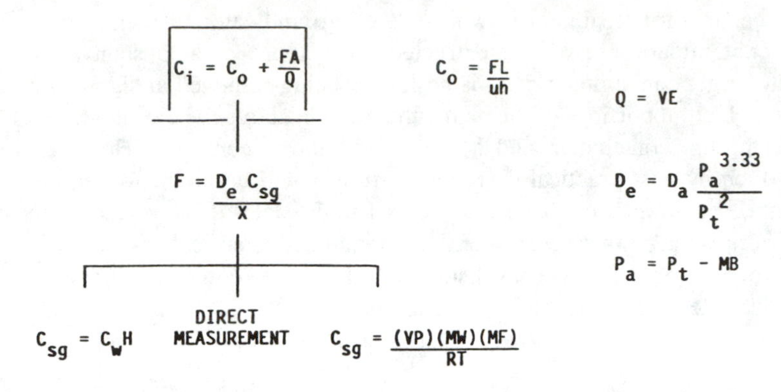

C_{sg} = Concentration in soil gas
C_w = Concentration in groundwater or soil pore water
C_i = Total concentration in indoor air
C_o = Concentration in outdoor air
VP = Vapor pressure
MW = Molecular weight
MF = Mole fraction
H = Henry's Law constant
D_e = Effective diffusion coefficient
D_a = Diffusion coefficient in air
P_a = Air filled porosity
R = Universal gas constant
T = Absolute temperature
P_t = Total porosity
B = Bulk density
M = Soil moisture
F = Flux
X = Distance between contamination and surface
A = Area through which flux occurs
Q = Ventilation rate
V = Volume of building
E = Exchange rate with outdoor air
L = Downwind length of contamination
u = Wind speed
h = Height of "box", e.g., height of doorway or ventilation system intake

Figure 2.1. A simplified vapor pathway evaluation.

and by temporal considerations, as the system may not have reached equilibrium. The next step is to calculate an effective diffusion coefficient based on estimates of porosity and soil moisture by the Millington-Quirk[74] equation. The effective diffusion coefficients and soil gas concentrations are used to calculate diffusive flux of benzene from the subsoils to the atmosphere. Finally, concentrations of benzene in outdoor and indoor air are estimated using box models that assume dynamic equilibrium.

Several assumptions are made. First, the model is for diffusive flux only. Pressure or convection driven flow is not considered. This assumption is probably reasonable for southern California, where most commercial and office buildings are under positive pressure and basements are not normally present. Second, benzene is present as a nondiminishing steady state source, a conservative assumption. Third, the system is at equilibrium. A related assumption is that exposure to benzene above equilibrium levels due to shutdown of the ventilation system is trivial in terms of lifetime exposure. Further, flux occurs only through infiltration areas such as cracks in the building slab and that flux through the slab itself is insignificant. Infiltration ratios, termed "specific leakage area," to estimate the effective floor infiltration area, were taken from Grimsrud et al.[75]

Table 2.2 gives input parameters for three different scenarios and concentrations of benzene in indoor air, calculated according to the scheme given in Figure 2.1. Scenario 1 is a worst case, while Scenarios 2 and 3 are perhaps more probable. In Table 2.3 the excess lifetime risk cancer risk was calculated by comparison to the California Department of Health Services health-based Applied Action Level for benzene in air (0.66 $\mu g/m^3$ at the 10^{-5} excess cancer risk level); by use of the EPA carcinogenic potency for benzene of 0.029 $(mg/kg\text{-}d)^{-1}$; and a time-weighted concentration comparison to the 20 $\mu g/d$ limit specified in the California Safe Drinking Water and Toxic Enforcement Act (CSDWTE) of 1986. Inhalation volume used was 23 m^3/d for the EPA carcinogenic potency factor and 20 m^3/d for the CSDWTE. Three different exposure conditions are given in Table 2.3: a total lifetime, and 45-year and 20-year working periods.

Inspection of Table 2.3 suggests that, for the situations modeled, additional cancer risk does not exceed the 1×10^{-5} level in either of more probable cases, or in the 20 year occupational exposure scenario when only flux through the foundation is considered. However, the 70 year lifetime exposure exceeds the 10^{-5} level for the worst case scenario and the 10^{-6} level for one of the more probable scenarios. (The 10^{-5} level is specified as the no significant risk level in the CSDWTE, while the 10^{-6} level is often used in environmental regulation.)

The results of this exposure modeling suggest that quantitative health-based risk assessments should be performed when residential construction is proposed over sites having subsurface hydrocarbon contamination. However, for commercial or office building as usually constructed in southern California, such risk assessments may be unnecessary. In the most common situation, relatively small structures are constructed over former service station sites. Construction is often single story, no basements, and areas outside the building are generally paved. In addition, the 20-year working period probably exceeds the average. Thus, a reasonable worst case for commercial development could be considered as Scenario 2 with the benzene flux through the foundation only, as permeation through the outside paving is expected to be small. It should be noted that Scenarios 1 and 2 may be considered to be conservative in that a relatively high mole fraction and small distance between the source of the contamination and the ground surface was assumed. However, use of a 70-kg body weight for adult males was less conservative, but in accordance with standard practice. Areas where basements are usually constructed must consider pressure-driven flow in addition to the diffusive flux considered herein. Consideration of pressure-driven flow may also appropriate for developments with subterranean parking structures.

This assessment of relative risk for commercial construction also seems reasonable on a relative risk basis. Shah and Singh[76] reported an average benzene

Table 2.2. Input Values for Benzene Vapor Diffusion Scenarios

Parameters	Scenario 1	Scenario 2	Scenario 3
Mole fraction	0.03	0.01	0.001
Vapor pressure (atm)	0.1	0.1	0.1
Molecular weight (mg/mole)	78,110	78,110	78,110
Universal gas constant (m^3-atm/mole °K)	8.2×10^{-5}	8.2×10^{-5}	8.2×10^{-5}
Temperature (°K)	293	293	293
Csg (mg/m^3)	9.6×10^3	3.2×10^3	3.2×10^2
Porosity total	0.40	0.40	0.35
Bulk density (g/cm^3)	1.3	1.5	1.6
Moisture	0.1	0.15	0.15
Porosity, air filled	0.27	0.18	0.11
Vapor diffusion (coefficient in air [m^2/h])	0.03	0.03	0.03
Vapor diffusion coefficient effective (m^2/h)	2.4×10^{-3}	6.2×10^{-4}	1.6×10^{-4}
Distance from source to surface (m)	3.05 (10 ft)	3.6 (12 ft)	4.6 (15 ft)
Flux (mg/m^2h)	7.6	5.5×10^{-1}	1.1×10^{-2}
Downwind length of site (m)	9.1 (30 ft)	10.7 (35 ft)	12.2 (40 ft)
Wind speed (m/h)	3,219 (2 mph)	8,047 (5 mph)	8,047 (5 mph)
Height of box (m)	1.8 (6 ft)	2.1 (7 ft)	2.4 (8 ft)
Ventilation exchange rate (exchanges/h)	0.5	0.5	1.0
Building foundation Area (m^2	232.6 (2,500 ft^2)	232.6 (2,500 ft^2)	232.6 (2,500 ft^2)
Height of building (m)	2.1 (7 ft)	2.5 (8 ft)	2.7 (9 ft)
Infiltration ratio (m^2/m^2)	1×10^{-3}	5×10^{-4}	2×10^{-4}
Building volume (m^3)	4.9×10^2	5.8×10^2	6.2×10^2
Effective flux area (m^2)	2.3×10^{-1}	1.2×10^{-1}	4.6×10^{-2}
C_o (mg/m^3) (ppbv)	1.2×10^{-2} (4)	3.5×10^{-4} (0.11)	6.7×10^{-6} (0.002)
FA/Q (mg/m^3) (ppbv)	6.9×10^{-3} (2)	2.3×10^{-4} (0.07)	8.2×10^{-7} (0.003)
Ci (mg/m^3) (ppbv)	1.9×10^{-2} (6)	5.8×10^{-4} (0.18)	7.5×10^{-6} (0.002)

concentration of 2.800 ppbv and a median of 1.665 ppbv for outdoor air from the EPA database of 122,820 records. They also reported an average benzene value of 5.162 ppbv and a median of 3.135 ppbv from 58,810 records from nonindustrial indoor air spaces. In addition, the South Coast Air Quality Management District[77] reported benzene values of 0 to 7 ppbv for various locations in southern California. Thus, background levels greatly exceed incremental increases caused by subsurface benzene contamination in the occupational scenarios.

It is always prudent to place a barrier beneath any building constructed over a contaminated site, but regulatory agencies should understand that installation

Table 2.3. Individual Excess Cancer Risk for Benzene Vapor Diffusion Scenarios

Conditions	Scenario 1	Scenario 2	Scenario 3
1. Total Ci—lifetime			
CSMDT[a]	3.0×10^{-4}	8.8×10^{-6}	1.1×10^{-7}
EPA[b]	1.8×10^{-4}	5.5×10^{-6}	7.1×10^{-8}
CSDWTE[c]	380	11.6	0.15
2. Flux through foundation (FA/Q) only			
CSMDT	1.0×10^{-4}	3.5×10^{-6}	1.2×10^{-8}
EPA	6.6×10^{-5}	2.2×10^{-6}	7.8×10^{-9}
CSDWTE	138	4.6	0.02
3. Total Ci—45 yr[d] working lifetime			
CSMDT	8.1×10^{-5}	2.3×10^{-6}	3.0×10^{-8}
EPA	4.9×10^{-5}	1.5×10^{-6}	1.9×10^{-8}
CSDWTE	103	3.1	0.04
4. Flux through foundation (FA/Q) only—45 yr working life			
CSMDT	2.7×10^{-5}	9.4×10^{-7}	3.2×10^{-9}
EPA	1.8×10^{-5}	5.9×10^{-7}	2.1×10^{-9}
CSDWTE	37	1.2	0.005
5. Total Ci—20 yr working life[e]			
CSMDT	2.1×10^{-5}	6.2×10^{-7}	7.7×10^{-9}
EPA	1.3×10^{-5}	3.8×10^{-7}	5.0×10^{-9}
CSDWTE	26	0.8	0.01
6. Flux through foundation (FA/Q) only—20 yr working life			
CSMDT	7.0×10^{-6}	2.4×10^{-7}	8.4×10^{-10}
EPA	4.6×10^{-6}	1.5×10^{-7}	5.5×10^{-10}
CSDWTE	9.7	0.3	0.001

[a]California Site Mitigation Decision Tree Applied Action Level for benzene (0.66 $\mu g/m^3$ at the 10^{-5} excess lifetime cancer risk level).
[b]EPA IRIS Carcinogenic Potency Factor of 0.029 $(mg/kg/d)^{-1}$.
[c]California Safe Drinking Water and Toxic Enforcement Act of 1986—values in the table are benzene concentration ($\mu g/m^3$) multiplied by 20 m^3/d for comparison to the acceptable daily intake of 20 $\mu g/d$.
[d]45 years, 6 days per week, 12 hours per day, or 27% of a 70 year lifetime.
[e]20 years, 5 days per week, 8 hours per day, or 7% of a 70 year lifetime.

of polyethylene flexible membrane liners as a remediation measure for vapor transport may be ineffective. In a closed system liner permeation experiment, Haxo[78] found no difference in toluene or o-xylene vapor concentrations above and below a flexible membrane liner. The source of the vapor was a dilute aqueous solution. The most effective barrier to diffusive flux, perhaps the only completely effective barrier, is an air gap that allows free circulation with outdoor air.

RESIDENTIAL DEVELOPMENT OVER HYDROCARBON CONTAMINATED SITES

The simplified analysis above should be modified for residential developments, even as a first approximation. The analysis is highly sensitive to the value of the area in the term FA/Q (see Figure 2.1). Area is taken as a ratio of infiltration area to total area from empirical data given in Grimsrud et al.[75] These ratios were taken to represent the ratio of the cracks or other infiltration zones in the foundation to the total foundation area. This line of reasoning assumes that the uncracked building slab acts as at least a partially effective barrier to benzene vapor. However, Zapalac[79] reported effective diffusion coefficients for concrete of the same order as calculated for soil in this study. Overall, the entire field of indoor air exposure is relatively new and any prediction of low level concentration is risky. Empirical verification of theory is needed with respect to vapor phase partitioning from volatile contaminants, vapor transport through porous media, and movement into and within buildings.

The exposure assessments for residential developments should differ from those for commercial structures in several important ways. First, a continuous exposure duration of 70 years should be assumed. Secondly, the area through which the flux occurs should be taken as the entire floor surface area. The slab could either be ignored, or an effective diffusion coefficient for concrete could be used and the concrete considered as a separate layer. Third, the assumption that there is no pressure-driven flow into the building should be examined more critically. Fourth, exposure to sensitive groups such as children should be given special consideration. All of the above could be considered as a first approximation to indicate the need for more sophisticated analyses. Bearing in mind the difficulty of detecting all contamination at a given site, the uncertainty of theoretical analysis and the probability that it will be unconvincing to the general public in any case, together with time-dependent and signal-to-noise ratio and difficulties inherent in direct measurement of low level contaminants with variable higher level ambient backgrounds (in the case of benzene, looking for tenths of ppb increments in perhaps tens of ppb background), the best alternative may be to avoid residential construction over previously contaminated sites. If residential development is nonetheless deemed to be the best use of contaminated sites, state-of-the-art-and-science risk assessments should be developed, perhaps supported by directed scientific research.

GUIDELINES FOR CLEANUP CRITERIA

In broad terms, the purpose of cleanup criteria for underground tank leaks is to protect exploitable groundwater from degradation, and to protect against excess cancer risk or other health hazards. Human health protection includes evaluation and minimization of long-term exposure to benzene vapors from contaminated soil and groundwater. The importance of thorough and accurate characterizations of contamination and site features cannot be overemphasized. Each site is unique to some degree. Nonetheless, some general guidelines for developing cleanup criteria may be stated on the basis of experience and erudition.

Numerical Cleanup Levels

Although simple generic numerical cleanup levels need not be used at all, they will probably continue to be used for risk management purposes due to the desire to provide definite direction for the large number of sites and limited resources available for investigation of sites. Simple generic numerical cleanup levels for soil should be applied only to the bottom of excavation or trench samples and not throughout the soil column (unless using the LUFT Manual General Risk Appraisal or similar methodology specifically designed to evaluate contaminant concentrations throughout the soil column). Numerical standards from the bottom of excavation samples should be used as screening level values only, i.e., concentrations of contaminants that indicate the need to conduct further sampling and more detailed site investigation.

Although numerical levels are most commonly applied to soil, they may also be applied to uppermost groundwater, especially if it is perched or located in a zone of low hydraulic conductivity. In fact, groundwater samples from a properly constructed and developed monitoring well should be much more representative than subsoil samples, and sampling and analytical methodologies should give more accurate results for groundwater than for subsoils.

The following are suggested as guidelines for generic numerical cleanup levels:

1. Consider the numbers to be flexible screening levels or cleanup goals. Numerical levels may be triggering levels to conduct further site assessment, initiate the need for treatment or monitoring, or to permit no action or closure on a practical risk management basis.
2. Include several categories or range of situations, such as low, medium, and high sensitivity conditions.
3. Start with a health-based number or accepted regulatory action level to increase defensibility.
4. Include both TPH and BTEX for soil. TPH is relatively cheap and can serve as a check in case more volatile constituents are lost in sampling or analysis.
5. Don't include too many factors or the system may be complicated beyond comprehensibility.
6. Develop the numbers conservatively and apply them strictly, as they are only screening levels that indicate the need for further investigation. In Orange County, for example, extensive contamination was found throughout the subsoil at several sites where the decision to conduct soil core sampling was based only on a few parts per million of xylene found in bottom of excavation samples.

Quantitative Site-Specific Health Risk Assessments

State-of-the-art-and-science risk assessments may be fine for large sites, but limited human resources appear to preclude their routine use for small sites unless standardized simplified methodologies are developed. Nonetheless, risk assessments are an ideal approach and necessary for certain situations; for example, where residential development is proposed over hydrocarbon contaminated sites.

1. Use for special situations, such as to determine excess lifetime cancer risk via the air exposure pathway for residential developments, day care centers, and other high sensitivity situations.
2. Use for relatively simple direct source to receptor analysis and relatively simple hydrogeology (e.g., contaminated zone to nearby well, rather than contaminated zone through several geologic strata to aquifer to well to water distribution system to human).
3. Use for situations where remedial action is difficult, such as sites where contamination has spread beneath existing buildings.
4. Develop adequate private and public sector human resources to develop and review the assessments. Public agencies should be prepared to develop staff expertise and conduct a thorough review of overall modeling concepts, assumptions, and input parameters.

Remediation Criteria

The American Petroleum Institute[10] has forwarded cleanup goals and criteria for the four hydrocarbon phases and Keely[72] has made the distinction between conventional, state-of-the-art and state-of-the-science remediation based on corresponding characterization scenarios. It may also be possible to better define the effects of biodegradation both qualitatively and quantitatively as criteria for passive remediation in the future. At present, the following may be reasonable technology-based criteria:

1. Goals are exposure minimization by maximum practical removal of contaminants and prevention of further spread of contamination.
2. The end point could be considered to be remediation to the upper asymptote of the cumulative hydrocarbon removal curve, along with constancy of composition of the hydrocarbon mixture and maximum loss of the more volatile and soluble "lower ends." This approach attempts to find the point of diminishing returns where continued operation results in a negligible decline in the concentration with stability in the relative proportions of contaminant constituents. It may be possible to test the system by shutting down and restarting, and operating it at various rates to check changes in concentrations and composition, or by confirming that the concentration times the flow rate is a constant over a range of values.

3. Include post-remediation monitoring or sampling and evaluation to determine if expansion of the system, or use of a different system, is indicated.
4. Recognize that remediation and monitoring may be necessary for an indefinite, long period of time.

AN ILLUSTRATION OF A REGULATORY APPROACH TO PETROLEUM HYDROCARBON CONTAMINATION FROM UNDERGROUND STORAGE TANKS

The following is provided to illustrate one possible scheme to integrate health and technology-based criteria. It is given as an illustrative example only and is focused on southern California.

Figures 2.2 through 2.5 illustrate an operational approach to regulatory site management. The first step (Figure 2.2) should be to take any indicated immediate corrective action and identify potential exposure pathways. The direct contact exposure route is not considered further as it is of limited importance in urban areas and is amenable to direct toxicological and exposure analysis.

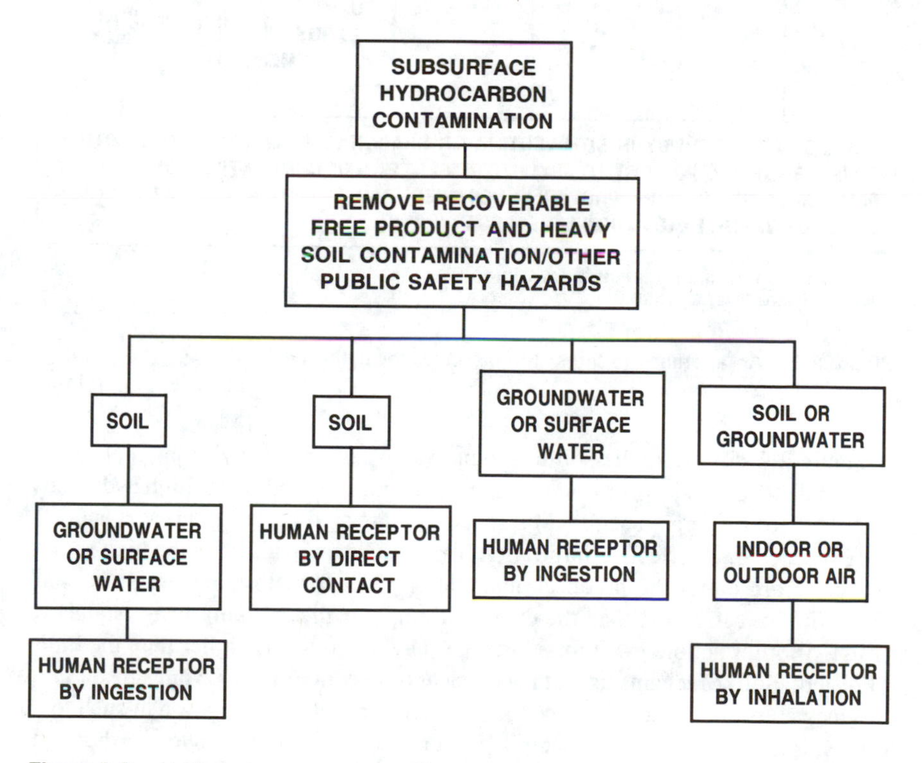

Figure 2.2. Hydrocarbon exposure pathways.

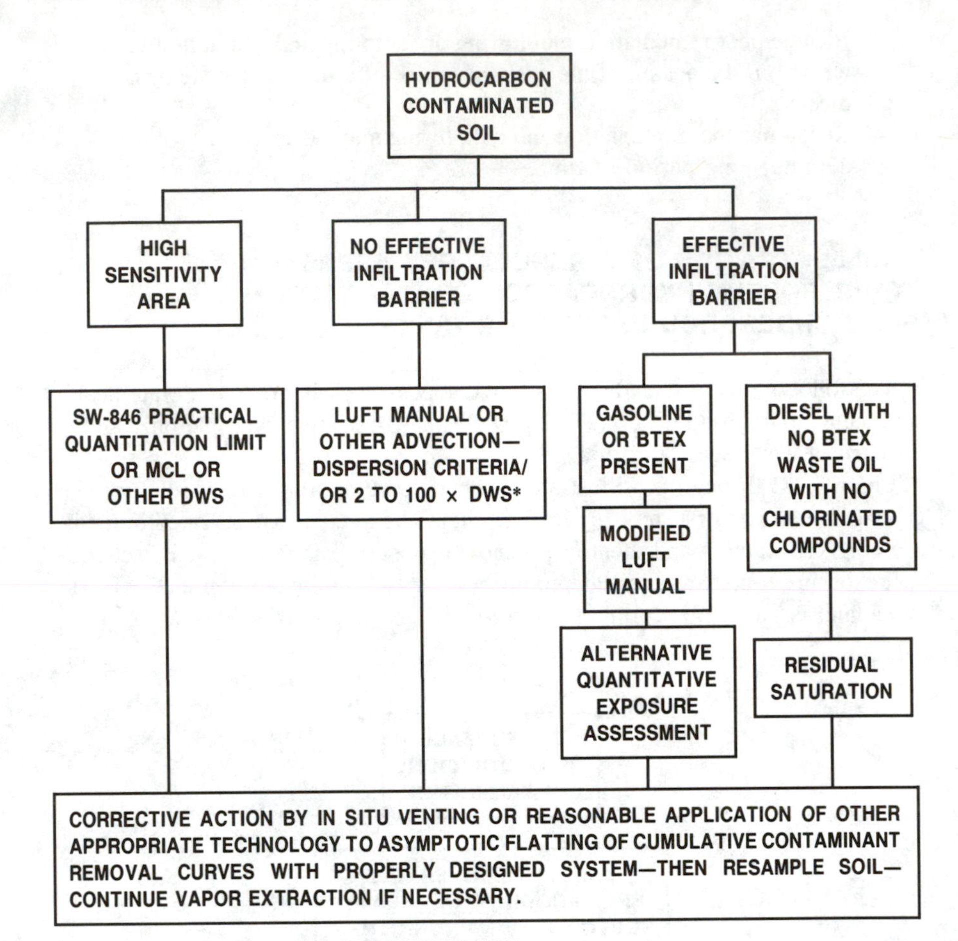

Figure 2.3. An approach to the soil to groundwater pathway.

Figure 2.3 considers the soil to groundwater pathway. Environmental sensitivity is divided into low, medium, and high categories based on (1) high hydrogeologic sensitivity (e.g., specified or specially managed recharge areas), and (2) the existence of an effective infiltration barrier. Cleanup goals for the high sensitivity site are either the practical quantitation limit for substances such as benzene with an MCL less than the detection limit, or the drinking water standard (MCL or other action level) for substances having an MCL higher than the limit of quantitation. Mechanisms that may concentrate contaminants from subsoil into groundwater are considered to be too unusual to merit consideration at such low concentrations. Further, the natural background of petroleum hydrocarbons in the subsoil at a depth of 10 to 15 feet is assumed to be negligible.

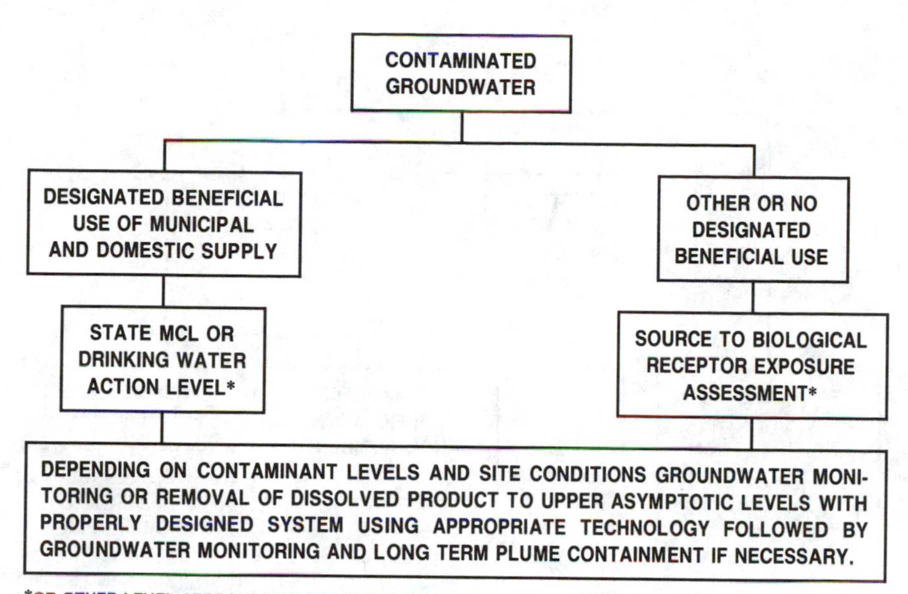

Figure 2.4. An approach to the groundwater pathway for California.

Sites with no effective infiltration barrier may use the LUFT manual. Other advection-dispersion analyses could be used, as well as a constant multiplier approach. The use of broadly applicable numerical levels is a risk management, not a risk assessment decision. It is important to understand that if simplified numerical levels are used, that they be used only for screening levels taken from the bottom of excavation samples. In this regard, a distinction must be made between the Leaching Potential Analysis part of the LUFT Manual, which is based on the bottom of excavation samples, and LUFT General Risk Appraisal, which uses state-of-the-art mathematical models to conservatively generate cleanup levels throughout the soil column.

The large range in multiplier values is due to the great range in the California MCLs, from benzene at 1 ppb to xylene at 1,750 ppb. A low multiplier is needed to keep xylene from getting too high and a high multiplier is needed so that benzene concentrations will make a real difference. Ideally a 5 to 10 foot "clean zone" should exist between the contaminated soil and groundwater.

Sites with effective infiltration barriers are further divided into sites where aromatic hydrocarbons are found or suspected and those where none are found. A constant multiplier or residual saturation approach is applied as appropriate. The criteria for an effective infiltration barrier is essentially a risk management decision. Standard paving does result in less infiltration over an urban watershed but, at any given site, cracks and depressions in paved surfaces could result in more of a "funnel effect" than a barrier. The conservative position is to require an

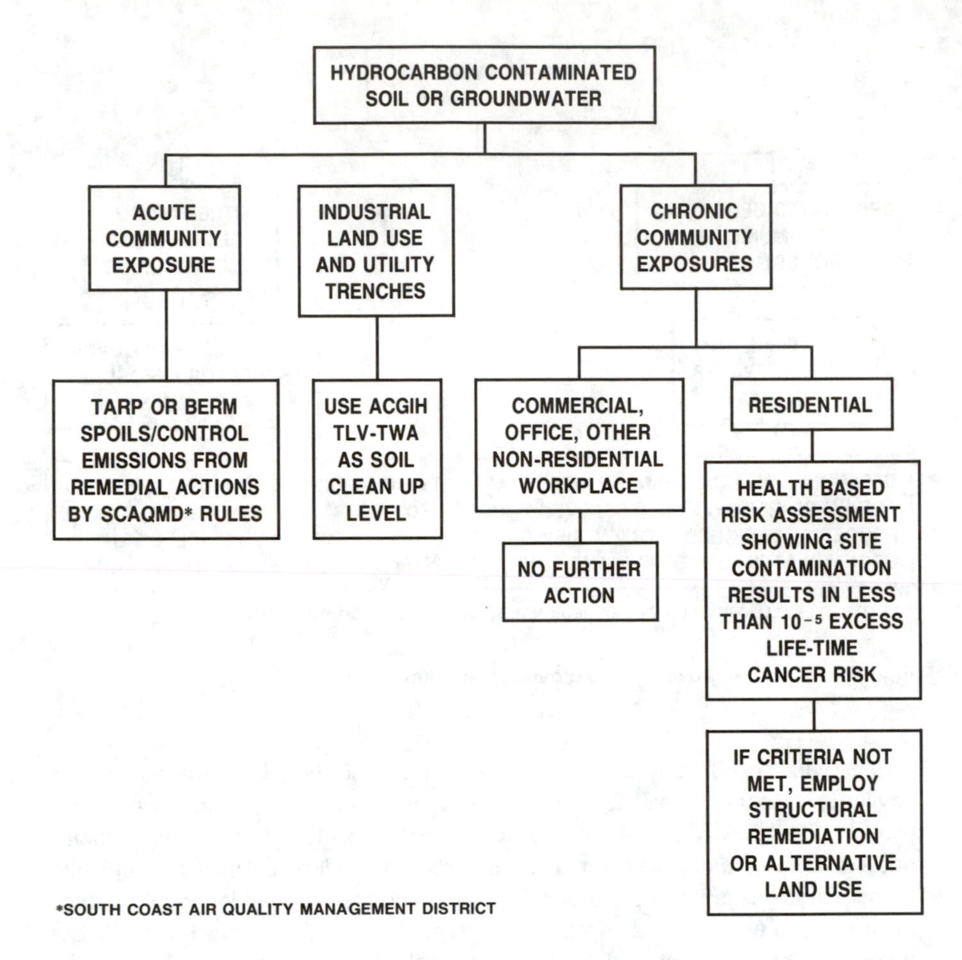

Figure 2.5. An approach to the soil/groundwater to air pathway for urban southern California.

engineered barrier and avoid locating water utility lines over contaminated areas. The modification to the LUFT manual for an effective infiltration barrier is an adjustment to the General Risk Appraisal tables that reduces the average annual rainfall by some factor. A reasonable reduction may be on the order of 50%. Thus the rainfall value could be reduced from 12 to 16 inches, to 6 to 8 inches for a site with an effective rainfall infiltration barrier in coastal southern California.

The essential point in each situation is, if these standards are not met, then in all cases proceed with application of corrective action and use technology-based criteria. A post-remediation site assessment should be performed to document remaining contamination and effectiveness of remediation. A reasonable application of appropriate technology may be considered to be more intensive in areas of high sensitivity.

Figure 2.4 shows an approach to evaluating existing groundwater contamination based on the California regulatory perspectives. The source to receptor assessment could be for nonhuman, as well as human receptors.

The soil, or groundwater, to air pathway is addressed in Figure 2.5. The criteria for industrial cleanups, designed for industrial occupational exposures, again postulates no concentrating mechanism. The more difficult problem of chronic exposures assumes that chronic effects of remediation and tank removals are not significant (acute exposures require immediate corrective action) based on the work of Johnson et al.[7], and that nonresidential exposure would be limited to a portion of total lifetime. In the case of chronic exposures, the 10^{-5} excess cancer risk level is selected because that level is specified as the no-significant risk level in the California Safe Drinking Water and Toxic Enforcement Act of 1986. In the case of residential exposures, a conservative lifetime quantitative risk assessment is suggested. If the health-based criteria are not strictly met, then structural remediation or an alternative land use is indicated.

CONCLUSIONS

Hydrocarbon releases to soil and groundwater are problematic because the effects of those releases are not yet known with certainty. This is especially true if the hydrocarbon contains the environmentally mobile carcinogen benzene, and is generally true with regard to the cumulative effects of many small and widely distributed hydrocarbon contaminated sites. It is not yet clear whether hydrocarbon contamination of soils and groundwater, especially from underground tanks storing refined petroleum products that contain benzene, is an overstated problem, or one that will have serious consequences. On one hand, analysis based on physical chemistry and advective-dispersive-diffusive transport indicates a cause for concern with respect to both the soil to groundwater to water supply exposure pathway, and the soil to air pathway, especially in indoor air if dwelling units are built over a contaminated site. On the other hand, relatively few drinking water supplies have been degraded due to underground storage tank and piping leaks. An important, yet unanswered question is where the substantial amount of benzene that has presumably leaked from underground tanks is now located. Both theoretical and empirical evidence suggests that bacterial biodegradation may act to hold small releases effectively in check and eventually transform hydrocarbons. However, the development of science-based mechanistic models to predict hydrocarbon fate and transport is complicated by micro and macro scale subsurface heterogeneities, the complexity of hydrocarbon mixtures, and a variety of difficult to estimate coefficients and other model input parameters, including biotransformation rates. These circumstances make the search for health-based soil cleanup criteria an uncertain and controversial process. The situation is similar for shallow groundwater in low permeability formations.

Even when health-based criteria are clear, as in drinking water aquifers, attainment of health-based standards is not always possible if liquid phase hydrocarbon is trapped in the subsurface, or if dissolved hydrocarbons are widely dispersed. The search for the single "correct" numerical cleanup level for soil and much of the groundwater is probably misguided from an assessment perspective. Although simple numerical concentration values can be considered goals or conservative screening levels, they are essentially risk management tools and should be considered to be flexible if they are used at all. Their absolute value is often less important than how they are used. Furthermore, there is no compelling need to set numerical cleanup levels prior to remediation.

At the operational level, a shift of emphasis away from the search for the "correct" health-based numerical criteria to a strategy of risk minimization by the rapid and reasonable application of corrective action technology seems appropriate, at least for smaller sites such as service stations. Cleanup criteria would then shift from the attainment of some acceptable no-significant-risk level in soil or uppermost groundwater zones, to a remediation technology-based criteria. The emphasis would shift from trying to define the correct soil or groundwater cleanup level to better defining reasonable application of "appropriate remedial technology." This approach recognizes that the state of our science and the state of public risk perception is not yet advanced to the point where we can achieve consensus on cleanup levels or on a standardized site evaluation process for the thousands of hydrocarbon contaminated sites. This approach also recognizes that some sites cannot reasonably be cleaned up to health-based standards, and that source reduction or containment is the most important element of exposure minimization. Emphasis should be placed on removing nonaqueous phase residual hydrocarbons, or failing that, on long-term remedies to prevent the spread of contamination. Post-remediation monitoring and sampling should also be included before the system is disassembled to determine if the remedial actions were effective.

In this chapter, the conservative health risk assessment process was used to estimate individual excess lifetime cancer risk due to long-term benzene vapor exposure by using a simplified generic diffusive transport model. I also tried to point out that unsaturated zone models for multiphase complex mixtures, such as petroleum hydrocarbons, do not yet predict contaminant fate with the confidence and detail useful for operational level evaluation of small sites. This makes risk assessment especially difficult for the vadose zone to groundwater to drinking water supply exposure pathway. The larger question for risk assessment, pointed out by Maxim[80] and others, is the difficulty of obtaining "reality checks" on low probability human health risk computations, and the dangers of misallocation of regulatory attention to trivial risks. Problems with high to low dose extrapolations and upper bound dose response assessments, as well as using a series of conservative assumptions for exposure assessments, are well known. Paustenbach[81] has pointed out that use of biological based cancer models and PB/PK (physiologically based pharmacokinetic) or disposition models, may result in more realistic risk assessments. Use of more realistic exposure scenarios and "plausibility audits"[80] would also make risk assessment a more useful regulatory tool.

In summary, the best apparent regulatory perspective seems to be a balance between health-based and technological-based criteria; that is, between the reasonable application of appropriate remedial technologies and public health evaluation. Health-based criteria should be considered as a goal or conservative level at which no corrective action is needed. If health-based criteria cannot be reached, then the emphasis should be shifted to achieving a reasonable reduction of contamination. However, in many, perhaps most cases, the public health impacts of subsoil and groundwater contamination at leaking underground storage tank sites cannot be accurately or precisely defined. Documents such as the LUFT Manual could be used as evaluative tools. Health-based risk assessments and environmental modeling can be used for certain situations, such as for estimating excess lifetime cancer risk for residential exposures, or used more generally if appropriate contaminant fate models are developed and validated.

At sites where reasonable remediation efforts will not result in reduction of contamination to established levels and risk assessment cannot demonstrate that the contamination is not significant, the apparent choices for further action are: (1) to contain and monitor contaminants until established numerical criteria are met; (2) restrict uses of the site by institutional controls; (3) close the site because application of appropriate remedial technology has been performed to or past the point of diminishing returns; or (4) close the site, based on a body of empirical evidence from monitoring programs or special studies. Dissolved phase contaminant plumes could be contained by pump-and-treat methods or slurry walls. Vapor migration could be controlled by in situ extraction. Groundwater or vapor monitoring networks could be established around unsaturated zone contamination to see if groundwater will be affected or if vapor migrates. Adequate groundwater monitoring should include establishing gradients, monitoring changes in gradients, locating wells up and down gradient inside and outside of contaminant plumes, and at least quarterly sampling for a minimum of one to two years.

Regulatory criteria could be modified if better methods can be developed to better predict, both qualitatively and quantitatively, the effects of natural containment mechanisms, such as biodegradation on a site-specific basis. Advances in understanding of dissolution and volatilization of liquid phase hydrocarbons (NAPL) in the subsurface may also have regulatory implications. Criteria could also be modified by advances in toxicological assessment and risk characterization. However, hydrocarbon contamination from underground storage tanks will most likely remain a regulatory problem for some time because scientific uncertainty and the large number of contaminated sites makes it difficult to put the problem in proper perspective without the development of a greater body of empirical knowledge. While management of petroleum hydrocarbons is more tractable than for chlorinated hydrocarbons, the public health concern will remain as long as environmentally mobile carcinogens, such as benzene, are present in refined petroleum products.

ADDENDUM

The vapor pathway evaluation presented in Table 2.1 is quite sensitive to soil moisture. Total porosity and water content determines the air-filled porosity, which is a determinant of the effective diffusion coefficient. The soil moisture values given in Table 2.2 were estimated from reported moisture content of soils disposed of at landfills and from soil science texts. In a recent paper, Jury et al.[82] used "standard values" of soil properties to simulate behavior of volatile organic chemicals in the subsurface. The standard values used by Jury et al. for water content were 0.18 for sandy soil and 0.375 for clayey soil. If sandy soil is assumed, a slightly lower air-filled porosity was used for this study (0.18 vs 0.22). An increase of 0.04 in air content results in doubling of the benzene flux and the excess lifetime cancer risk. Nonetheless, only the total lifetime exposure risk of Scenario 2 changed from less than to greater than the 1 in 100,000 excess cancer risk level. Therefore, implications regarding the needs for vapor risk assessments for commercial and residential developments are unchanged even if the Jury et al. air content value for sandy soils is used.

REFERENCES

1. Kostecki, P. T., E. J. Calabrese, and E. Garnick. "Regulatory Policies for Petroleum-Contaminated Soils: How States Have Traditionally Dealt with the Problem," in *Soils Contaminated by Petroleum: Environmental and Public Health Effects,* E. J. Calabrese and P. T. Kostecki, Eds. (New York: John Wiley & Sons, Inc., 1988), pp. 415–433.
2. Bell, C. E., P. T. Kostecki, and E. J. Calabrese. "State of Research and Regulatory Approach for Cleanup of Petroleum Contaminated Soils," in *Petroleum Contaminated Soils,* Volume 2, E. J. Calabrese and P. T. Kostecki, Eds. (Chelsea, MI: Lewis Publishers, Inc., 1989), pp. 73–94.
3. "Risk Assessment Guidance for Superfund, Volume 1, Human Health Evaluation Manual," U.S. Environmental Protection Agency, EPA/540/1-89/002, (1989).
4. "Superfund Exposure Assessment Manual," U.S. Environmental Protection Agency, EPA/540/1-88/001 (1988).
5. "The California Site Mitigation Decision Tree Manual," California Department of Health Services (1986).
6. "Leaking Underground Fuel Tank Field Manual: Guidelines for Site Assessment, Cleanup and Underground Storage Tank Closure," California State Water Resources Control Board (1989).
7. Johnson, P. C., M. B. Hertz, and D. L. Byers. "Estimates for Hydrocarbon Vapor Emissions Resulting from Service Station Remediations and Buried Gasoline-Contaminated Soils," in *Petroleum Contaminated Soils,* Volume 3, (Chelsea, MI: Lewis Publishers, Inc., 1990).
8. Potter, T. L. "Analysis of Petroleum Contaminated Soil and Water: An Overview," in *Petroleum Contaminated Soils,* Volume 2, E. J. Calabrese and P. T. Kostecki, Eds. (Chelsea, MI: Lewis Publishers, Inc., 1989), pp. 97–109.

9. Mackay, D. ''The Chemistry and Modeling of Soil Contamination with Petroleum,'' in *Soils Contaminated by Petroleum: Environmental and Public Health Effects,* E. J. Calabrese and P. T. Kostecki, Eds. (New York: John Wiley & Sons, Inc., 1989), pp. 5–17.
10. ''A Guide to the Assessment and Remediation of Underground Petroleum Releases,'' American Petroleum Institute Publ. 1628, Washington, D.C. (1989).
11. Freeze, R. A., and J. A. Cherry. *Groundwater* (Englewood Cliffs, NJ: Prentice-Hall, Inc., 1979).
12. Mackay, D. M., P. V. Roberts, and J. A. Cherry. ''Transport of Organic Contaminants in Groundwater,'' *Environ. Sci. Technol.* 19(5): 384–392 (1985).
13. Dragun, J. *The Soil Chemistry of Hazardous Materials* (Silver Spring, MD: Hazardous Materials Control Institute, 1988).
14. ''IARC Monographs on the Evaluation of Carcinogenic Risks of Chemicals to Humans: Some Industrial Chemicals and Dyestuffs,'' Vol. 29. WHO International Agency for Research on Cancer (1982).
15. ''Documentation of Applied Action Level for Benzene,'' California Department of Health Services (1986).
16. Calabrese, E. J., P. T. Kostecki, and D. A. Leonard. ''Public Health Implications of Soils Contaminated with Petroleum Products,'' in *Soils Contaminated by Petroleum: Environmental and Public Health Effects,* E. J. Calabrese and P. T. Kostecki, Eds. (New York: John Wiley & Sons, Inc., 1988), pp. 191–229.
17. Bauman, B. J. ''Soils Contaminated by Motor Fuels: Research Activities and Perspectives of the American Petroleum Institute:, in *Petroleum Contaminated Soils,* Volume 1, P. T. Kostecki and E. J. Calabrese, Eds. (Chelsea, MI: Lewis Publishers, Inc., 1989), pp. 3–19.
18. Bauman, B. ''Current Issues in Management of Motor Fuel Contaminated Soils,'' in *Petroleum Contaminated Soils,* Volume 2, E. J. Calabrese and P. T. Kostecki, Eds. (Chelsea, MI: Lewis Publishers, Inc., 1989) pp. 31–42.
19. Slovic, P. ''Perception of Risk,'' *Science* 236: 280–285 (1987).
20. Cohrssen, J. J., and V. T. Covello. ''Risk analysis: A Guide to Principles and Methods for Analyzing Health and Environmental Risks,'' Council of Environmental Quality, NTIS (1989).
21. ''Literature Survey: Hydrocarbon Solubilities and Attenuation Mechanisms,'' American Petroleum Institute Publ. 4414 (1985).
22. ''Literature Survey: Unassisted Natural Mechanisms to Reduce Concentrations of Soluble Gasoline Components,'' American Petroleum Institute Publ. 4415 (1985).
23. Bonazountas, M. ''Mathematical Pollutant Fate Modeling of Petroleum Products in Soil Systems,'' in *Soils Contaminated by Petroleum: Environmental and Public Health Effects,* E. J. Calabrese and P. T. Kostecki, Eds. (New York: John Wiley & Sons, Inc., 1988), pp. 31–111.
24. McKee, J. E., F. B. Laverty, and R. M. Hertel. ''Gasoline in Groundwater,'' *Jour. Water Pollut. Control Fed.* 44(2):293–302 (1972).
25. Hoag, G. E., and M. C. Marley. ''Gasoline Residual Saturation in Unsaturated Uniform Aquifer Materials,'' *J. Environ. Eng.* 112(3):586–604 (1986).
26. Hoag, G. E. Personal communication, 1990.
27. Mackay, D. M., and J. A. Cherry. ''Groundwater Contamination: Pump-and-Treat Remediation,'' *Environ. Sci. Technol.* 23(6):630–636 (1989).

28. Hunt, J. R., N. Sitar, and K. S. Udell. ''Nonaqueous Phase Liquid Transport and Cleanup, 1. Analysis of Mechanisms,'' *Water Resour. Res.* 24(8):1247–1258 (1988).
29. Javandel, I., R. Falta, and H. Holman. ''Hydrocarbon Contaminants in Shallow Soil Systems,'' Second Berkeley Symposium on Topics in Petroleum Engineering, Berkeley, CA, 1988, pp. 65–68.
30. Baehr, A. L. ''Selective Transport of Hydrocarbons in the Unsaturated Zone Due to Aqueous and Vapor Phase Partitioning,'' *Water Resour. Res.* 23(10):1926–1938 (1987).
31. Falta, R. W., I. Javandel, K. Pruess, and P. A. Witherspoon. ''Density-Driven Flow of Gas in the Unsaturated Zone Due to the Evaporation of Volatile Organic Compounds,'' *Water Resour. Res.* 25(10):2159–2169 (1989).
32. Roy, W. R., and R. A. Griffin. ''Mobility of Organic Solvents in Water-Saturated Soil Materials,'' *Environ. Geol. Water Sci.* 7(4):241–247 (1985).
33. Freeze, R. A., and J. A. Cherry. ''What Has Gone Wrong,'' *Ground Water* 27(4):458–464 (1989).
34. ''Ground Water Handbook,'' U.S. Environmental Protection Agency, EPA/625/6-87/016 (1987).
35. Borden, R. C., and P. B. Bedient. ''Transport of Dissolved Hydrocarbon Influenced by Oxygen-Limited Biodegradation 1. Theoretical Development,'' *Water Resour. Res.* 22(13):1973–1982 (1986).
36. Wilson, B. H., G. B. Smith, and J. F. Rees. ''Biotransformations of Selected Alkylbenzenes and Halogenated Aliphatic Hydrocarbons in Methanogenic Aquifer Material: A Microcosm Study,'' *Environ. Sci. Technol.* 20(10):997–1002 (1986).
37. Gibson, T. L., A. S. Abdul, and R. H. Olsen. ''Microbial Degradation of Aromatic Hydrocarbons in Hydrogeologic Material: Microcosm Studies,'' in *Proceedings of the Second National Outdoor Action Conference on Aquifer Restoration, Ground Water Monitoring and Geophysical Methods* (Dublin, OH: National Well Water Association, 1988) pp. 53–69.
38. Kindred, J. S., and M. A. Celia. ''Contaminant Transport and Biodegradation 2. Conceptual Model and Test Simulations,'' *Water Resour. Res.* 25(6):1149–1159 (1986).
39. Barker, J. F., G. C. Patrick, and D. Major. ''Natural Attenuation of Aromatic Hydrocarbons in a Shallow Sand Aquifer,'' *Ground Water Monitoring Review* 7(1):64–71 (1987).
40. Chiang, J. P., J. P. Salanitro, E. Y. Chai, J. D. Colthart, and C. L.Klein. ''Aerobic Biodegradation of Benzene, Toluene and Xylene in a Sandy Aquifer-Data Analysis and Computer Modeling,'' *Ground Water* 26(6):823–834 (1989).
41. Hunt, J. R., J. T. Geller, N. Sitar, and K. S. Udell. ''Subsurface Transport Processes for Gasoline Components,'' *Proceedings of the* 1988 *Joint CSCE-ASCE National Environmental Engineering Conference,* Vancouver, B.C., 1988, pp. 1–8.
42. Borden, R. C., P. B. Bedient, M. D. Lee, C. H. Ward, and J. T. Wilson. ''Transport of Dissolved Hydrocarbons Influenced by Oxygen-Limited Biodegradation 2. Field Application,'' *Water Resour. Res.* 22(13):1983–1990 (1986).
43. Wilson, J. T., J. F. McNabb, J. W. Cochran, T. H. Wang, M. B. Tomson, and P. B. Bedient. ''Influence of Microbial Adaptation on the Fate of Organic Pollutants in Groundwater,'' *Environ. Toxicol. Chem.* 4:721–726 (1985).
44. Abriola, L. M., and G. F. Pinder. ''A Multiphase Approach to the Modeling of Porous Media Contamination by Organic Compounds 1. Equation Development,'' *Water Resour. Res.* 21(1):11–18 (1985).

45. Corapcioglu, M. Y. and A. L. Baehr. "A Compositional Multiphase Model for Groundwater Contamination by Petroleum Products 1. Theoretical Considerations," *Water Resour. Res.* 23(1):191–200 (1987).
46. Sleep, B. E., and J. F. Sykes. "Modeling the Transport of Volatile Organics in Variably Saturated Media," *Water Resour. Res.* 25(1):81–91 (1989).
47. Baveye, P., and A. Valocchi. "An Evaluation of Mathematical Models of the Transport of Biologically Reacting Solutes in Saturated Soils and Aquifers," *Water Resour. Res.* (1986).
48. Sitar, N., J. R. Hunt and K. S. Udell. "Movement of Nonaqueous Liquids in Groundwater," Proceedings of Geotechnical Practice for Waste Disposal, 87/GT Div. ASCE, Ann Arbor, MI, 1987, pp. 205–222.
49. Skopp, J. "Analysis of Time-Dependent Chemical Processes in Soils," *J. Environ. Qual.* 15(3)205–213 (1986).
50. Hillel, D. "Movement and Retention of Organics in Soil: A Review and a Critique of Modeling," in *Petroleum Contaminated Soils,* Volume 1, P. T. Kostecki and E. J. Calabrese, Eds. (Chelsea, MI: Lewis Publishers, Inc. 1988), pp. 81–86.
51. Callahan, M. A. "The Utility of Environmental Fate Models to Regulatory Programs," in *Petroleum Contaminated Soils,* Volume 1, P. T. Kostecki and E. J. Calabrese, Eds. (Chelsea, MI: Lewis Publishers, Inc. 1988), pp. 105–112.
52. Bonazountas, M., Personal communication, 1990.
53. Rudy, R. J., and S. J. Bedosky. "Segregation and Distribution of Gasoline Compounds in the Floridan Aquifer at Wacissa, Florida," in *Proceedings of Petroleum Hydrocarbons and Organic Chemicals, in Ground Water: Prevention, Detection and Restoration* (Dublin, OH: National Well Water Association, 1988), pp. 897–913.
54. Peterec, L. J. "A Case Study in Petroleum Contamination: The North Babylon, Long Island Experience," in *Soils Contaminated by Petroleum: Environmental and Public Health Effects,* E. J. Calabrese and P. T. Kostecki, Eds. (New York: John Wiley & Sons, Inc., 1988), pp. 231–255.
55. "Organic Chemical Contamination of Small Public Water Systems in California," California Department of Health Services, 1990.
56. "Organic Chemical Contamination of Large Public Water Systems in California," California Department of Health Services, 1986.
57. "Leaking Underground Storage Tank Information System (LUSTIS) Report," California State Water Resources Control Board, 1990.
58. Brown, H. S. "A Critical Review of Current Approaches to Determining 'How Clean is Clean' at Hazardous Waste Sites," *Hazard. Wastes & Hazard. Materials* 3(3):233–259 (1986).
59. Kostecki, P. T., E. J. Calabrese, and H. M. Horton. "Review of Present Risk Assessment Models for Petroleum Contaminated Soils," in *Petroleum Contaminated Soils,* Volume 1, P. T. Kostecki and E. J. Calabrese, Eds. (Chelsea, MI: Lewis Publishers, Inc., 1989), pp. 263–300.
60. Schiffman, A. "Cleanup of Polluted Ground Water—A Solution in Search of the Right Problem," *Ground Water Monitoring Review* 9(1):5–10 (1989).
61. Rubel, F. N., B. J. Burgher, and E. C. McGriff, Jr. "An Overview and Suggested Methodology to Determine the Adequacy of Cleanup of Contaminated Soils," in *Petroleum Contaminated Soils,* Volume 2, E. J. Calabrese and P. T. Kostecki, Eds. (Chelsea, MI: Lewis Publishers, Inc., 1989), pp. 409–416.
62. Daugherty, S. J. "The California Leaking Underground Fuel Tank Field Manual: A Guidance Document for Assessment of Underground Fuel Leaks," in *Petroleum*

Contaminated Soils, Volume 2, E. J. Calabrese and P. T. Kostecki, Eds. (Chelsea, MI: Lewis Publishers, Inc. 1989), pp. 453–469.
63. Henry, E. C., and M. F. Hansen. "Letting the Sleeping Dog Lie: A Case Study in the No-Action Remediation Alternative for Petroleum Contaminated Soils," in *Petroleum Contaminated Soils,* Volume 2, E. J. Calabrese and P. T. Kostecki, Eds. (Chelsea, MI: Lewis Publishers, Inc. 1989), pp. 471–483.
64. "Superfund Public Health Evaluation Manual," U.S. Environmental Protection Agency, EPA/540/1-86/060, 1986.
65. "Initial Statement of Reasons for Proposed Regulations: Criteria for Identification of Hazardous and Extremely Hazardous Wastes," California Department of Health Services (Undated).
66. "Guidelines for Addressing Fuel Leaks," San Francisco Regional Water Quality Control Board, 1985.
67. "Regional Board Staff Recommendations for Initial Evaluation and Investigation of Underground Tanks: Tri-Regional Recommendations," North Coast, San Francisco and Central Valley Regional Water Quality Control Boards, 1988.
68. Marshack, J. B. "Waste Classification and Cleanup Level Determination-Draft Guidance Document," Central Valley Regional Water Quality Control Board (Sacramento, CA, 1985).
69. Ames, B. N., R. Magaw, and L. S. Gold. "Ranking Possible Carcinogenic Hazards," *Science* 236:271–280 (1987).
70. Lehr, J. N. "Toxicological Risk Assessment Distortions: Part 2." *Ground Water* 28(2):170–175 (1990).
71. Haley, J. L., D. L. Lang, and L. Herrinton. "EPA's Approach to Evaluating and Cleaning Up Ground Water Contamination at Superfund Sites," *Ground Water Monitoring Review* 9(4):177–183 (1989).
72. Keely, J. F. "Performance Evaluations of Pump-and-Treat Remediations," U.S. Environmental Protection Agency Ground Water Issue, EPA/540/4-89/005, 1989.
73. Bastian, B. N., and P. C. Johnson. "Subsurface Vapor Migration Assessment of Indoor Air Quality," Shell Oil Company Report to the Orange County Health Care Agency, 1989.
74. Millington, R. J., and J. M. Quirk. "Permeability of Porous Solids," *Trans. Faraday Soc.* 57:1200–1207 (1961).
75. Grimsrud, D. T., M. H. Sherman, and R. C. Sonderegger. "Calculating Infiltration: Implications for a Construction Quality Standard," *Proceedings of Thermal Performance of External Envelopes of Buildings, American Society of Heating, Refrigeration and Air Conditioning Engineers,* 1983, pp. 422–452.
76. Shah, J. J., and H. B. Singh. "Distribution of Volatile Organic Chemicals in Outdoor and Indoor Air," *Environ. Sci. Technol.* 22(12):1381–1388 (1988).
77. "Analysis of Ambient Data from Potential Toxics 'Hot Spots' in the South Coast Air Basin," South Coast Air Quality Management District (1988).
78. Haxo, H. E., Jr. "Transport of Dissolved Organics from Dilute Aqueous Solutions Through Flexible Membrane Liners," *Proceedings of the Fourteenth Annual Research Symposium on Land Disposal, Remedial Action, Incineration and Treatment of Hazardous Waste,* U.S. EPA, Cincinnati, OH (1988), pp. 145–166.
79. Zapalac, G. H. "A Time-Dependent Method for Characterizing the Diffusion of ^{222}Rn in Concrete," *Health Physics* 45(2):377–383 (1983).

80. Maxim, D. L. ''Problems Associated With the Use of Conservative Assumptions in Exposure and Risk Analysis,'' in *The Risk Assessment of Environmental Hazards: A Textbook of Case Studies,* D. J. Paustenbach, Ed. (New York: John Wiley & Sons, Inc., 1989), pp. 526–560.
81. Paustenbach, D. J., and R. E. Keenan. ''Health Risk Assessment in the 1990s,'' *Hazmat World* 2(10):44–56 (1989).
82. Jury, W. A., D. Russo, G. Streile, and H. El Abd. ''Evaluation of Volatilization by Organic Chemicals Residing Below the Soil Surface,'' *Water Resour. Res.* 26(1):13–20 (1990).

CHAPTER 3

Private Sector Perspectives on Hydrocarbon Contamination

John J. Hills, Public Utilities Department, Environmental Services Division, Anaheim, California

In order to effectively resolve a problem, one should attempt to gain a comprehensive understanding of all facets of the problem. With regard to hydrocarbon contamination, it is essential for regulators, consultants, attorneys, bankers, business owners, and others involved with this issue to develop a comprehensive understanding of the major perspectives associated with hydrocarbon contamination in order to address and effectively resolve the specific hydrocarbon contamination issues these groups are, or will be, facing.

Other chapters present the issues associated with hydrocarbon contamination from regulatory perspectives. The purpose of this chapter is to identify the hydrocarbon contamination issues from private and public sector perspectives. In doing so, I hope to present another important perspective that will provide those involved with hydrocarbon contamination a more complete picture of the scope of the problem.

PRIVATE SECTOR ISSUES AND CONCERNS

As a result of my involvement with hydrocarbon contamination issues over the last six years and discussions with representatives from private sector companies during and since that time, I have identified five main issues with which I believe most companies with hydrocarbon contamination have been confronted. These issues are: financial implications; technological limitations; availability of remediation resources; short- and long-term liability; and environmental considerations.

FINANCIAL IMPLICATIONS

There are numerous financial implications associated with hydrocarbon contamination. However, the major implications I would like to identify are as follows: remediation costs; loss of income; fines/penalties; adverse publicity; and property devaluation.

Remediation Costs

Costs to clean up hydrocarbon contamination can range from a few thousand to millions of dollars. The lower end of the range is often associated with remediating a minor amount of low level soil contamination which is often encountered when there has been a relatively recent small volume hydrocarbon release. The cost of remediating this type of contamination is relatively low due to the fact that the contamination may be easily removed and quickly disposed of. This type of remediation approach is often referred to as "scoop and run."

The upper end of the range would be associated with remediating sites with extensive soil and groundwater contamination. These types of sites have usually become contaminated as a result of hydrocarbon releases occurring over long periods of time. The costs of remediating these types of sites may be very high due to the large volumes of soil and groundwater contamination that must be removed or treated in place, and the associated treatment expenses. Remediations of this type, because of the technical complexities, require the services of experienced professionals to ensure a proper, expedient, and cost-effective solution.

Loss of Income

The remediation efforts described above usually entail a considerable amount of equipment and contractor activity at the contaminated site. This is especially true if large volumes of contaminated soil are being removed. Needless to say, this type of situation may result in some amount of disruption to business, a decrease in productivity, and a potential loss of income. Examples of this are the numerous service stations across the country that are in the process of replacing their underground storage tanks. Hydrocarbon contamination is often encountered during the tank replacement process. If there is even a moderate amount of contamination requiring soil or groundwater treatment or removal, the remediation process may require that the station be closed anywhere from a couple of weeks to several months. During this time, the facility owner is not only paying a contractor for his services for remediating the site, but because of the closure, no business income is being generated.

In addition to the loss of income, remediating a petroleum contaminated site may, depending upon the location of the contamination, result in a considerable disruption to business operations. Public or employee access to and/or within a facility may be limited during the remediation process, discouraging customers from doing business at the facility or hampering employees' performance of their duties.

Fines/Penalties

Local ordinances, state regulations, and federal laws have been established that require property owners to clean up contaminated sites in a "timely manner." Many of these same statutes allow fines to be levied against property owners who fail to clean up contaminated sites within the time frames specified by the regulatory agencies. While it is true that most property owners wish to resolve their soil contamination problems as soon as possible, there are property owners who, for any number of reasons, are reluctant to execute a timely cleanup. As a result, fines into the millions of dollars have been assessed against these "nonresponsible parties" for noncompliance with cleanup directives.

Adverse Publicity

Most businesses are concerned with their credibility in their communities, for their credibility often affects the success of their business. Hydrocarbon contamination of the environment, whether accidental, intentional, or out of negligence, will often result in the business owner being branded as a polluter, regardless of the cause. The net effect, as in the recent Alaska and California oil spills, may be a significant loss of business due to the loss of credibility. Depending upon the type and size of the business, this loss of credibility/income may far exceed the cleanup costs associated with remediating the spill.

Loss of Property Value

There are states now having real estate disclosure laws that require sellers of real property to notify potential buyers of chemical contamination present on the property. Needless to say, many buyers, upon receiving such a notification, would be extremely cautious to purchase real property contaminated with hazardous chemicals such as benzene, toluene, or xylene. In this situation, it could be said that demand for the property may be appreciably diminished, resulting in decreased saleability and a lower property value.

In addition, many buyers for their own protection, have environmental assessments made of the properties they are intending to purchase. These assessments, through public records, historical data, and soil samples taken from the property will often identify prior hydrocarbon contamination. This often raises a red flag to conservative buyers who are reluctant to become involved with properties having past environmental contamination problems. In these situations, the number of potential buyers may diminish due to the uncertainties associated with the effectiveness of prior cleanup activities at the site and the new property owner's potential for long-term liability.

TECHNOLOGICAL UNCERTAINTIES

There are a number of technological uncertainties that property owners often encounter during the remediation of sites with hydrocarbon contamination. The following are a number of the uncertainties often encountered:

- What remediation method would be most effective in cleaning up the site?
- From an exposure standpoint, will the remediation method selected allow the contamination to be transferred from a relatively controlled medium (i.e., soil), to a relatively uncontrolled medium (i.e., air)?
- What is the lowest cost remediation method which could be employed that would effectively clean up the site?
- Will the remediation method selected be capable of attaining the cleanup levels established for the site?
- What remediation method will create the least disruption to the business and operations?

These are a few of the technological issues which confront owners of hydrocarbon contaminated property. To provide an adequate response to each of these questions would require a full chapter in a technical publication. However, it is important for these concerns to be identified in order to understand this aspect of the problem facing owners of contaminated sites.

AVAILABILITY OF RESOURCES

Even after hydrocarbon contamination has been identified and a commitment is made to mitigate the problem, property owners are often faced with yet another issue—securing the services in a *timely* manner of *qualified* remediation companies. Until the implementation of the state and federal underground storage tank regulatory programs, there were very few companies specializing in remediating hydrocarbon contaminated sites. There were even fewer companies with proven track records. As the state and federal programs developed, it appeared that every contractor (licensed or not) who owned a backhoe became a site remediation "expert" overnight. Because of this, there have been numerous instances where significant dollar amounts have been paid to "remediation experts" who were not only unable to resolve the problem, but often created more problems than originally existed.

Today, there are many qualified companies that have developed an expertise in remediating hydrocarbon contaminated sites. These companies have also developed reputations that, due to the current workload, keep them in demand. As a result of this, it may often be difficult to secure the services of these companies in a timely manner, which is critical in order to minimize the spread of contamination, subsequent adverse environmental impacts that may be caused by the contamination, and the associated remediation costs.

It is to everyone's advantage to initiate a well-organized, comprehensive remediation plan as soon as possible after contamination is discovered. However, because of the aforementioned reasons, it is often not possible to do so.

SHORT- AND LONG-TERM LIABILITY

A considerable amount of short- and long-term liability may be created by owners of hydrocarbon contaminated sites, depending upon the levels of hydrocarbon constituents that remain after remediation is completed and the methodology employed to remediate the site.

Based upon the current and/or future land use proposed for a contaminated site, a property owner may remediate a site to the satisfaction of all regulatory agencies involved with the cleanup. However, over time, the land use of the site may change significantly from what was proposed when the cleanup levels were established for the site. Consider the following scenario. Cleanup levels are established for a contaminated site based on a proposed usage for the site as a parking lot. However, sometime later the site becomes a day care center with a "tot lot." Children at the center begin complaining of nausea, and a subsequent investigation reveals that low levels of hydrocarbons are present throughout the site. The regulatory agency overseeing the investigation requires the current property owner to implement mitigation measures that, depending upon the costs of the additional attenuation/remediation measures the current property owner will "request" that the former property owner(s) "financially assist" in the efforts to resolve this problem.

Significant liability may also be created by the act of remediating a contaminated site. As an example, excavation of hydrocarbon contaminated soil and disposal of the soil at a hazardous waste landfill may be the most expeditious method of remediating a site. However, a method such as this may create both short- and long-term liability for the property owner. The short-term liability may be created by contaminates escaping into the atmosphere during the excavation process, resulting in real or perceived adverse health effects to "exposed" individuals. In order to resolve a problem such as this, a property owner may be required to compensate "exposed" individuals or become involved in costly litigation. Additional liability may be introduced by the safety hazards created as a result of the extensive excavations resulting from the soil removal process. While certain measures may be taken to reduce this risk, a certain amount of risk associated with cave-in or fall-in accidents remains. The long-term liability is created by the act of disposing of contaminated soil at a hazardous waste landfill. Individuals remediating contaminated property that they own are, by law, eternally responsible for the contaminated soil generated during the remediation of the site. This means that, although the soil was disposed of at a permitted landfill in accordance with all laws and regulations, the property owner would be required to financially assist future remediation of the landfill, should the landfill ever be found to be a threat to public health or the environment.

ENVIRONMENTAL CONSIDERATIONS

Although a substantial portion of adverse publicity is given (and often due) to companies that have willfully, or through negligence, contaminated the environment, there are many companies that are genuinely committed to preserving our environment. These companies are determined not to allow conditions or situations within their control to exist that may adversely impact our natural resources, and will allocate the resources required to mitigate any adverse public health and/or environmental impacts they have created. The problem that arises in these situations is that it is often unclear as to what level of remediation is realistically necessary to mitigate "adverse public health and/or environmental impacts." This concern, which may be paraphrased "how clean is clean?" is an issue that the regulatory sectors also struggle with in their efforts to protect public health and the environment. As stated earlier, many companies are committed to allocating whatever resources are required to mitigate an environmental problem, as long as there is some assurance that their mitigation efforts are indeed necessary.

CONCLUSION

It is my hope that this chapter has provided the reader with a more comprehensive understanding of the perspectives, issues, and concerns facing property owners with hydrocarbon contamination. This understanding, in conjunction with the chapters addressing other perspectives, will perhaps allow the reader to more effectively deal with the various aspects of hydrocarbon contamination the reader is, or will be, facing.

CHAPTER 4

Unique Problems of Hydrocarbon Contamination for Ports

Donald W. Rice, Los Angeles Harbor Department, Los Angeles, California

INTRODUCTION

Since the early 1900s, port facilities in the United States have been involved in the import and export of petroleum products. A typical marine oil terminal consists of a berth for tanker vessels, a tank farm storage area consisting of aboveground storage tanks, and a pipeline system for moving products to or from the terminal facility. Some facilities also receive or ship products by rail or tank truck.

The WORLDPORT L.A. is a 7,000 acre land and water area that is administered by the Department of the City of Los Angeles under a tidelands grant from the State of California for the purposes of commerce, navigation, and fisheries.

Over half of the oil-refining of California lies within 20 miles of WORLDPORT L.A. It is therefore not surprising that the port is a major hub for the handling of crude oil and petroleum products, including gasoline, aviation gas/jet fuel, and marine fuels. It is also not surprising that port facilities, given their long history of handling petroleum products, contain areas where soils and groundwater are contaminated with hydrocarbons. This contamination is localized but can be extensive.

Petroleum and petrochemical products are handled at terminal facilities that are leased to oil companies. In 1911 the port handled one million barrels of oil. The total liquid hydrocarbon throughput at the port during fiscal year 1987–1988 reached 175 million barrels. Facilities range in size from small marine fueling stations to major petroleum terminals. The port has 19 berths handling crude oil or petroleum products for 15 separate terminal operators. There are 20 separate active tank farms and 3 decommissioned former tank farm sites.

Among the facilities in the Port of Los Angeles there are approximately 500 aboveground storage tanks with 500 million gallons of storage capacity (see Figure 4.1).

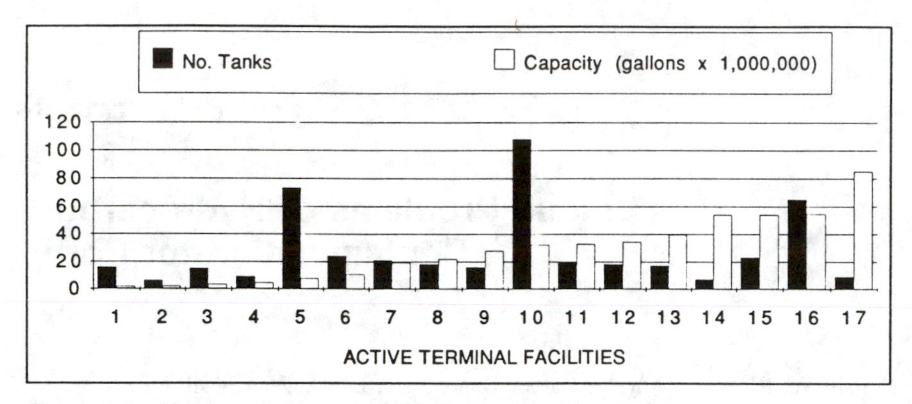

Figure 4.1. Number of storage tanks and capacity at terminals.

The purpose of this chapter is to describe the nature of the problem that the Port of Los Angeles faces with hydrocarbon contaminated soils, the policy and administrative context in which solutions to the problem are being sought, and some of the unique problems or opportunities that the port has in managing the issue.

PROBLEM

The problem that the port faces with petroleum hydrocarbon contaminated soils is being able to develop port lands in a timely and cost-effective manner where contamination is discovered. The port is presently planning a 5-year Capital Development Program consisting of 150 projects at an estimated cost of $592,240,000. Several of these projects, some of which are now under construction, are being built on former marine oil terminal facility sites. The need to manage hydrocarbon contaminated soils and groundwater is a critical present and future need for the port in order to achieve its development plans.

PHYSICAL CONTEXT OF HYDROCARBON CONTAMINATION

One of the unique aspects of marine terminal hydrogeology is the shallow depth to groundwater, coupled with tidal influences/perturbations, due to the low topographic setting proximal to open water. Such shallow groundwater conditions enable monitoring wells to be drilled and constructed with small track-mounted drilling rigs or, in some cases, even by hand methods. In the case of WORLDPORT L.A., this factor is invaluable since the high retaining walls of the bulk terminals often allow for very limited working space.

Past studies of the port area have shown that the shallow aquifer generally consists of approximately 10 feet of dredged harbor fill which sits on natural sediments of marine clay and silt.[1]

The natural sediments have been found in one study to be of much lower permeability than the overlying fill.[2]Tidal influence has been found to be limited due to the relatively low permeability of the fill that comprises the uppermost shallow aquifer. Furthermore, contamination derived from surface spills or releases is retained, or its vertical movement retarded, by this low permeability surficial layer. Sampling programs for facilities, therefore, must be designed for shallow groundwater conditions, low permeability surficial materials, and hydrogeologically complex sediments within the saturated zone.

ADMINISTRATIVE AND POLICY CONSIDERATIONS

Under the policy and guidance of the Board of Harbor Commissioners a program has been developed and is continuing to evolve that will provide solutions to the contamination problems at the port's liquid bulk terminals.

In 1986 the port developed a strategic plan to periodically review the organization's purpose for being, the goals or results it wants to achieve in order to carry out that purpose, and policies and strategies that will govern the use of departmental resources to achieve these goals. A key environmental issue of this plan is soil and groundwater contamination. The continued use and future development of facilities on existing port lands require that sites evidencing soil and groundwater contamination be evaluated and, if necessary, be restored.

On April 13, 1988 the Board of Harbor Commissioners adopted a Hazardous Materials Management Policy that addressed the port's responsibilities and interests as a "property owner" in preventing releases of hazardous materials and implementing effective cleanup work in the event of releases. Regulatory compliance, cooperation, and coordination with regulatory agencies, and development of an ongoing program to address these concerns were also subjects of the policy.

Permits and leases with tenants have been modified over the last several years to assure closer environmental monitoring of facilities and to increase the cleanup obligations of tenants at their leased premises. Under current and prior agreements with the port, tenants are required to characterize soil or groundwater contamination problems on their leased premises and remediate the property.

The Deputy Executive Director for Development chairs a Liquid Bulk Task Force comprised of representatives from the Property, Engineering, and Environmental Divisions of the port. The Task Force reviews information on characterization, remediation, and lease renewal activity pertaining to liquid bulk terminals, and develops policy alternatives and implementation programs for consideration by Executive Management and the Board of Harbor Commissioners.

PROGRAM IMPLEMENTATION AND RESULTS

WORLDPORT L.A. has taken the approach of establishing and maintaining a key management role in the cooperative resolution of the contamination problems the port and its tenants are facing. In implementing its program, the port coordinates all remediation and compliance activities with the appropriate regulatory agencies. These agencies include the California Department of Health Services, Regional Water Quality Control Board, Los Angeles County Department of Health Services, and the Los Angeles City Fire Department, as dictated by specific site conditions.

A guidance document has been prepared and provided to tenants in order to standardize the site characterization process among the marine oil terminal tenants, to ensure consistent and comparable investigations are conducted at each of the facilities. This guidance document addresses project goals and objectives, the sampling and analysis plan, including soil and groundwater investigation methods, analytical methods, and quality assurance and reporting protocols.

Soil and groundwater characterization results are now available for 14 of the port's marine oil terminal tank farm areas. Three additional sites have characterization plans in the preparation stage. The status of characterization and remediation activity at the ports marine terminals is shown in Table 4.1.

Table 4.1. Los Angeles Harbor Marine Oil Terminal Sites Characterization/ Remediation Status

Activity	No. of Terminals
Characterization results for soils/groundwater	14
Characterization plans in progress	3
Remediation underway	
Product recovery	5
Groundwater remediation	2 (4 sites no product on groundwater)
Soil remediation	3

Product recovery or groundwater and soil remediation are in progress at several terminals. At seven terminal sites product or product contaminated groundwater is being removed and treated. Soil remediation is underway at three sites. Two of these are tenant bioremediation projects, while the third is a steam stripping operation at a former chemical storage tank farm. The port will soon establish a new bioremediation area for its own use.

Key issues to be addressed and resolved as the program continues include: (1) identification of available and applicable remediation technologies and an evaluation of present bioremediation projects; (2) the setting of site-specific cleanup goals; (3) an evaluation of the technical feasibility of remediation of soils at operating terminal facilities; and (4) specifications for modernization and upgrading of existing facilities to achieve contamination source control and development of design and operational guidelines and inspection protocols.

REFERENCES

1. Harding Lawson Associates. ''Site Assessment Report,'' Port of Los Angeles, California, June 1986.
2. Harding Lawson Associates. ''Geotechnical Investigation Berths 218–221,'' Port of Los Angeles, California, September 1983.

CHAPTER 5

Review of State Cleanup Levels for Hydrocarbon Contaminated Soils

Charles E. Bell, Paul T. Kostecki, and **Edward J. Calabrese,**
School of Public Health, University of Massachusetts, Amherst

BACKGROUND

The information in this chapter has been condensed from data collected as part of an ongoing national survey to investigate the regulatory approaches of states to site investigation and cleanup of hydrocarbon contaminated soils (HCS), primarily resulting from leaking underground storage tanks. Regulatory representatives have been asked to respond to a series of questions concerning: the current state of research relating to HCS; emergency response measures; site notification requirements; site assessment protocols, including field and analytical methods used; corrective action requirements; pre- and post-treatment disposal options; pending legislation, rule making, or policy decisions affecting corrective actions; and classification or treatment of HCS as a hazardous waste.

Of particular concern to both regulatory and regulated communities is the development and application of cleanup levels to be used both as a tool for guidance and the establishment of cleanup objectives. Limited guidance at the federal (Environmental Protection Agency) level, a lack of appropriate analytical methods specifically designed to sample petroleum hydrocarbons in soils, and variation in site-specific factors affecting fate and transport processes all contribute to the variation of acceptable cleanup goals.

The key to any cleanup effort is that the adopted criteria are risk based. Essentially, this has been adopted as the foundation for development of soil cleanup criteria for other agents such as PCBs, dioxin, and lead. In order to be consistent within the current regulatory context, the parameters used to determine regulatory compliance should also be risk based. This is not always the case. The nature of terms such as total petroleum hydrocarbons (TPH) preclude the utility of the value directly into the risk assessment process. This is due to variable composition between and among petroleum products, site-specific characteristics, and the degree to which the chemical composition of the TPH value changes over time. However, the TPH value can function as a cost-effective screening mechanism for site assessment, and provide guidance for further corrective action. Consequently, the TPH value used in conjunction with other measures such as benzene may provide a basis for relative ranking among selected sites, as well as an estimation of actual risk for the development of remediation goals.

RESULTS AND INTERPRETATION

In order to understand the functional role of cleanup numbers it is necessary to define some terminology commonly used. Cleanup numbers may be classified into four categories based on their application (Table 5.1). Soil cleanup standards (CS) are fixed standards comparable to air or water quality standards that have been established by law as a rule and are normally compound-specific. Corrective action levels (CAL) are used at sites to be remediated. These are site-specific contaminant concentrations which must be achieved in order for corrective action to be deemed complete by the agency. Guidance or action levels (GL) function to direct or dictate the response by the regulatory agency and serve to advise the responsible party if further corrective action is required. A remediation goal (RG) is an acceptable endpoint contaminant concentration, the value of which may be determined on a site-by-site basis.

Only two states surveyed (Georgia and Tennessee) reported having cleanup standards for petroleum in soils using laboratory analyses. Three other states (Florida, Maine, and North Dakota) do not require laboratory analysis of soil samples, but rather have opted to utilize results from field screening methods as a basis for determining cleanup requirements. Seven states have developed corrective action levels as part of their programs, while 21 other states have established guidance levels, and another 16 states have set remediation goals.

The surveying of states has also revealed that TPH is most widely employed as a guidance level for cleanup or as a site-specific remediation goal. Forty-two out of 50 states indicate that their agencies utilize TPH in regulatory oversight and that specific operation levels range from 10 to 10,000 ppm. Thirty-five of the 42 states that use TPH also require BTEX sampling or some other measurement of volatiles.

Few states have adopted rigid cleanup requirements for soils as demonstrated by the number of states that have incorporated cleanup standards or corrective

Table 5.1. States Classified by Categories of Cleanup Levels

State	Parameters	Method	Term	Level	Related Information	References
AK	TPH[a]	EPA 418.1[b] using Infrared Spectroscopy (IR)	GL	100 ppm for diesel 50 ppm for gasoline	These are recently proposed soil sampling guidelines developed to provide an enforceable policy framework. HCID method used to identify hydrocarbon type. Samples need only be analyzed for TPH if HCID test indicates the presence of diesel or other non-gasoline fractions.	Petroleum Contaminated Soil Sampling Guidelines. (Draft) AK Dept. of Environmental Conservation. March, 1990.
	BTEX[c]	EPA 8020 and EPA 5030 or EPA 8240 by GC/MS	RG	total BTEX < =10 ppm		
	HCID[d]	EPA 8015 by GC				
AL	TPH[a]	EPA 9071[e], Standard Method 503[f]	CAL	100 ppm TPH	No futher action required if TPH < =100 ppm and groundwater >5 feet below the surface or TPH values are < =10 ppm for all soil samples.	AL Dept. of Env. Mgt. Admin. Code R., Water Quality Program. Sections 335-6-15.26 to 335-6-15.34.
	BTEX[d]	EPA 5030 or EPA 3810, EPA 8020 or EPA 8240				
	Lead	EPA 239.2				
AR	TPH BTEX	EPA 418.1 EPA 8020	RG	None	Remediation goals determined on a case-by-case basis as a function of water quality considerations.	None.
AZ	TPH	Modified EPA 8015 for gasoline, modified 418.1 for diesel	GL	100 ppm TPH	Nonenforceable, health based guidance levels derived from AZ Dept. of Health recommended groundwater cleanup and action levels. These values represent ten times allowable limits for groundwater.	None.
	BTEX	EPA 8020 or modified EPA 8015 for TEX		130 ppm benzene 200 ppm toluene 68 ppm ethylbenzene 44 ppm xylenes		
CA	TPH	Modified EPA 8015 EPA 5020 for gas EPA 5030 for diesel	RG	10–1000 ppm TPH for gas, 100–10,000 ppm TPH for diesel, 0.3–1 ppm benzene, 0.3–50 ppm toluene, 1–50 ppm xylenes, 1–50 ppm ethylbenzene	TPH analytical methods used based on procedure recommended by CA Dept. of Health Services. The analyses performed based on the nature of the contaminant. LUFT Field Manual intended to be used as a tool for developing site-specific cleanup levels as opposed to state-wide remediation goals.	CA LUFT Field Manual, State Water Resources Control Board, Sacramento, CA. November, 1989.
	TRPH[g]	EPA 418.1 for diesel				
	BTEX	EPA 8020				
	HVOs[h]	EPA 8010				

continued

Table 5.1. Continued

State	Parameters	Method	Term	Level	Related Information	References
CO	TPH BTEX	EPA 418.1 or modified EPA 8015 EPA 8020	RG	None	Site-specific remediation goals often to detection limits or level deemed necessary to protect groundwater and minimize potential offsite impacts.	CO House Bill No. 1299, Amendment to Sec. 1, Article 20, Tit. 8, CO Revised Statutes, Underground Storage Tanks.
CT	BTEX HVOs	EPA 8020 EPA 8010	GL	50 ppm TPH for gasoline	If the TPH concentration exceeds 50 ppm then EP Toxicity testing for heavy metals must be performed. Soils exceeding EP Toxicity for metals are disposed of as hazardous material. Soils contaminated with fuel oils may be mixed with clean sand to prevent free draining of product and landfilled.	None.
DE	TPH BTEX	EPA 418.1 EPA 8020	RG	100 ppm TPH and 10 ppm BTEX for gasoline	DE Division of Air and Waste Management reportedly in the process of developing guidance levels which may be incorporated into existing corrective action process for UST sites.	None.
FL	Not required		GL	500 ppm TPH	Soils with TPH readings exceeding 500 ppm on organic vapor analysis instruments with flame ionization detector require remediation. Soils with vapor readings between 10–500 ppm may require cleanup depending on site-specific factors. Lab analysis of soils performed for verification purposes only.	Guidelines for Assessment and Remediation of Petroleum Contaminated Soils. FL Dept. of Environmental Regulation. January, 1989.
GA	TPH BTEX	EPA 418.1 EPA 602	CS	100–500 ppm TPH total BTEX 20–100 ppm	Soils >100 ppm TPH or 20 ppm total BTEX must be remediated at sites within 3 miles of public drinking water wells or 1/2 mile from private wells. Remediation of soils in excess of 500 ppm TPH or total BTEX >100 ppm is required at all other UST corrective action sites.	Rules of GA Dept. of Natural Resources Environmental Protection Division, Chapter 391-3-15, UST Management. December, 1989.
HI	TPH	EPA 3550 for diesel, EPA 5030 and EPA 8015 or EPA 8020 for gas	RG	50 ppm TPH	A remediation goal of 50 ppm TPH may be applied at sites limited to soil contamination. The agency utilizes state water quality standards to evaluate site-specific environmental and public health risks and determine cleanup requirements at sites where groundwater may be impacted.	None.

continued

Table 5.1. Continued

State	Parameters	Method	Term	Level	Related Information	References
IA	TPH BTEX	EPA 418.1, EPA 8250 EPA 8020, EPA 8240	RG	100 ppm TPH	Soils greater than 100 ppm total volatile or total extractable hydrocarbons (depending on the nature of the contaminant) are considered contaminated and must be remediated. Remediation goals determined on a case-by-case basis.	None.
ID	TPH BTEX EDB[i]	Modified EPA 8015 EPA 503.1, EPA 602, EPA 624, EPA 8020, EPA 8240 EPA 8010	RG	100 ppm TPH for gasoline, 1000 ppm for diesel, individual BTEX values of 1 ppm	Soils with TPH >100 ppm for gasoline spills and 1000 ppm TPH at diesel sites or individual BTEX concentrations >1 ppm should be remediated. More stringent goals may be applied in cases where "beneficial" groundwaters may be impacted.	Recommended Cleanup Criteria for Sites Contaminated with Petroleum Released from USTs. ID Water Quality Bureau. June, 1989.
IL	BTEX	EPA 5030, EPA 8240	CAL	0.025 ppm benzene and total BTEX <=16.025 ppm	Soil cleanup objectives may only be applied at sites where groundwater is not impacted. Soils meeting these criteria are no longer considered a potential source of groundwater contamination. Soils exhibiting petroleum odors or visibly contaminated must also be removed.	Guidance Manual for Petroleum-Related LUST Cleanups in Illinois. IL Env. Protection Agency, Div. of Pollution Control. May, 1990.
IN	TPH	EPA 418.1	GL	100 ppm TPH	Cleanup levels are determined on a site-by-site basis. The agency is considering the use of a 100 ppm TPH action level as a screening tool for site investigations.	None.
KS	TPH BTEX EDC[j]	Appropriate EPA SW846 methods	CAL	100 ppm TPH 1.4 ppm benzene 0.8 ppm EDC	TPH is defined as the sum of individual concentrations of toluene, xylenes, ethylbenzene, and MTBE for the purpose of determining cleanup requirements. Soil contaminant levels greater than 1.4 ppm benzene, 0.8 ppm EDC and/or 100 ppm TPH require remediation.	Corrective Action Policy Manual. KS Dept. of Health and Environment, UST Program. October, 1989.
KY	TPH BTEX PAHs[k]	EPA 9070 or EPA 9071 for waste oil EPA 8020 for gasoline EPA 8100 for diesel	RG	Background	Collection of both background and suspected contaminated soil samples are required at all tank removals. The agency normally require soils be remediated to background or detection limit.	Guidelines for Site Investigations of Leaking Underground Storage Tank Sites in Kentucky. KY Dept. of Environmental Protection, UST Program. January, 1990.

continued

Table 5.1. Continued

State	Parameters	Method	Term	Level	Related Information	References
LA	BTEX TPH	EPA 8020 for gasoline Modified EPA 8015 (CA)	RG	None	Remediation goals are determined on a site-by-site basis but are comparable to those applied in other states.	Proposed Underground Storage Tank Rules and Regulations. LA Dept. of Environmental Quality, UST Division. May, 1990.
MA	TOVs[l]	Field screening method for gasoline, Standard Methods 503B/E or EPA 418.1	RG	10 ppm total volatiles 100 ppm TPH	Gasoline contaminated soils >1800 ppm TOVs or oil residuals with a weight/weight concentration >300 ppm TPH are regulated as hazardous waste. TOVs are measured as tot. organic headspace vapors expressed as benzene utilizing a portable PID or FID.	Management Procedures for Excavated Soils Contaminated with Virgin TPH Petroleum Oils. Policy #WSC-89-001 MA Dept. of Environmental Protection. June, 1989.
MD	TPH BTEX[m] MBTE	Modified EPA 418.1 EPA 601, 602, 624 or 625 based on field screening	RG	None	Free-phased product must be removed to background or non-measurable levels. Dissolved contamination may also require removal to an endpoint determined both feasible and cost-beneficial.	None.
ME	TPH	Field screening method	RG	20–50 ppm TPH	Remediation goals are determined on a site-by-site basis; however, 20–50 ppm TPH as measured by field instrumentation (H-Nu) are often considered acceptable.	None.
MI	BTEX PAHs TPH	EPA 602, 8020 or 8240 EPA 8270, 8250 Standard Methods 503B, 503D, 503E	CAL	background, detection limit or risk-based value	One of three options may be selected by RP as cleanup objective: cleanup to background or detection limit; levels that do not pose an unacceptable risk based on standard exposure assumptions or based on a site-specific assessment of risk.	MI Dept. of Natural Resources. Environmental Contamination Response Activity, Admin. Rules for 1982 PA 307 as passed. May, 1990.
MN	BTEX TPH Lead	EPA 8015, 8020 EPA 418.1 Appropriate EPA test method	RG	50 ppm TPH	Additional action levels of 10 ppm TPH for gasoline and less than 1 ppm for heavier fuels are used for field screening (using PID) at sites where soils are to be excavated. Laboratory analysis of remaining soils is used to determine if further remedial investigation is required.	Excavation of Petroleum Contaminated Soil. MN Pollution Control Agency, Tanks and Spills Section. April, 1990.

continued

Table 5.1. Continued

State	Parameters	Method	Term	Level	Related Information	References
MO	VOA[n] TPH	EPA 8020 or 8240 (VOA scan) Modified EPA 418.1	GL	10 ppm TPH and 10 ppm total BTEX or VOA for petroleum	Residual contamination levels are reviewed as part of tank closure reporting requirements. Further corrective action may be required depending on site-specific conditions.	UST Guide for Tank Closure. MO Dept. of Natural Resources. June, 1989.
MS	BTEX TPH	Appropriate EPA test method Standard Methods 503	RG	100 ppm total BTEX for gasoline, 100 ppm TPH for diesel	These cleanup levels are typically applied at sites in which there are no "sensitive environmental receptors" such as public or private wells and geologic recharge areas.	None.
MT	BTEX TPH	EPA 8020 for gasoline EPA 8015 or EPA 418.1 or modified EPA 8015 (CA)	GL	100 ppm TPH 10 ppm total BTEX	Contaminant concentrations exceeding these values may trigger further investigation (health based risk assessment) to determine site-specific remediation goals.	1989 Admin. Rules of Montana, Sub-Chapters 5–7, Sections 16.45.401–16.45.701.
NC	VOs SVOs TPH	EPA 5030 with modified CA method EPA 5030 and 3550 with modified CA method	CAL	10 ppm TPH	Additional analytical methods may be required based on the nature of contaminant. Remediation of soils not required if TPH <10 ppm. A Site Sensitivity Evaluation (SSE) is performed using a numerical scoring procedure to determine site-specific cleanup levels for sites with contaminated soils >10 ppm TPH.	Guidelines for Remediation of Soil Contaminated by Petroleum. NC Division of Environmental Management. July, 1990.
ND	Not required			None	Laboratory analysis of soil samples not normally required unless overexcavation fails to remove residual contamination. Guidance levels under development.	None.
NE	VOA TRPH	EPA 624, 601, 602, 8020 or 8024 for gas EPA 418.1 or 3540	RG	None.	Remediation goals established on a site-by-site basis. Proposed rules currently under development may include a matrix of cleanup levels for soils based on site-specific criteria.	UST System Site Assessment Protocol for Permanent Closure and Change In Service. NE Dept. of Environmental Control. September, 1989.

continued

Table 5.1. Continued

State	Parameters	Method	Term	Level	Related Information	References
NH	BTEX TPH	Modified EPA 8015 EPA 418.1 for diesel	RG	1 ppm BTEX and 10 ppm TPH for gas, 1 ppm BTEX and 100 ppm TPH for all others	More stringent remediation goals may be set depending on site-specific factors. Current goals established using the CA LUFT Field Manual leaching potential analysis for gasoline and diesel using data from NH case histories.	Policy for Management of Soils Contaminated from Spills/Releases of Virgin Petroleum Products. NH Dept. of Environmental Services. June, 1989.
NJ	VOs[o] SVOs[p] TPH	Appropriate EPA SW846 method	GL	1 ppm VOs 10 SVOs 100 ppm TPH	Three percent (30,000 ppm) petroleum contaminated soil is classified as hazardous and requires disposal (if not treated) in a hazardous waste facility. PCS exceeding 100 TPH may require cleanup based on the extent of VOs and SVOs present.	Environmental Cleanup Responsibility Act, NJ Dept. of Environmental Protection
NM	VOs TPH	EPA 8240 or other EID approved Method EPA 418.1	GL	10 ppm benzene 50 ppm aromatics 100 ppm TPH	Soils contaminated with gasoline or lighter hydrocarbons require remediation if benzene concentrations exceed 10 ppm, all detected aromatics exceed 50 ppm, or field measurements (headspace method) are greater than 100 ppm. For diesel and heavier hydrocarbons, TPH levels greater than 100 ppm indicate remediation is necessary.	Draft NM UST Bureau Soils Policy, NM Health and Environment Department, Environmental Improvement Division (EID), July, 1990.
NV	TPH	Modified EPA 8015	GL	100 ppm TPH	Excavation of soils contaminated in excess of 100 ppm TPH may be required depending on site-specific conditions including depth to ground water, land use and hydrocarbon product type.	Hydrocarbon Spill and Remedial Action Policy. NV Dept. of Conservation and Natural Resources. October, 1987.
NY	VOs SVOs	EPA 8020 EPA 8270	GL	compound specific	Recently proposed soil cleanup rules call for soil guidance values to be calculated for individual compounds using the water partition theory equation (Cs = f*Koc*Cw), where Cw is the more stringent NY Dept. of Health groundwater/drinking water standard in ppb, Koc the contaminant-specific partition coefficient, and an assumed organic content (f) of 2.5 percent.	Methodology for Soil Guidance Module (Draft). NY State Dept. of Env. Conservation, Bureau of Spill Response. September, 1989.

continued

Table 5.1. Continued

State	Parameters	Method	Term	Level	Related Information	References
OH	VOs SVOs TPH	EPA 8020 EPA 8270 EPA 418.1 (IR method)	RG	Background	All contaminated soils must be remediated to background or detection limit. RPs may propose alternative, site-specific cleanup standards if it can be demonstrated they would result in no adverse impact on any environmental media or pose a threat to human health or local fauna.	Corrective Action Policy and Procedures. Appendix C1, LUST Trust Fund Cooperative Agreement, OH Dept. of Commerce. September 20, 1988.
OK	BTEX TPH	Appropriate EPA SW846 method	GL	10 ppm total BTEX 50 ppm TPH	These are temporary action levels established for tank closures and sites where contamination is limited to soils. They are also used to advise RPs if further corrective action may be required.	None.
OR	BTEX TPH	EPA 5030 and EPA 8020 or 8240 Modified EPA 418.1	CS	40–130 ppm TPH for gasoline 100–1000 ppm TPH for diesel	Site-specific target cleanup level is derived from a numerical matrix score based on five evaluation parameters: depth to groundwater; mean annual precipitation; soil type; sensitivity of the uppermost aquifer; potential receptors. These rules are currently undergoing revisions regarding analytical and sampling methods, and reporting requirements.	Numeric Soil Cleanup Levels for Motor Fuel and Heating Oil (Draft). OAR 340-122-305 to 340-122-360. July, 1990.
PA	BTEX TPH	Appropriate EPA SW846 method		None	A computer modeling system is currently under development named Risk Assessment/Fate and Transport (RAFT) intended for use as a tool in the determination of site-specific cleanup levels based on the principles of environmental fate and exposure, exposure assessment and risk assessment.	User's Manual for Risk Assessment/Fate and Transport (RAFT) Modeling System. PA Dept. of Environmental Regulation, Bureau of Waste Management. October, 1989.
RI	VOCs	Appropriate EPA SW846 method	GL	50 ppm TPH	Site-specific cleanup goals based on the nature of the contaminant and whether the site is located on commercial or residential property. Laboratory analysis of soils required for VOCs in the case of gasoline spills, naphthalene for diesel, and PAHs for heavier fuels.	None.

continued

Table 5.1. Continued

State	Parameters	Method	Term	Level	Related Information	References
SC	BTEX TPH	EPA 602 or EPA 8020 EPA 418.1		None	Cleanup goals determined on a site-by-site basis. Contaminated soils may not be stockpiled on-site. Many landfill operators will not accept soils in excess of 10 ppm total BTEX so much of the material requires pre-treatment (incineration) prior to disposal.	None.
SD	VOs SVOs	EPA 8200 or 602 for gas EPA 8100 or 610 for diesel	RG	10 ppm TPH	Regulations require removal of all visibly contaminated soils. Site-specific cleanup levels may be negotiated or the RP can opt to clean to the 10 ppm TPH level.	SD Administrative Rules, Chapter 74:03:28, Underground Storage Tanks.
TN	BTX TPH	Appropriate EPA SW846 method Modified EPA 418.1 (CA) method	CS	10–500 ppm total BTX 100–1000 ppm TPH	The contamination plume must be clearly defined in order to apply soil cleanup levels. A site-specific cleanup level is determined based on the highest level of soil permeability observed within the plume and whether groundwater present below the site is classified as a drinking or non-drinking water supply.	Petroleum UST Program Rules 1200-1-15.01 to 1200-1-15.07, Technical Standards and Corrective Action Requirements for Owners and Operators of Petroleum USTs TN Dept. of Health and Environment. October, 1989.
TX	BTEX TPH	EPA 8020, 602, EPA 624 EPA 418.1 ASTM[q] D3328-78B	GL	30 ppm total BTEX 100 ppm TPH	Soils do not require remediation if TPH is less than 100 ppm or total BTEX is less than 30 ppm at sites where groundwater is not threatened. Sites involving groundwater contamination are handled on a case-by-case basis.	None.
UT	BTEX TPH	EPA 8020, 602 Modified EPA 8015 (CA)	GL	50 ppm TPH	Soil sampling for BTEX is required at gasoline contaminated sites where initial TPH measurements are in excess of 50 ppm. Remediation goals are determined on a site-by-site basis. Soil cleanup levels based on a scoring system of site-specific criteria are under development.	None.

continued

Table 5.1. Continued

State	Parameters	Method	Term	Level	Related Information	References
VA	TPH	Modified EPA 418.1	GL	100 ppm TPH	The 100 ppm TPH guidance level is normally applied at UST site closures where it has been determined there is no potential environmental (groundwater) or health-related impact. Sites in excess of 100 ppm TPH require further sampling using appropriate EPA approved methods based on the nature of the contaminant. Additional testing procedures and cleanup criteria (BTEX) are required for soil disposal.	Petroleum Contaminated Soil Disposal Guidelines. VA Dept. of Waste Management. April, 1990.
VT	TPH	Modified EPA 418.1	GL	20 ppm volatiles	An action or guidance level of 20 ppm total volatile constituents, as measured using a photo ionization device (PID), is used to determine if soils require remediation at sites where groundwater contamination is not a factor. Final remediation goals are determined on a site-by-site basis. Treated soils less than 100 ppm volatiles may be disposed in a landfill.	None.
WA	BTEX TPH	EPA 8020 EPA 418.1	RG	0.1 ppm benzene 4.0 ppm toluene 3.0 ppm ethylbenzene 2.0 ppm xylenes 100 ppm TPH for gas 200 ppm TPH for diesel	Proposed rules would allow for application of these compliance levels or the establishment of alternative site-specific levels using a leaching potential analysis and risk appraisal approach (similar to California's) to estimate the levels of BTEX and TPH that may be left in place.	Interim Underground Petroleum Storage Tank Removal and Remediation Guidelines. (Draft) WA Dept. of Ecology. January, 1990.
WI	VOCs TPH	Appropriate EPA SW846 method Modified EPA 418.1 (CA)	 CAL	 10 ppm TPH	Wisconsin uses the 10 ppm TPH detection limit for closure of UST sites. Sites with soils exceeding 10 ppm TPH require further analyses (BTEX) to characterize the degree and extent of contamination. Sampling for total lead may also be required depending on the nature of the contaminant.	None.

continued

Table 5.1. Continued

State	Parameters	Method	Term	Level	Related Information	References
WV	BTEX TPH	Appropriate EPA SW846 method Modified EPA 418.1 (CA)	GL	Detection limit for BTEX 100 ppm BTEX, 100 ppm TPH	Contaminated soils in excess of 100 ppm TPH require corrective action while those less than 100 ppm TPH may be left on site. Additional requirements call for total BTEX to be less than detectable limits in order for the material to be left on site. Sites in which groundwater used as drinking water has been impacted are handled on a case-by-case basis.	None.
WY	BTEX TPH	Appropriate EPA methods	RG	10–100 ppm TPH	Sites are considered clean when contaminant concentrations are <10 ppm TPH for soils located in areas where depth to groundwater is <50 feet, and <100 ppm TPH where depth to groundwater is >50 feet.	None.

[a] Total Petroleum Hydrocarbons.
[b] EPA Method 418.1 from the Manual of Methods for Chemical Analysis of Water and Waste, EPA 600/4-79-020. "Modified" refers to the use of an extraction procedure provided in the manual. "(CA)" refers to an alternative analytical extraction procedure recommended by the California Dept. of Health Services.
[c] Benzene, toluene, ethylbenzene, and xylenes.
[d] Qualitative hydrocarbon identification by a chromatographic method.
[e] U.S. EPA Test Methods for Evaluating Solid Waste Physical/Chemical Methods, SW846.
[f] Standard Methods for the Examination of Water and Wastewater. Published jointly by the American Public Health Association, American Water Works Association and Water Pollution Control Federation.
[g] Total Recoverable Petroleum Hydrocarbons, synonomous with TPH.
[h] Halogenated Volatile Organics including 1,2-dibromoethane and 1,2-dichloroethane.
[i] 1,2-dibromoethane.
[j] 1,2-dichloroethane.
[k] Polynuclear Aromatic Hydrocarbons.
[l] Total Organic Volatiles.
[m] Methyl Tertiary Butyl Ether.
[n] Volatile Organic Aromatics.
[o] Volatile Organics.
[p] Semivolatile Organic.
[q] American Society for Testing and Materials.

action levels into their respective programs. Those that do, frequently provide an alternative methodology for deriving site-specific, risk-based cleanup levels. Nearly all states with rules or guidelines that establish remediation goals also contain language that allows for modification of the level as the situation demands. Overall, states have resisted using stringent cleanup numbers or adopting regulations that would limit their ability to derive acceptable cleanup levels on a case by case basis. This trend is unlikely to change until the factors affecting the fate and transport of these chemicals in the soil matrix, subsequent exposure pathways, and the risk to the public health are more completely understood.

CHAPTER 6

Sharing Responsibilities: A New Approach to Groundwater Remediation Involving Multiple Potentially Responsible Parties

Jean B. Kulla, McLaren/Hart, Irvine, California
Angelo J. Bellomo, McLaren/Hart, Burbank, California

INTRODUCTION

The benefits available to Potentially Responsible Parties (PRPs) in working cooperatively on the remediation of a groundwater problem are not well recognized. This is partially due to the reality that the need to proceed on a site is initially driven by one party only, with the other PRPs often lacking the motivation to be involved in what might become a very costly undertaking.

As an example, a developer wishing to aggressively proceed with a site investigation may find no such interest on the part of a prior owner-operator or adjoining property tenants, who would prefer to avoid the recognition of a problem to which they may all have contributed. For this reason most PRPs will defer participating in an investigation and cleanup until forced to by reason of expediting a property transaction or development, responding to a government order, or preparing for defense in a cost-recovery action.

Alternatively, a number of PRPs are motivated to become involved in a ground-water problem, but with each perceiving that they are dealing with distinctly separate problems. In such cases, multiple PRPs are participating, but their individual and uncoordinated efforts are in fact complicating a problem requiring a unified approach to remediate. In either situation, the lack of aggressive cooperation among PRPs has resulted in substantial and continuing litigation between them. It appears however, that this trend will soon be changing.

THE BENEFITS OF PROACTIVE COOPERATION

There are two important reasons why PRPs are beginning to consider a more proactive and cooperative relationship in multiparty cleanups. First, a number of cost-recovery actions have proceeded on the basis of site investigations and cleanup that were designed and funded by a single party. Obviously there is a distinct advantage to this party in defining the magnitude and contributing factors to the problem, as well as the approach taken to resolve it. It is reasonable to expect that the single party may have designed the investigation to favorably influence the findings, and selected a remedial approach that was most responsive to their needs. Ultimately, the other PRPs will find themselves arguing against the findings of a remedial investigation they had no part in designing. For this reason, many PRPs, regardless of how minor their exposure may seem, are participating to the extent necessary to influence the design of investigations and the selection of remedies.

Secondly, there is a growing recognition among PRPs of the need to proactively initiate work on a site before the work becomes government-driven. When this occurs, the lead regulatory agency will normally assign a project manager who may at times consult with as many as five to ten other agency staff who are technical experts in various fields. A company that has historically delayed work on a site to the point where the work becomes agency-driven, may well find itself paying substantially more, including the costs of the many government employees assigned to oversee the work. In some states there is a backlog of sites which although known to the state agency, have not been the subject of a government order due to their relatively low priority. The state regulatory agencies which otherwise direct or oversee site remediation have begun to develop policies of encouraging PRPs to work with local government on these sites without state agency involvement. Working under such a proactive arrangement, outside of an agency order, clearly provides PRPs with an opportunity to collaborate among themselves in planning a cost-effective and equitably-shared remediation.

Beyond the avoidance of delays in an agency-driven process and the need to influence the work of a single party, less obvious benefits are also available through proactive cooperation among PRPs. A cooperative approach will necessarily place an appropriate emphasis on planning. This is important in assuring that the collective efforts of the group are tied to a unified set of project goals and objectives. Interim and individual actions can be reviewed for consistency with each other and with what is projected to be the final remedy. Early planning and collaboration between the parties also allows for their individual and collective needs to be identified from the beginning, and this will ensure that critical issues or "fatal flaws" in the plans of any single party are known at that time.

Additionally, when a cooperative approach is agreed to from the beginning, it improves the prospects for the group to develop a set of cost-allocation criteria *before* the investigation is undertaken. It is much less likely that the group would accept criteria for allocating responsibility and costs associated with the problem *after* the remedial investigation is completed and the facts are known by all. When the allocation criteria are developed up-front and accepted by the group at that time, the necessary investigations can proceed to gather the facts to which these criteria will be applied. In this way, the chances for an equitable allocation of costs are optimized.

Finally, a proactive and cooperative relationship among the parties will minimize extensive negotiations and litigation between them, and with the governmental agencies, thus reducing the overall costs of remediation.

With respect to groundwater remediation, the importance of cooperation can not be overstated. Most groundwater remediation is accomplished by pumping or extracting the contaminated groundwater, passing the contaminated water through some type of treatment process, and either reinjecting the 'clean' water or releasing the water to the sewer system.

The extraction process is designed to 'capture' contaminated water in the groundwater plume; in doing so, the localized flow patterns are altered. Ground-water from the local surrounding area and potentially under neighboring properties is drawn toward the extraction wells. This drawing of groundwater flow toward extraction wells on a property may achieve the property owners' objective to capture the plume for which they are responsible, but what does it do to the contaminant plume on the neighboring or upgradient properties? What does it do to the respective extraction and remediation systems on those properties? The consequence may be to make them inefficient or ineffective.

The problem of uncoordinated groundwater remediation efforts by multiple PRPs is comparable to multiple parties holding oil rights to a large reservoir and each party trying to produce their share of the oil independently. The consequences are the inefficient and costly extraction of the oil. A number of years ago the oil industry recognized this folly and today multiple parties join together in a production unit to extract oil more efficiently and cost-effectively. The same philosophy should be used, where applicable, to the extraction of groundwater contamination plumes.

DEFINING THE PROBLEM

Apportioning liability among multiple PRPs for an environmental investigation and remediation necessitates that the technical factors and the criteria on which they are based are largely defined from the beginning. Many environmental attorneys, regulators, engineers, and scientists hold that every site is unique and therefore the technical factors relative to that site are particular to it. In detail that may be true; however, a basic set of physical and chemical factors govern the occurrence, movement, and extraction of contaminants, and these factors can be used to quantify the contamination, the costs of cleanup, and apportion the liability equitably among the PRPs.

In any environmental investigation, quantification of the nature and extent of the contamination is the goal. The situation becomes more complicated when multiple PRPs are involved with multiple degrees of responsibility for the type and amount of contamination. With only one PRP involved, the investigation should focus on identification of the source of the extent of contamination. With a multi-PRP investigation, the source and extent of contamination also have to be identified, but with the additional burden of delineating and differentiating the percent of the contaminant from each source and its relative extent. With these added burdens, additional science must be used from the beginning and the investigation should be geared to address such factors as:

- Specific contaminants which are historically associated with each PRP source, and the chemical differences between each of the original contaminants
- Historic operations and practices which may have affected the release of contaminants from each source
- Estimated amounts of specific contaminants which potentially may have been released from each source
- Contaminant migration pathways and distance to which contaminants may have migrated from the original sources
- Chemical alterations to the original contaminants which may have occurred over time
- Heterogeneity of the contaminant plume and the concentrations and total amount of each contaminant in the plume.

DETERMINING PROPORTIONATE CLEANUP COSTS

Once the relative amounts of each contaminant and the magnitude of the source contribution to the plume or plumes have been defined, the alternative cleanup technologies and costs can be addressed. As in the case of only a single PRP, the alternative cleanup technologies must be evaluated considering the most feasible, cost-effective, and agency-acceptable solutions for the whole problem. In addition, "contaminant clean-up proportionate costs" must also be considered. The contaminant proportionate costs are the proportionate amount of the total cleanup costs that are due to a specific source or party. If the contaminants are different and are from different sources, assessing the proportionate costs may be relatively straightforward. If the contaminants are the same or similar and from similar sources, assessing the proportionate costs is more difficult.

Scenario No. 1: Proportionate Costs from Different Chemical Sources

It has been found during the investigation phase that two companies have sources that have contributed to a heterogeneous complex plume located downgradient of both companies. Company A, a bulk storage terminal facility, has contributed chlorobenzene, trichloroethylene (TCE), and tetrachloroethylene (PCE). Company B, a petroleum refinery, has contributed unrefined crude oil, gasoline, diesel, and heating fuel (see Figure 6.1). In this particular case the chemical contaminants in the plume are clearly distinguishable as to their relative sources: halogenated organic—Company A; hydrocarbon—Company B. The alteration products of the contaminants are also relatively distinguishable as to their origin. However, the cleanup criteria and the cleanup technologies relative to the different chemical contaminants are different, more difficult, and more costly due to their occurrence together.

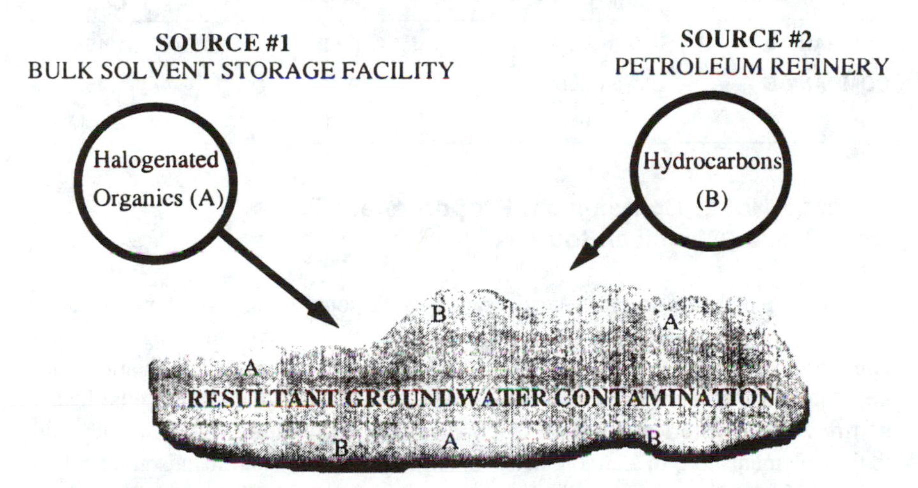

Figure 6.1. Hypothetical Case No. 1: groundwater contamination from distinctly different chemical sources.

It has been determined through a feasibility study that the best technologies to clean up the plume downgradient of the companies is to extract the groundwater and pass the contaminated water through a steam stripper. The rough-order-estimate of the initial costs is $2.5 M. It is recognized that if the chlorinated contaminants occurred in a separate plume, extraction costs would be about the same. An air stripper would be employed to treat the extracted water. The rough-order-estimate of the initial costs for extraction and treatment using an air stripper would be about $1.75 M. It is recognized that if the nonhalogenated hydrocarbons occurred in a separate plume, extraction costs would be about the same. The treatment system would employ aboveground bioremediation. The rough-order-estimate of the initial costs for extraction and treatment using bioremediation would be about $1.25 M.

The above analysis shows that if the contaminants had occurred as separate contaminant plumes, the cost to remediate each separately would have been less than the cost to remediate the complex plume. This is because of the more expensive technology necessary, the steam stripper, to treat the combination of chlorinated hydrocarbons, volatiles, semivolatiles, and polynuclear aromatic hydrocarbons. Proportionately, remediating the chlorinated hydrocarbons from the combined contaminant plume is more costly than remediating the petroleum hydrocarbons. That is because the costs are driven by the relative expense of the technology which must be used for the cleanup as well as the relative quantities of the contaminants. The proportionate costs, based on the above example, for Company A are $1.46 M and for Company B are $1.04 M, as shown in Table 6.1.

Table 6.1. Calculation of Proportionate Cleanup Costs

Scenario Number 1			
	Cost if Separate Plumes Existed	Proportionate Share	Actual Proportionate Cost
COMPANY A	$1.75M	0.58	$1.45M
COMPANY B	$1.25M	0.42	$1.05M
	$3.0M	1.00	$2.50M

Scenario No. 2: Contaminant Proportionate Costs from Similar Chemical Sources

Assessing the proportionate costs from similar sources may be very difficult. The key to proportioning the costs lies in the documentation of the magnitude and comparative contribution of the respective PRP. In part, the documentation may be company and agency records of the magnitude and frequency of spills, leaks, or fires. Other historical research, such air photographic interpretation, may add to the documentation. In addition, determining the relative contaminant source potential considering the transport distance of the chemical migration from the release point, the potential timing of the release, and the migration pathways will place boundaries on the amount of a PRP's contribution to a contaminant plume.

For example, it has been found through the investigation phase that two companies have sources that have contributed to a hydrocarbon plume located downgradient from both companies, as shown in Figure 6.2. Company A and Company B are both petroleum refining companies and many of their products are similar. Documentation exists to show leaks and spills of both crude oil and refined products. A careful geochemical study must be done to try to distinguish between the crudes used by the two refineries and the refined products. Chemical fingerprint typing, stable isotopes, trace metal indicators, and other techniques may be warranted, depending on the circumstances, to correlate the contaminant plumes to their respective sources. Transport modeling which includes mixing and reaction terms may be used to help delineate the relative contribution of contamination to the plume. Once the relative amount of contaminant contribution from

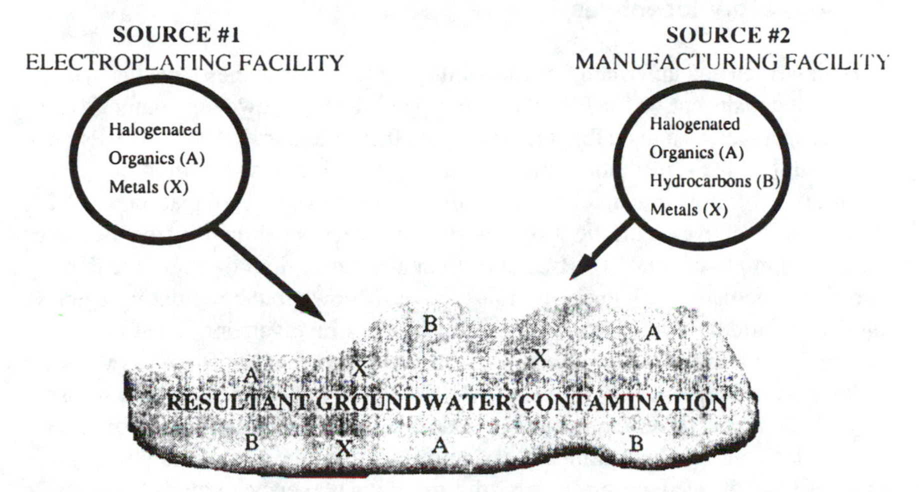

Figure 6.2. Hypothetical Case No. 2: groundwater contamination from similar chemical sources.

the two companies is determined, an expanded feasibility study is done to determine cleanup alternatives and costs.

THE NEED FOR CHANGE IN REGULATORY POLICY

For almost a decade, an inordinate emphasis has been placed upon the importance of government funding for site remediation. Presently there is a growing recognition that the nation's ability to clean up hazardous waste sites cannot be measured against the size of state and federal Superfunds. The role of government in this process needs to be reevaluated and much greater emphasis be placed upon the efforts of PRPs. Regulatory policy should be further developed to encourage proactive cooperations among PRPs. Following are three specific needs in this area.

1. Stronger Requirement for Prenotification of PRPs

The extent to which individual parties are associated with a groundwater contamination problem often remains vague until the time of cost recovery. Although the cost recovery provisions of state and federal laws require some degree of PRP notification, greater emphasis should be placed on the early identification and notification of PRPs. In the cases where a single party is proceeding with the assessment and remediation of site, the early identification of other PRPs by that party should be strongly tied to the party's ability to later recover costs.

2. Regulatory Incentives

Both written and unwritten policies of the regulatory agencies have historically provided a disincentive for PRPs to participate. Today, however, many governmental agencies recognize the important role that voluntary efforts of PRPs have in expediting site remediation and conserving public funds. It would now seem appropriate for government officials to offer incentives for participation by PRPs. One such incentive would be less government process and involvement of sites where voluntary effort of PRPs, rather than an agency directive, are the driving force in remediation. Historically, if a site were known to the regulatory agency, cleanup would have to be accompanied by agency involvement. The process by which this involvement took place was the same, whether the PRP was working voluntarily or pursuant to an order, or whether or not public funds were to be used.

With the recent development and availability of technical standards for site investigations and cleanup, a much less active role on the part of the government is necessary. In recognition of this, the regulatory agencies could direct PRPs to follow established guidance, with minimal regulatory oversight. Additionally, in those cases where some level of government involvement is required, a greater commitment to reduce review times and the waiving of government oversight fees could be offered where PRPs are participating voluntarily.

3. Technical Guidance and Criteria

The need for technical guidance in the conduct of site investigations and cleanups is widely recognized. Often, the uncertainties associated with site remediation have resulted in an inordinate degree of discretion on the part of both consultants and governmental officials, which has unnecessarily complicated and prolonged remedial actions. With a greater focus on voluntary cleanups, regulatory agencies should devote more effort to the development of guidance.

Beyond technical guidance for the conventional issues of remediation, there is a need for objective criteria by which the allocation of responsibility and costs among multiple parties could be derived. If such criteria were available, they would facilitate better working relationships among PRPs already committed to proceed with a site, as well as encourage the proactive participation of those parties who are not otherwise sufficiently motivated to act.

CHAPTER 7

Challenges Encountered in Hydrocarbon Contaminated Soil Cleanup

Andrew C. Lazzaretto, Department of Environmental Management, Santa Fe Springs, California

Much of my experience relating to the cleanup of hydrocarbon contaminated soils has been garnered from serving the city of Santa Fe Springs, California as a redevelopment consultant and project manager. Therefore, my comments will be centered on that community. To set the stage I believe it might be helpful for me to relate some of the history and background of Santa Fe Springs (SFS).

THE SETTING

The community was first founded as an agricultural settlement in the latter part of the nineteenth century, with virtually all of the farms and ranches either planted in orchards or engaged in raising cattle and livestock. The Southern Pacific Railroad had a line running through the area primarily to serve the needs of the ranchers and farmers.

The community at the time was known as Fulton Wells in honor of a large hotel complex which had been erected around a well-known mineral spring touted for its curative value. The local population had been aware for some time of the presence of brackish water in shallow wells and of the peculiar odor which permeated much of the surrounding area. The problem with the local water supply did, in fact, affect the agricultural prosperity of the area, but did not stop the founding of a number of impressive ranchos.

Between 1915 and early 1917 a number of petroleum engineers who recognized the telltale land forms and other physical clues, including the presence of large quantities of shallow methane (natural) gas, began exploring for oil. The discovery well for the Santa Fe Springs field was drilled by Union Oil of California on February 12, 1917. In October 1919, Union Oil began drilling "Bell 1," and after a little over two years of drilling, the well began producing over 2,500 barrels per day. Bell 1 spawned the subsequent growth and activity of the Santa Fe Springs field.

The field contains 10 producing zones, extending from a depth of 3,700 feet to approximately 10,000. The cumulative production of the oil field is approximately 600 million barrels of oil, of an estimated 1.6 billion barrels of oil in place. To date, only 37% of the oil has been recovered. The field was unitized by Mobil, and went into secondary recovery in 1971.

After the first successful wells began producing, literally hordes of companies raced to the community and began tying up property with oil leases. As soon as leases were secured, and sometimes even before, the drillers began erecting their rigs and went to the task of trying to remove the oil from the earth.

The petroleum specialists of the day believed that the oil was located in "pools" and the drilling strategy of the time called for the placing of as many wells into the pool as possible in order to draw out the petroleum. Because they feared being excluded when the pool ran dry, speculators rushed to drill as many wells as possible in the shortest possible time.

The California State Division of Oil and Gas (CDOG) did exist at the time and did require drillers to secure the standard permits. It is clear from the state records, however, that many permits were never pulled or inadequate processing took place which resulted in some serious gaps in the public record.

Other conditions impacted the Santa Fe Springs oil field. I mentioned earlier the presence of a great deal of methane gas in the area. In the very early days of the field, producers relied on the natural pressure within the formation to bring the oil to the surface. Pumping technology had not yet come into general use, which added to the frenzied approach many producers took. Again, they wanted to get their share before the "pool" went dry. In addition, blowout prevention (BOP) devices had not come into general use or were very crude at best. This situation led to a number of large well blowouts which lasted for weeks, and to many smaller accidents which were far less spectacular, but nonetheless had lasting effects on the environment and the soil in the area.

The last circumstance influencing the environmental legacy of the SFS oil field concerns the way virtually all the wells were drilled and the methods used in handling the crude oil once it was brought to the surface. During the 'discovery days' of the Santa Fe Springs field, the "tool and cable" method of drilling was employed. This was an effective means of well drilling for the time, but limited the depth to which wells could be drilled.

Soon, the oil industry developed the "rotary" method of drilling; the same method which is still in use today.

With the rotary method, drillers used a mixture of ingredients often containing chemical additives, in a clay-based slurry called 'drilling mud,' which was circulated in and out of the well during drilling. Some of the chemicals and additives used, like lead, are now known to be harmful. Today the industry uses Baker tanks to contain the drilling mud and to capture the cuttings.

In the early days of the SFS field, the drillers simply gouged a large excavation near the well and went to work. Similarly, when the well began producing, another pit was created onsite and the oil was allowed to flow into the hole, from which it was removed by hand "by the barrel." When the decision was made to abandon the well and the site, the crews simply covered over the sumps, usually after tossing in whatever spent equipment, trash, concrete, and miscellaneous debris they could find, and they left.

Needless to say, current operating practices are a far cry from what they were; for example, lead type additives are not used or permitted. However, the problems created by past mistakes must be identified and solved today. The legacy of the SFS field is awesome.

PUBLIC HEALTH CONCERNS

When faced with the prospect of approving a development plan, we attempt to answer a few very basic questions in an effort to assess the public health implications:

1. What is the current use on site?
2. What is the proposed use? What type of construction will be employed?
3. What previous use was the site subjected to?
4. What is the present condition of the site?
5. What are the surrounding uses?
6. Is there evidence of methane gas? If so, at what levels? Are mitigation measures warranted?

There are many others—but these are representative of the questions we ask developers and ourselves. While attempting to answer these questions we are faced with some very formidable problems.

Formal guidelines have yet to be adopted by the regulatory agencies for determining when the concentrations of petroleum hydrocarbons in soil will be considered "hazardous" or a threat to the waters of the state. The most serious recurring problem facing Santa Fe Springs is the presence of property contaminated primarily by petroleum hydrocarbons with concentrations between 100 plus and over 50,000 parts per million (ppm). The contamination is typically a result of the previous oil-drilling operations described earlier.

Because there are no clear-cut regulatory guidelines to follow, the city has had to develop an administrative methodology to deal with this ongoing problem. SFS has determined that when it is suspected that property may be contaminated with petroleum hydrocarbon wastes, in all such circumstances a site characterization study must be conducted by a qualified environmental engineer to determine the characteristics of the waste, and the lateral and vertical extent of the contamination. Based on this study, the city requires assurances from the developer that:

(1) the contamination is not ''hazardous waste'' subject to regulation by the California Department of Health Services (''DHS''),
(2) that the contamination poses no threat to the quality of surface and groundwater and therefore is not subject to regulation by the Regional Water Quality Control Board (''RWQCB''), and
(3) that the resulting development will not pose a threat to the eventual occupants of the premises.

Characterization studies most often follow a pattern containing three elements:

PHASE I —An investigation of available records; Fairchild Collection at Whittier College, the photo collection at the University of California at Santa Barbara, the Munger maps, Division of Oil and Gas records, city records, regulatory agency records and lists, etc.

PHASE II —Initial field investigation, which will include soil samples, trenching, boring, etc.

PHASE III—Additional boring and sampling to determine the lateral and vertical extent of the contamination found in Phase II. After site remediation has begun, it may be necessary to conduct additional tests to refine the findings of Phase III or to determine the parameters of contamination found after the site work is under way.

In the absence of clearer guidelines, the notification of regulatory agencies and remediation requirements for contaminated properties has had to involve a case-by-case assessment. However, through the school of experience and by working regularly with agencies like DHS and RWQCB, Santa Fe Springs has propagated practical and reasonable standards for development. For example, if total petroleum hydrocarbons (TPH) is 1,000 ppm or below and poses no threat to groundwater, it would be considered acceptable and would be permitted to be left in place. The same would be true if gasoline or diesel fuel did not exceed 100 ppm. We have adopted similar guidelines for the presence of methane gas. Based on the knowledge we've acquired to date, the city is contemplating the adoption of an ordinance mandating tests for gas. In the case of metals, the state DHS standards are followed.

CASE STUDIES

During the recent past the city of Santa Fe Springs has taken on the role of lead agency in the cleanup of a number of properties. These cases involved land which was owned by the city as redevelopment agency, and was being prepared for sale and/or development.

Let me share a few of our more interesting experiences:

Bloomfield East

This 3+ acre parcel presented us with a number of challenges:

- over $100,000 site characterization study;
- including magnometer to discover anomalies to help to pinpoint problems. Could not establish correlation;
- two large, deep cesspools (discovered when the grading equipment fractured the dome);
- one unknown, previously unabandoned well;
- many pockets of discarded trash and debris;
- Bell 1—converted to a water well. Almost six (6) months to abandon at a cost of over $500,000.

Bloomfield West

- same initial problems with site characterization, $150,000 or more in costs;
- leaking pipelines;
- difficulty in locating wells;
- eventually found four unknown wells, one deserted and three previously abandoned. One of the three was full of cement from surface to bottom; DOG required the abandonment go all out.
- excavations reached over 40 ft chasing contaminated soil.
- ended up with 14,000 cubic yards of soil which had to be remediated. Bioremediation now underway.
- cost of clean up $1,100,000. Approximately 152,460 square feet = $7.22 per sq ft.
- Methodology:
  a. historical investigation.
  b. perimeter trenching, removal of pipelines, surface samples, borings, diagonal trenching, and potholing.
  c. vertical and lateral testing.
  d. excavation and grading.
  e. remediation.

LESSONS LEARNED

A. Do a comprehensive site characterization study, but don't go to extremes. Put limited faith in the results because you will almost always be surprised once you start the remediation process. If faced with the choice of doing more tests or starting the excavation, go with the excavation. You'll get more mileage for your money, and even if you perform more tests, you'll never have all the answers anyway.

B. If you have to go to the regulatory agencies, be well prepared. The agencies are getting more sophisticated and more knowledgeable, and are better prepared to assist you. There is still too much delay and bureaucracy, but things are improving. Some of the agencies are beginning to realize which are the more threatening situations versus those that are more routine and less of an environmental threat.

C. Nothing can replace experience and good sound judgment; e.g., degraded crude oil vs refined product.

D. Mobility is always an issue with the water board.

E. Having ongoing legal counsel is very important. Don't get paranoid, but much of what you face requires judgment. It is good to have a well centered attorney who knows environmental law helping you to brainstorm critical issues.

F. New technology is needed to speed the remediation process. Now, it takes many months and a lot of area to handle oil field waste "in situ." The only other alternative is expensive "dig and haul," and not practical for the future.

G. Regulatory agencies need to adopt a more routine approach to approving requests to clean up crude oil derived contamination. These are not "Superfund" sites we're dealing with, and the agencies should recognize the difference.

H. Local agencies, e.g., cities, should adopt the use of indemnity agreements to protect themselves and to save property owners and developers the unnecessary time and expense of going through a complete site assessment.

CHAPTER 8

Sampling and Analysis of Soils for Gasoline Range Organics

Jerry L. Parr, Gary Walters, and **Mike Hoffman,** Enseco Incorporated, Arvada, Colorado

This chapter presents findings from a laboratory study sponsored by the American Petroleum Institute (API) designed to establish a reliable method for sampling and analyzing gasoline range organics in soil. The basis for this project was a recognition by professionals in the petroleum industry that the current approaches were inadequate due to the following concerns:

- loss of volatile organics during sampling and sample handling, resulting in significant bias,
- wide variety of laboratory techniques used to measure "gasoline," providing data of varying quality,
- the generally poor documentation of the performance of methods used to measure "gasoline."

The approach developed to remedy this situation consisted of three phases. Phase I was a survey of existing technology to select procedures for evaluation. This phase included a workshop from which a consensus approach of field and laboratory professionals was achieved. This approach defined the experimental design for Phases II and III.

Phase II was a Single Laboratory Method Validation study of the methods selected in Phase I. (Note: The term validate is used to indicate that the performance characteristics of the method are known and the method has been standardized.) The preferred method validated in Phase II was then further evaluated as part of a field study in Phase III. This phase also included a comparison to other existing sampling and analysis technologies used to measure petroleum hydrocarbons.

PHASE I

The Phase I study accomplished two goals. First, the objectives of the analytical method to be implemented were defined. Second, based on these objectives, sampling and analytical procedures were selected for evaluation.

The objectives were as follows:

- develop a protocol which fully integrates sampling, sample handling, and the laboratory procedure,
- rely upon existing technology,
- focus on gasoline range organics, defined as compounds in the C_6 to C_{10} boiling point range, at environmental (ppb-ppm) levels,
- consider regulatory guidelines, and
- document the performance of the method.

Given these basic objectives, a general approach was defined.

This general approach is based on gas chromatography with flame ionization detection (GC/FID). This technology provides the following advantages:

- ability to quantitatively measure all organic compounds in the gasoline range,
- ability to generate a ''finger print'' which can be used to provide supplemental information,
- ability to configure a photoionization detector (PID) in conjunction to measure volatile aromatics, and
- sensitivity to hydrocarbons in the nanogram range.

In addition to selecting GC/FID as the technology, another basic concept was defined. A synthetic mixture of aliphatic and aromatic compounds present in gasoline was defined. (See Table 8.1.) This synthetic mixture was selected to eliminate bias resulting from the use of commercial ''gasolines'' as calibration standards. Chromatograms of the synthetic mixture (method standard) and gasoline are shown in Figures 8.1 and 8.2.

PHASE II

With these two basic concepts, a variety of options were selected for evaluation. First, two options were selected to evaluate for the sample introduction into the GC/FID, headspace, and purge and trap. For purge and trap, both direct purge

Table 8.1. Synthetic Reference (Method Standard)

Compound	Weight Percent
2-Methylpentane	15
2,2,4-Trimethylpentane	15
Heptane	5
Benzene	5
Toluene	15
Ethyl benzene	5
m-Xylene	10
p-Xylene	10
o-Xylene	10
1,2,4-Trimethylbenzene	10

and methanol extraction were evaluated. For each of the laboratory options, integrated sampling and sample handling options were evaluated.

Work performed as part of the Phase II method validation consisted of determining the performance characteristics of the headspace and purge and trap GC/FID methods. The performance characteristics determined were:

- linear range,
- method detection limits
- bias/interference from petroleum products, and
- matrix effects by soil type

The results from the Phase II laboratory studies are shown in Tables 8.2 and 8.3. As shown in these tables, all three techniques are viable methods for measuring gasoline range organics in soil. Each technique offers advantages and disadvantages as summarized below.

Table 8.2. Detection Limit/Dynamic Range/Bias

			Accuracy (as Percent Recovery)		
Method Option	PQL, μg/g	Upper Limit, μg/g	Sand	Clay	Loam
Headspace	40	500 ±	83	66	50
Methanol Extraction, Purge and Trap	5	100[a]	70	68	60
Direct Purge and Trap	0.1	2	84	19	18

[a]Can be extended via dilutions.

Table 8.3. Method Bias to Petroleum Products

	Accuracy (as Percent Recovery)	
Product	Headspace	Purge and Trap
Gasoline	84	50–70
Jet Fuel (JP-4)	80	72
Kerosene	8	2
Diesel	4	4

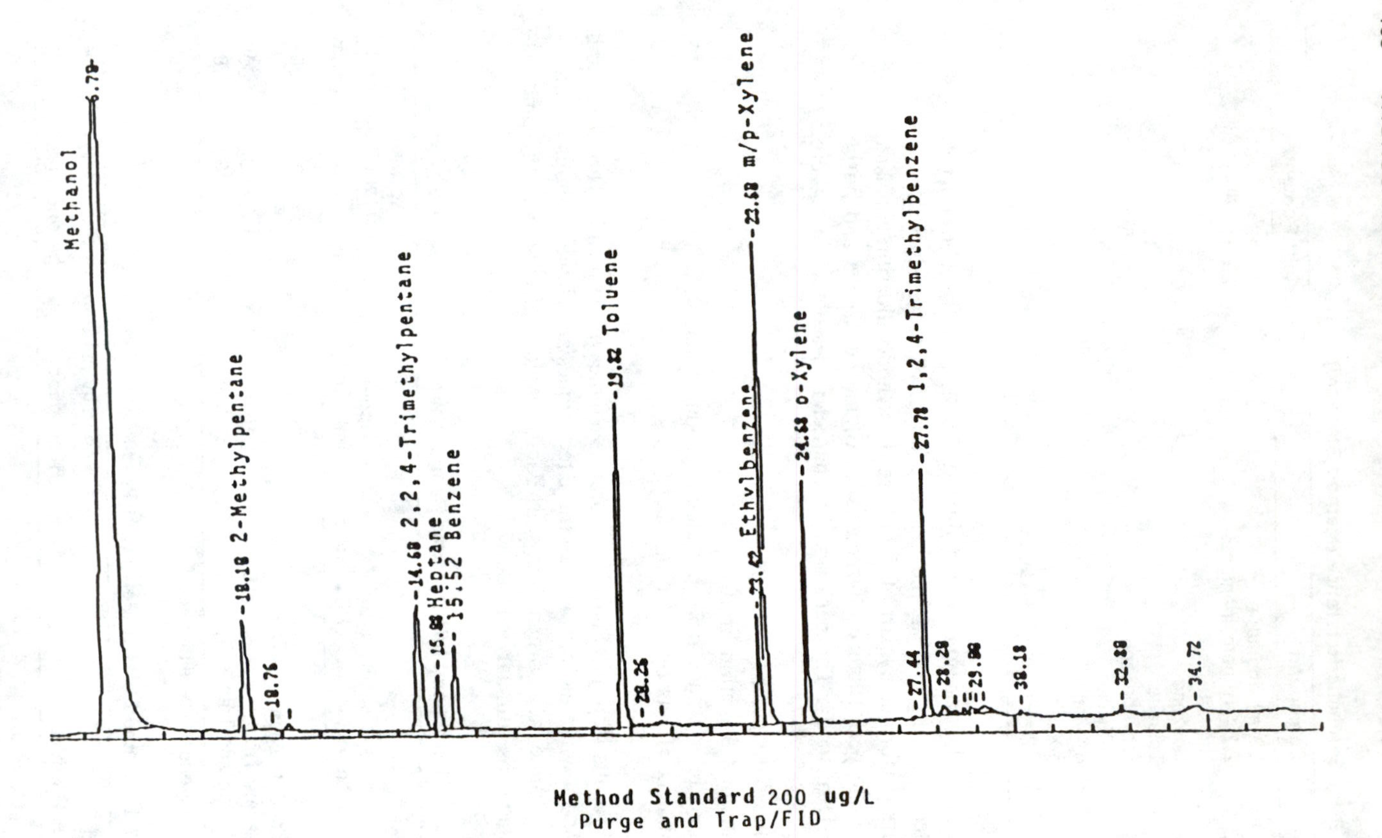

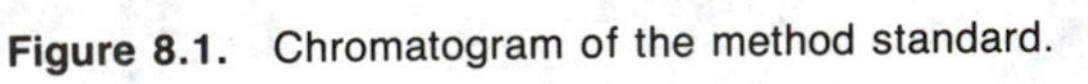
Figure 8.1. Chromatogram of the method standard.

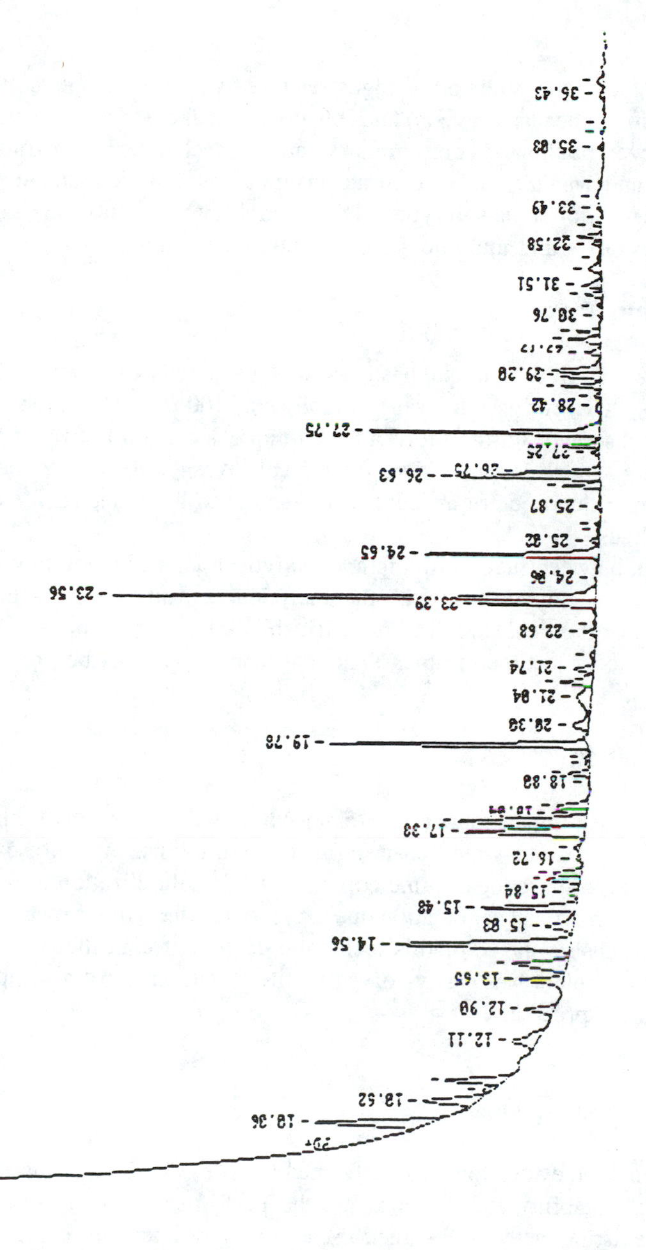

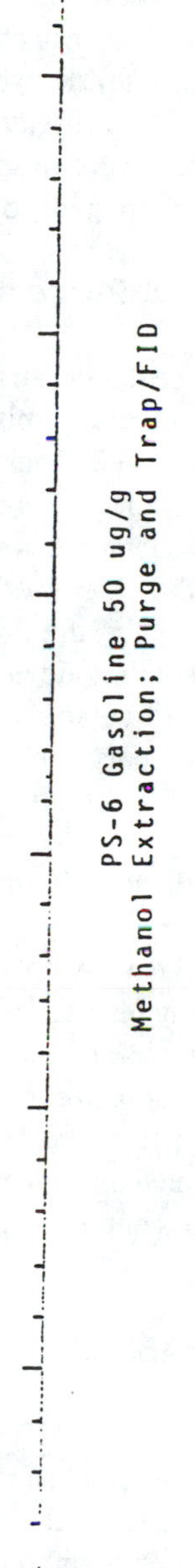

Figure 8.2. Chromatogram of gasoline.

Headspace

The headspace technique offers advantages relative to ease of sample collection (field collection in headspace vials) and low operational cost (as a screening technique). However, as shown in the summary data, the technique is constrained in both sensitivity and dynamic range. Accurate quantitation is laborious and matrix effects were observed for some soil types. The technology is not in wide use by commercial laboratories and multiple samples must be collected.

Direct Purge and Trap

The direct purge and trap technique has the lowest sensitivity, with a practical quantitation limit (PQL) for gasoline range organics of 100 ppb. However, significant bias was observed. Furthermore, the technique is constrained by an upper limit of 2 μg/g, which is impractical for typical investigations. The use of the technique would thus require an additional sample to be collected for one of the other methods.

Finally, the link between field sampling and analytical determination requires a sampling device that can interface with the analytical instrumentation with no loss of volatiles. For the field study in Phase III, this link was accomplished by using purge tubes to collect the samples. This approach would not be practical in most situations.

Methanol Extraction

The methanol extraction/purge and trap technique has moderate sensitivity, little if any bias on soil type, and a wide dynamic range via dilutions. Field preservation with methanol stabilizes the organic components via solubilization and furthermore minimizes microbial degradation due to the toxic effects of the methanol.

In fact, the only disadvantage of this technique centers around the field use of methanol. This problem was easily solved by the development of a shipping and sample handling procedure.

PHASE III

The Phase III method evaluation was performed to (1) evaluate sampling techniques plus sample stability and (2) compare the performance of the selected method to other existing methods designed to measure gasoline. This evaluation was performed by conducting a stability study plus sampling and analyzing soil containing gasoline in both laboratory and field conditions. The laboratory study was performed by spiking gasoline onto soil and collecting multiple samples. The field study was performed by collecting samples from a site known to be contaminated with gasoline.

The sampling techniques and analytical procedures shown in Table 8.4 were evaluated. Comparison of the "other methods" to the methanol addition, purge and trap method are in Table 8.5.

Table 8.4. Evaluation of Sampling Techniques and Analytical Procedures

Sampling Technique	Analytical Method
GRO Method	
Core sampler	Purge and trap GC/FID
Sealed purge tube	Purge and trap GC/FID
Conventional bottle	Purge and trap GC/FID
Methanol addition	Purge and trap GC/FID
Other Methods	
Conventional bottle	8020 (BTEX)
Conventional bottle	8240 (GC/MS)
Conventional bottle	TPH-IR
Conventional bottle	California LUFT Manual

Table 8.5. Comparison of Other Methods to the Methanol Addition, Purge and Trap Method

Method	Measured Concentration, µg/g	
	Field Sample	Lab Spike
GRO-Methanol Extraction	130[a]	8.3[b]
GRO-Direct Purge	NA	6.6[b]
LUFT	64[a]	0.96[a]
BTEX (total)	5.9[a]	1.8[a]
8240	51	0.82
TPH-IR	ND (50)	ND (50)

[a]Average of duplicate runs
[b]Average of triplicate runs
NA (Not Analyzed) Field sample exceeds linear range of Direct Purge and Trap.

Field samples were collected from an abandoned service station site in Laramie, Wyoming. For the Gasoline Range Organics (GRO) analyses, soil was sampled directly into 10 mL of methanol in a 40 mL vial for the methanol extraction technique, and a second sample for the direct purge analysis was collected using a 3/4-inch capped purge tube. All other techniques were sampled by traditional sampling methods (filling a 7 oz jar in the field, then subsampling smaller vials for each test). The lab spike was prepared at 50 µg/g and sampled in the same manner as the field sample.

The GRO tests have higher concentrations than the other evaluated methods. Since the TPH-IR Method has poor recovery for gasoline, Not Detected (ND) at 50 µg/g is expected. The BTEX analysis only measures selected components of gasoline. The lower values from LUFT and 8240 for gasoline are at least in part related to the sampling technique. As shown in Table 8.6, BTEX measured by Method 8020 in the field and lab GRO samples is substantially higher.

A separate study indicated that soil samples collected in methanol or capped purge tubes are stable up to 28 days for the measurement of Gasoline Range Organics and BTEX. An evaluation of traditional sampling techniques and a core sampler demonstrated losses greater than 90% after one month.

Table 8.6. Comparison of Sampling Technique for BTEX Results

	BTEX Concentration, μg/g	
Sampling Technique	**Field Sample**	**Lab Spike**
Conventional Bottle	5.9	1.8
Methanol Stabilization	19	2.9
Sealed Purge Tube	NA	2.6

CONCLUSIONS

Based on the results of this work, the preferred technique for measuring gasoline range organics in soil consists of field preservation with methanol followed by purge and trap GC/FID. A recommended method is contained in this chapter's Appendix. As compared to the other sampling and analytical methods studied, this technique has the following advantages:

- field stabilization/preservation of volatiles,
- reasonable sensitivity,
- large dynamic range (via dilutions),
- widespread availability,
- appropriate for gasoline range, and
- minimal matrix effects.

This work will be described in more detail in an API research report scheduled to be completed in late 1990.

APPENDIX A

Method for Determination of Gasoline Range Organics

1. Scope and Application

1.1 This method is used to determine the concentration of gasoline range organics in water and soil. Gasoline or other specific petroleum products are identified by the use of pattern recognition techniques.

1.2 The Practical Quantitation Limit (PQL) of this method for gasoline range organics is approximately 5 μg/g for the methanol extraction method (soils), and 100 μg/L for ground water.

1.3 This method is based on a purge-and-trap, Gas Chromatography (GC) procedure. This method should be used by, or under supervision of analysts experienced in the use of purge-and-trap systems and gas chromatographs. The analysts should be skilled in the interpretation of gas chromatograms and their use as a quantitative tool.

1.4 With the optional PID detector, this method can be extended for the specific determination of volatile aromatics (BTEX) as specified in EPA Method 8020.

2. Summary of Method

2.1 This method provides gas chromatographic conditions for the detection of certain volatile petroleum fractions such as gasoline. Samples are analyzed utilizing purge-and-trap sample concentration. The gas chromatograph is temperature programmed to facilitate separation of organic compounds. Detection is achieved by a flame ionization detector (FID) or FID with photoionization detector (PID) in series (photoionization detector first in the series).

2.1.1 Identification of various petroleum products is performed by comparison of the chromatograms of samples and commercial products. Quantitation is based on FID detector response to an external standard (10 component method standard).

2.2 Water samples can be analyzed directly for gasoline range organics by purge-and-trap extraction and gas chromatography. Soil or solid samples are dispersed in methanol to dissolve the volatile organic constituents. A portion of the methanolic solution is analyzed by purge-and-trap GC following the normal water method.

2.3 Special field sampling techniques are recommended to minimize the loss of volatiles from soil by using conventional sampling and sample handling techniques. Collection of small volume soil core samples is considered to be the more reliable means of minimizing VOC losses from the samples when compared to placing soil in larger jars which require later subsampling, and which will be subject to the resultant volatile losses during handling.

2.4 This method is based on USEPA SW-846 methods 5030, 8000, 8020, 8015, and a single laboratory method evaluation study conducted by the American Petroleum Institute.

3. Definitions

3.1 Gasoline Range Organics (GRO): All chromatographic peaks eluting between 2-methyl pentane and 1,2,4-trimethylbenzene. Quantitation is based on a direct comparison of the area within this range to the total area of the 10 components in the method standard.

3.2 Method Standard: A 10 component blend of typical gasoline compounds (Table 8.7). This standard serves as a quantitation standard and a retention time window-defining mix for gasoline range organics. It may also be used as the PID calibration standard for the optional determination of BTEX by Method 8020.

3.3 Gasoline Control Standard: A commercial gasoline used by the laboratory as a quality control check.

3.4 Surrogate Control Sample: A reagent water or method blank sample spiked with the surrogate compounds used in the method. The surrogate recovery is used to evaluate method control.

3.5 Laboratory Control Sample: A reagent water or method blank sample spiked with the gasoline control standard. The spike recovery is used to evaluate method control and must be greater than 50%.

3.6 Pattern Recognition Standards: Various commercial gasolines and other petroleum products used by the laboratory to identify petroleum products.

3.7 Other terms are as defined in SW-846.

Table 8.7. Method Standard Components and Concentrations

Component	Relative Concentration Weight%
2-Methylpentane	15
2,2,4-Trimethylpentane	15
Heptane	5
Benzene	5
Toluene	15
Ethylbenzene	5
m-Xylene	10
p-Xylene	10
o-Xylene	10
1,2,4-Trimethylbenzene	10

4. Interferences

4.1 High levels of heavier petroleum products such as diesel fuel may contain some volatile components producing a response within the retention time range for gasoline. Other organic compounds, including chlorinated solvents, ketones, and ethers are measurable. As defined in the method, the GRO results include these compounds.

4.2 Samples can become contaminated by diffusion of volatile organics through the sample container septum during shipment and storage. A trip blank prepared from reagent water and carried through sampling and subsequent storage and handling can serve as a check on such contamination.

4.3 Contamination by carryover can occur whenever high-level and low-level samples are sequentially analyzed. To reduce carryover, the sample syringe and/or purging device must be rinsed between samples with reagent water or solvent. Whenever an unusually concentrated sample is encountered, it should be followed by an analysis of a solvent blank of reagent water to check for crosscontamination. For volatile samples containing high concentrations of water-soluble materials, suspended solids, high boiling compounds or organohalides, it may be necessary to wash the syringe or purging device with a detergent solution, rinse with distilled water, and then dry in a 105°C oven between analyses. The trap and other parts of the system are also subject to contamination; therefore, frequent bake-out and purging of the entire system may be required. A screening step is recommended to protect analytical instrumentation.

5. Safety Issues

5.1 The toxicity or carcinogenity of each reagent used in this method has not been precisely defined; each chemical compound should be treated as a potential health hazard. Exposure to these chemicals must be reduced to the lowest possible level by whatever means available. The laboratory is responsible for maintaining a current awareness file of OSHA regulations regarding the safe handling of the chemicals specified in this method. A reference file of material safety data sheets should also be made available to all personnel involved in the chemical analysis. Additional references to laboratory safety are available and have been identified for the information of the analyst.

6. Apparatus and Materials

6.1 Gas Chromatograph

6.1.1 Gas Chromatograph: Analytical system complete with gas chromatograph suitable for purge-and-trap sample introduction and all required accessories, including detectors, column supplies, recorder, gases and syringes. A data system capable of determining peak areas is recommended.

6.1.2 Columns:

6.1.2.1 Column 1: 105-m × 0.53 mm I.D. Restek RTX 502.2 0.3 micron film thickness, or equivalent.

6.1.2.2 Other columns may be used—capillary columns are recommended to achieve necessary resolution. At a minimum, the column should resolve 2-methylpentane from the methanol solvent front in a 25 μg/g LCS standard and should resolve ethylbenzene from m/p-xylene ($< 25\%$ valley). Some columns may require subambient cooling to achieve these guidelines.

6.1.3 Detector: Flame ionization (FID), or FID in series with a Photoionization detector (PID).

6.2 Syringes: 5-mL Luerlock glass hypodermic and a 5-mL gas-tight syringe with shutoff valve.

6.2.1 For purging large sample volumes for low detection limit analysis of aqueous samples for petroleum products, 25- or 50-mL syringes may be used. Subsequently substitute the appropriate volume in the method wherever 5-mL is stated when low detection limits are required.

6.3 Volumetric flask: 10-, 50-, 100-, 500-, and 1,000-mL with a ground-glass stopper.

6.4 Microsyringes: 1-μL, 5-μL, 10-μL, 25-μL, 100-μL, 250-μL, 500-μL, and 1,000-μL.

6.5 Syringe valve: Two-way, with luer ends (three each), if applicable to the purging device.

6.6 Balance: Analytical, capable of accurately weighing to the nearest 0.0001 g, and a top-loading balance capable of weighing to the nearest 0.1 g.

6.7 Glass scintillation vials: 20-mL, with screw-caps/crimp caps and Teflon® liners or glass culture tubes with a screw-cap and Teflon® liner, or equivalent.

6.8 Spatula: Stainless Steel.

6.9 Disposable pipets: Pasteur.

6.10 Purge-and-trap device: The purge-and-trap device consists of three separate pieces of equipment: the sample purger, the trap, and the desorber. Several complete devices are commercially available.

6.10.1 The recommended purging chamber is designed to accept 5-mL samples with a water column at least 3-cm deep. The gaseous headspace between the water column and the trap must have a total volume of less than 15 mL. The purge gas must pass through the water column as finely divided bubbles with a diameter of less than 3 mm at the origin. The purge gas must be introduced no more than 5-mm from the base of the water column. The sample purger, illustrated in Figure 8.3 meets these design criteria. Alternate sample purge devices may be used, provided equivalent performance is demonstrated.

6.10.2 The trap must be at least 25 cm long and have an inside diameter of at least 0.105 in. Starting from the inlet, the trap must be packed with the following adsorbents: 1/3 of 2,6-diphenylene oxide polymer, 1/3 of silica gel, and 1/3 of coconut charcoal. It is recommended that 1.0 cm of methyl silicone-coated packing be inserted at the inlet to extend the life of the trap (see Figures 8.4 and 8.5). If the determination of dichlorodifluoromethane or other fluorocarbons of similar volatility is not necessary, the charcoal can be eliminated and the polymer increased to fill 2/3 of the trap. If only compounds boiling above 35°C are to be analyzed, both the silica gel and charcoal can be eliminated and the polymer increased to fill the entire trap. Prior to initial use, the trap should be conditioned overnight at 180°C by backflushing with an inert gas flow of at least 20 mL/min. Vent the trap effluent to the hood, not to the analytical column. Prior to daily use, the trap should be conditioned for 10 min. at 180°C with backflushing. The trap may be vented to the analytical column during daily conditioning; however, the column must be run through the temperature program prior to analysis of samples.

6.10.3 Another alternate trap uses 7.6-cm Carbopack B and 1.3-cm Carbosieve S-III (Supelco Cat# 2-0321R). This trap should be desorbed at 240°C and baked to 300°C.

6.10.4 The desorber should be capable of rapidly heating the trap to 180°C for desorption. The polymer section of the trap should not be heated higher than

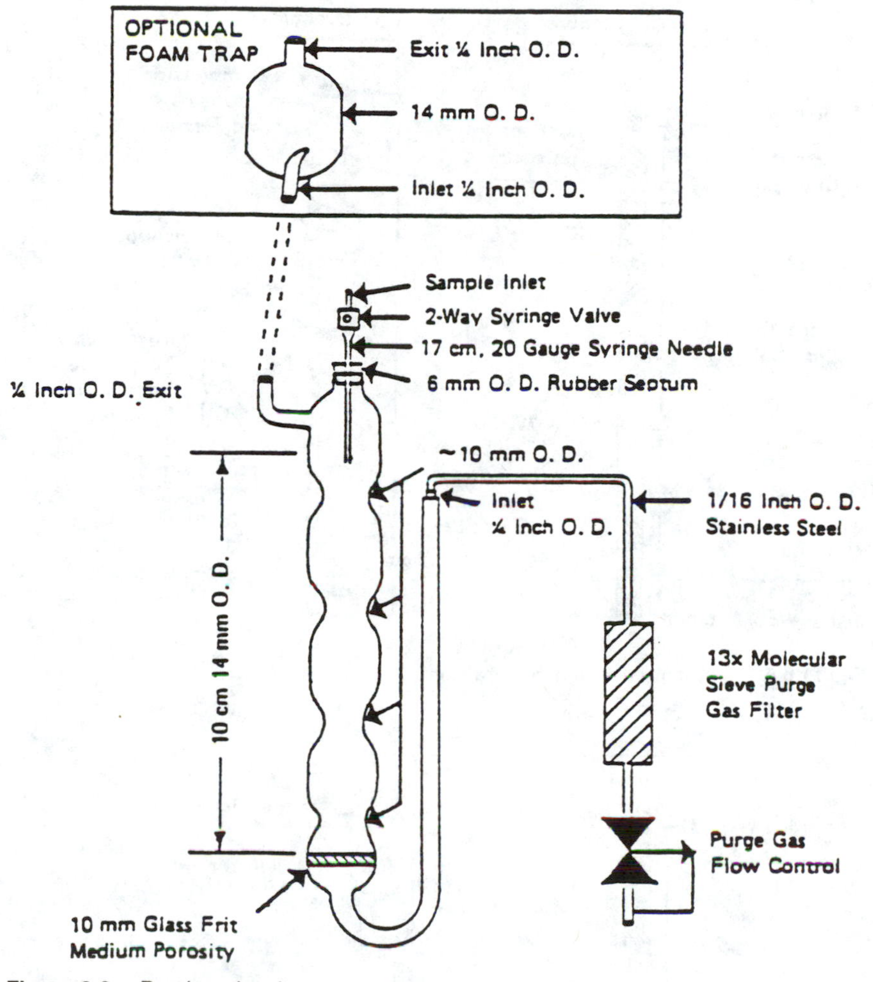

Figure 8.3. Purging chamber.

180°C, and the remaining sections should not exceed 220°C during bake-out mode. The desorber design illustrated in Figures 8.4 and 8.5 meet these criteria.

6.10.5 The purge-and-trap device may be assembled as a separate unit or may be coupled to a gas chromatograph, as shown in Figures 8.6 and 8.7.

6.10.6 Trap Packing Materials

6.10.6.1 2,6-Diphenylene oxide polymer: 60/80 mesh, chromatographic grade (Tenax GC or equivalent).

6.10.6.2 Methyl silicone packing: OV-1 (3%) on chromosorb-W,60/80 mesh or equivalent.

6.10.6.3 Silica Gel: 35/60 mesh, Davison, grade 15 or equivalent.

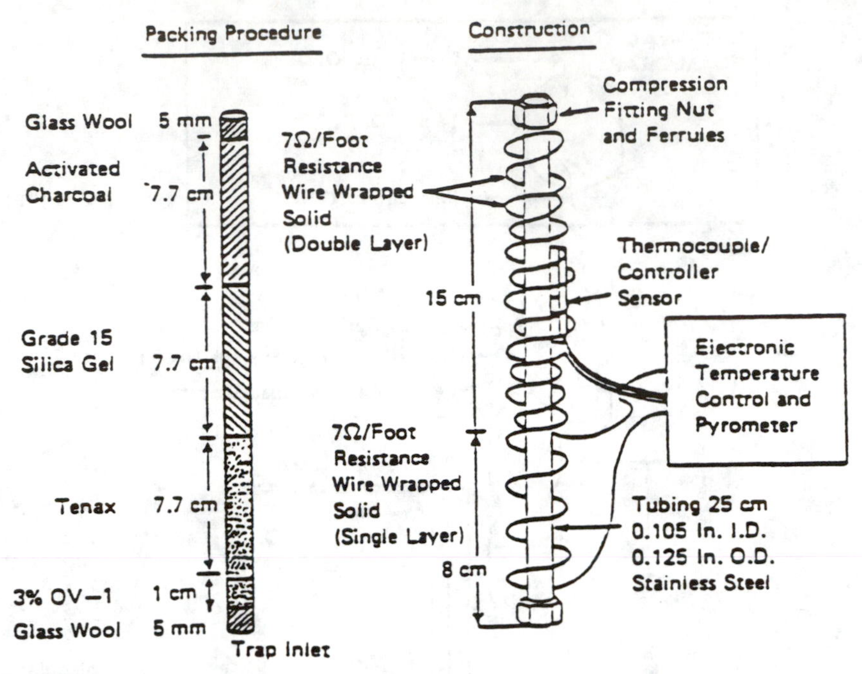

Figure 8.4. Trap packings and construction.

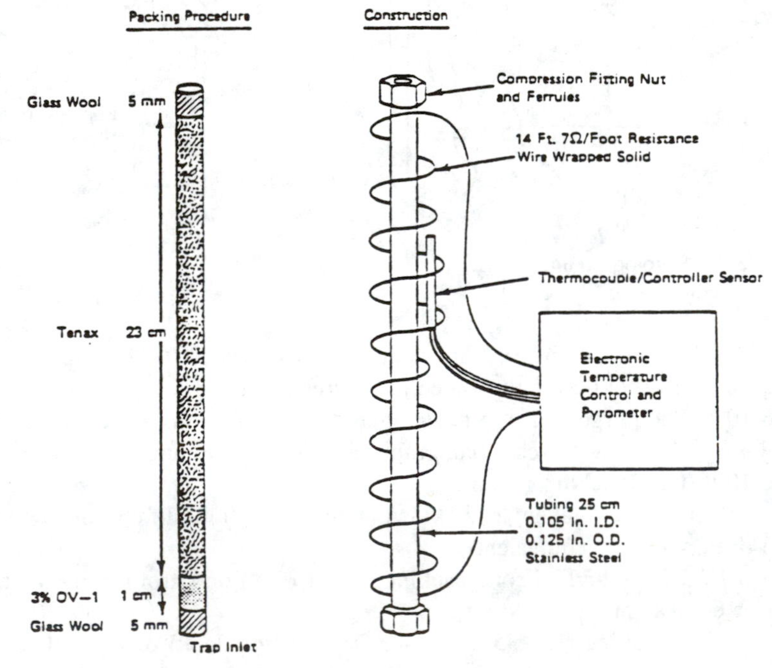

Figure 8.5. Trap packing and construction.

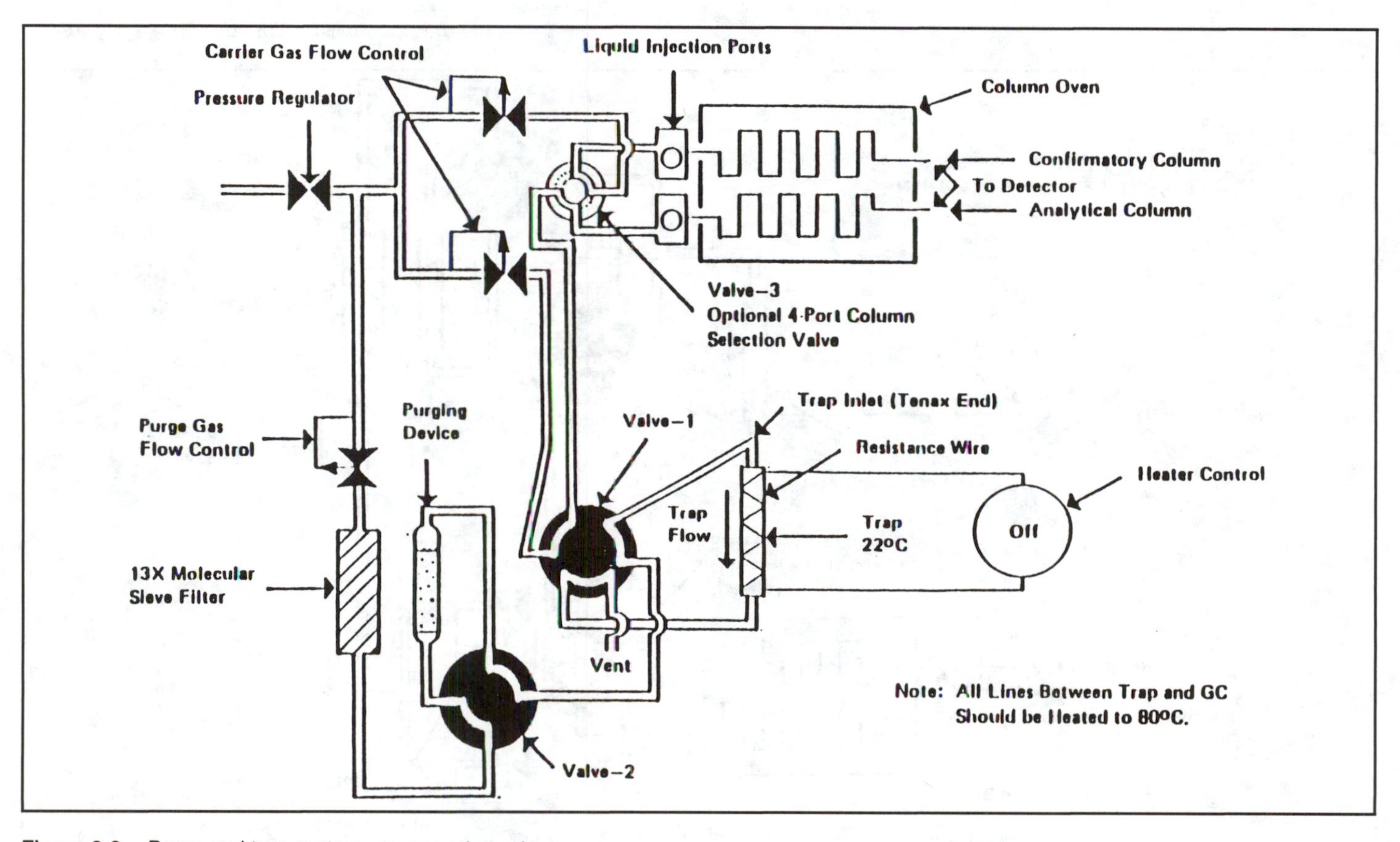

Figure 8.6. Purge-and-trap system, purge-sorb mode.

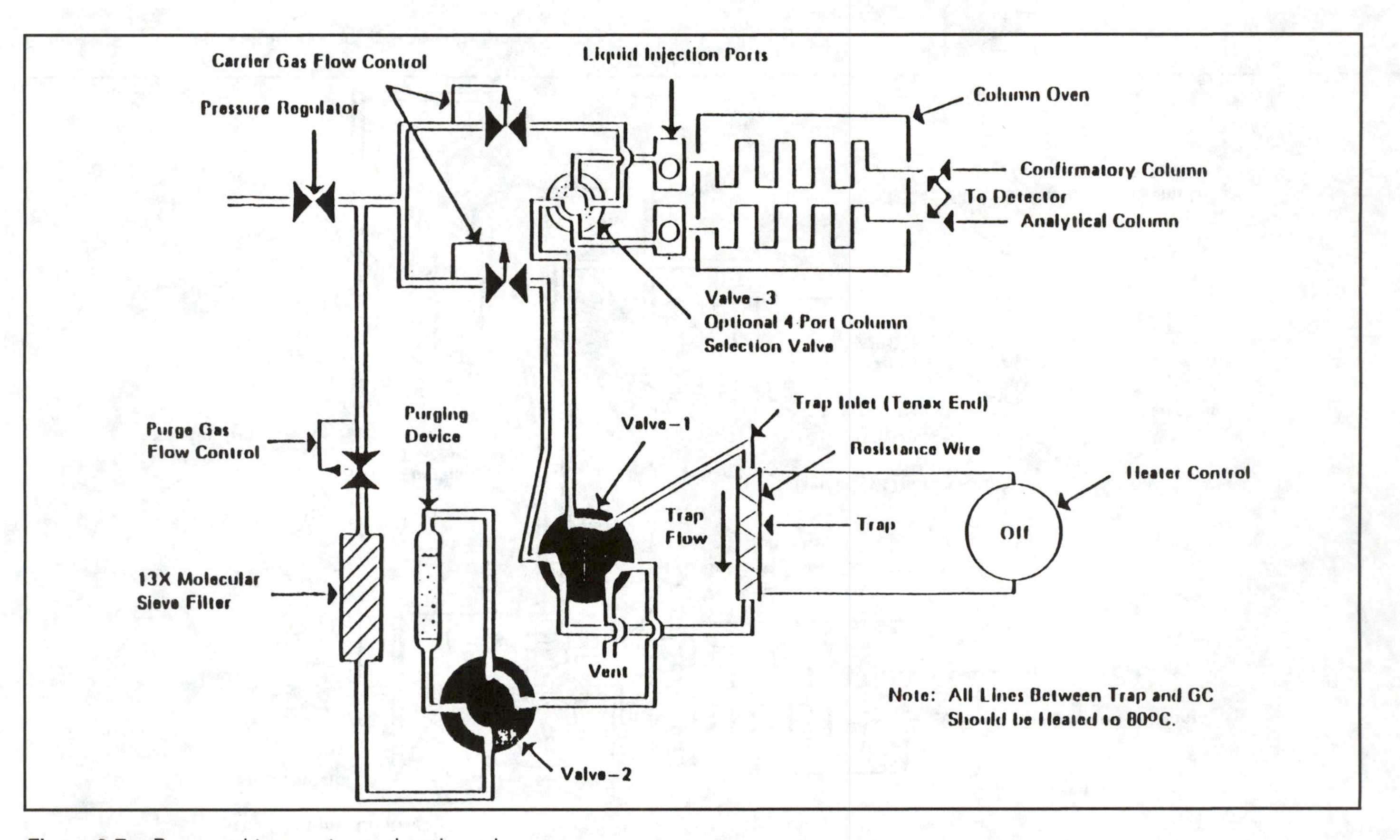

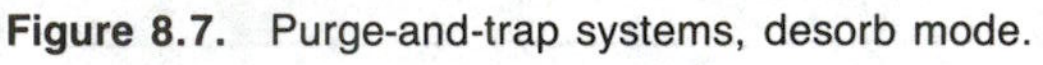

Figure 8.7. Purge-and-trap systems, desorb mode.

6.10.6.4 Coconut charcoal: Prepare from Barnebey Cheney, CA-580-26 lot #M-2649, by crushing through 26 mesh screen.

7. Reagents

7.1 Reagent Water: Carbon-filtered water purged with helium prior to use.

7.2 Gasoline Control Standards: One reference standard is API PS-6 gasoline, a characterized gasoline used in petroleum research. (Major components in Table 8.8). Other gasolines can be used if they are thoroughly evaluated by the laboratory. At a minimum, the laboratory should generate precision and accuracy data to establish control limits as specified in Method 8000.

Table 8.8. Major Components of API PS-6 Gasoline

Compound	Percent Weight
2-Methylbutane	8.72
m-Xylene	5.66
2,2,4-Trimethylpentane	5.22
Toluene	4.73
2-Methylpentane	3.93
n-Butane	3.83
1,2,4-Trimethylbenzene	3.26
n-Pentane	3.11
2,3,4-Trimethylpentane	2.99
2,3,3-Trimethylpentane	2.85
3-Methylpentane	2.36
o-Xylene	2.27
Ethylbenzene	2.00
Benzene	1.94
p-Xylene	1.72
2,3-Dimethylbutane	1.66
n-Hexane	1.58
1-Methyl, 3-ethylbenzene	1.54
1-Methyl, 4-ethylbenzene	1.54
3-Methylhexane	1.30

Source: Reference #6.

7.3 Method Standard: The 10 component quantification standard which also serves as the quantitation range (retention time window defining mix) standard. The components and concentration (weight %) are in Table 8.7. The standard is prepared by adding the appropriate volume of each component to yield the desired weight %.

7.4 Stock Standards: Prepare stock standards for the gasoline and method standards in methanol.

7.4.1 Place about 9.8 mL of methanol in a 10-mL tared ground-glass stoppered volumetric flask. Allow the flask to stand, unstoppered, for about 10 min. or until all alcohol-wetted surfaces have dried. Weigh the flask to the nearest 0.1 mg.

7.4.2 Using a 100-μL syringe, immediately add two or more drops of gasoline or method standard to the flask; then reweigh. The liquid must fall directly into the alcohol without contacting the neck of the flask.

7.4.3 Reweigh, dilute to volume, stopper, and then mix by inverting the flask three times. Calculate the concentration in micro grams per microliter (μg/μL) from the net gain in weight. When compound purity is assayed to be 96% or greater, the weight may be used without correction to calculate the concentration of the stock standard. Commercially prepared stock standards may be used at any concentration if they are certified by the manufacturer or by an independent source.

7.4.4 Transfer the stock standard solution into a Teflon®-sealed screw-cap/crimp cap bottle. Store, with minimal headspace, at −10°C to −20°C and protect from light.

7.4.5 Standards must be replaced after 6 months, or sooner if comparison with check standards indicates a problem.

7.5 Secondary Dilution Standards: Using stock standard solutions, prepare secondary dilution standards in methanol, as needed. The secondary dilution standards should be prepared at concentrations such that the aqueous calibration standards prepared in Section 7.6 will bracket the working range of the analytical system. Secondary dilution standards should be stored with minimal headspace for volatiles and should be checked frequently for signs of degradation or evaporation, especially just prior to preparing calibration standards from them.

7.6 Calibration Standards: Calibration standards at a minimum of three concentration levels are prepared in reagent water from the secondary dilution of the stock standards. One of the concentration levels should be at a concentration near, but above, the method detection limit. The remaining concentration levels should correspond to the expected range of concentrations found in real samples or should define the working range of the GC.

7.7 Internal Standard: An internal standard (1-chloro-4-fluorobenzene) is recommended for 602/8020 quantitation on the PID detector. Due to potential interferences, the internal standard is not recommended for FID quantitation.

7.8 Surrogate Control Standard (SCS): The analyst should monitor both the performance of the analytical system and the effectiveness of the method in dealing with each sample matrix by spiking each sample, standard, and reagent water blank with one or two surrogate compounds, bromofluorobenzene, or trifluorotoluene. From stock standard solutions prepared as in Section 7.4 prepare a surrogate spiking solution at 50 μg/mL of each surrogate in methanol. Add 5.0 μL of this surrogate spiking solution directly into the 5-mL syringe with every sample and reference standard analyzed.

7.9 Laboratory Control Sample (LCS) Standard: From the stock PS-6 gasoline standard or other appropriate gasoline control standards (Section 7.4) prepare a secondary dilution standard at 500 μg/mL in methanol. Addition of the following amounts yields the indicated concentrations:

0.005 mL added to 5 mL water: 500 μg/L
0.5 mL added to 10 g soil (methanol extraction): 25 μg/g

7.10 Methanol: pesticide quality or equivalent. Store away from other solvents.

8. Sample Collection, Preservation and Handling

8.1 Aqueous samples should be collected in triplicate without agitation and without headspace in contaminant-free glass 40-mL vials with Teflon®-lined septa in the caps. The Teflon® layer must contact the sample. Sample vials should contain 200 μL of 50% HCL as a preservative for aromatic analytes. Refrigerate samples at 4°C after collection.

8.2 Soil samples require special procedures to minimize the loss of volatiles during transit from the field to laboratory. Samples for the methanol extraction method should be collected in duplicate tared 40-mL vials that contain 10 mL methanol (includes 0.5 mL of surrogate solution at 50 μg/mL). A reagent methanol blank should be prepared in the same manner as the sample vials. Soil for the vials can be collected using a 10-mL plastic syringe with the end sliced off. A sufficient number of vials (2 recommended) should be collected to provide for backup analyses in the event of breakage. A soil volume of 6–8 mL corresponds to about 10 g. In addition, soil should be collected in a wide mouth glass jar with a Teflon®-lined lid for soil screening analysis prior to analysis of vial samples. The soil should be disturbed as little as possible and the containers filled as full as possible. Refrigerate all samples at 4°C after collection.

8.3 According to SW-846(2), soils for volatile organic analysis must be held at 4°C for a maximum of 14 days.

8.4 For reference, an API study (4) has indicated that samples sampled (preserved) in methanol can be held for up to 28 days at 4°C with no apparent losses. Samples taken by conventional techniques are subject to volatile losses throughout their storage period. These losses may exceed 90% after 28 days.

9. Procedure

9.1 Volatile compounds are introduced into the gas chromatograph by purge-and-trap. Purge-and-trap may be used directly on ground water samples. Soils and solids should be analyzed by methanol extraction. It is highly recommended that all samples be screened prior to analysis. This screening step may be analysis of a solid sample's methanol extract (diluted), the headspace method (SW-846 Method 3810), or the hexadecane extraction and screening method (SW-846 Method 3820). See Table 8.9.

9.2 Gas Chromatography conditions (recommended)

9.2.1 Column 1: Set helium column pressure to 20#. Set column temperature to 40°C for 1 min, then 5°C/min to 100°C, then 8°C/min to 240°C and

Table 8.9. Quantity of Methanol Extract Required for Analysis of Soils/Sediments

Approximate Concentration, GRO	Volume of Methanol Extract[a]
5–100 μg/g	100 μL
200 μg/g	50 μL
1000 μg/g	10 μL
5000 μg/g	100 μL of 1/50 dilution[b]

Calculate appropriate dilution factor for concentrations exceeding this table.
[a]The volume of methanol added to 5-mL of water being purged should be kept constant. Therefore, add to the 5-mL syringe whatever volume of methanol is necessary to maintain a volume of 100 μL added to the syringe.
[b]Dilute an aliquot of the methanol extract and then take 100 μL for analysis.

hold for 7.5 min. Conditions may be altered to improve resolution of gasoline range organics.

9.2.2 Other columns-set GC conditions to meet the criteria in 6.1.2.2.

9.3 Calibration:

9.3.1 Prepare final solutions containing required concentrations of calibration standards, including surrogate standards, directly in the 5-mL glass syringe. Add the aliquot of calibration solution directly to the reagent water in the glass syringe by inserting the needle through the syringe end. When discharging the contents of the microsyringe, be sure that the end of the syringe needle is well beneath the surface of the reagent water. Similarly, add 5.0-μL of the surrogate standard solution. Attach the two-way syringe valve to the syringe and then inject the standard into the purge vessel through the two-way valve. Proceed with purge-and-trap analysis procedure.

9.3.2 Run the method standard at a minimum of three concentration levels above the detection limits and covering the expected range of samples or the linear range of the instrument. The recommended calibration range (and corresponding method amounts are shown in Table 8.10.

Table 8.10. Method Standard

Nanograms to Detector	Water μL	Soil-MeOH extraction μg/g
250 (12.5 to 37.5)[a]	50	2.5
1000 (50 to 150)	200	10
2500 (125 to 375)	500	25

[a]Nanograms per individual component in parentheses.

An additional low point at 10 μg/L (0.5 to 1.5 μg/L for individual aromatics) is recommended for the optional PID quantitation. For the FID quantitation of a multicomponent product like gasoline, the linear range is related to the areas of individual components. Individual components in the method standard are 3 to 5 times the concentration of the same components in PS-6 gasoline. Therefore, considering the calibration curve, 500 μg/L of the method standard is the high point on the curve; but gasoline at 2000 μg/L (100 μg/g in soil) is within the range of the calibration curve.

9.3.3 External Standard Calibration

9.3.3.1 For quantification utilizing the method standard response, prepare calibration standards at a minimum of three concentration levels by adding appropriate volumes of the stock standards and surrogate standards to a 5-mL glass syringe. One of the external standards should be at a concentration near, but above, the method detection limit. The other concentrations should correspond to the expected range of concentrations found in samples or should define the working range of the detector. Due to potential carry over, do not purge more than 10 μg of gasoline in 5 mL of water (2000 μg/L).

9.3.3.2 Inject each calibration standard utilizing the purge-and-trap. Tabulate area response for the ten components against mass injected. The results can be used to prepare a calibration curve for the detector. Alternatively, the ratio of the response to the amount injected, defined as the calibration factor (CF), can be calculated for each analyte at each standard concentration. If the percent relative standard deviation (% RSD) of the calibration factor is less than 25% over the working range, linearity through the origin can be assumed, and the average calibration factor can be used in place of a calibration curve.

$$\text{Calibration Factor} = \frac{\text{Total area of peak}}{\text{Standard amount } (\mu g/mL)}$$

9.3.3.3 The working calibration curve or calibration factor must be verified on each working day by the injection of a continuing calibration standard. If the response for the method standard varies from the predicted response by more than 25% a new calibration curve must be prepared.

$$\text{Percent Difference} = \frac{R1 - R2}{\text{Ravg.}} \times 100$$

where:

R1 = Average calibration from the calibration curve.
R2 = Calibration factor from continuing calibration
Ravg. = (R1+R2) /2

9.4 Retention Time Window and Pattern Recognition

9.4.1 Before establishing windows, be certain that the GC system is within optimum operating conditions. Make three injections of the method standard throughout the course of a 72-hour period. Serial injections over less than a 72-hour period result in retention time windows that are too tight.

9.4.2 Calculate the standard deviation of the three absolute retention times for each method standard component.

9.4.2.1 The retention time window for individual peaks is defined as plus or minus three times the standard deviation of the absolute retention time for each component. For multiresponse petroleum products, the analyst may use the retention time window but should primarily rely on pattern recognition.

9.4.2.2 In those cases where the standard deviation for a particular analyte is zero, the laboratory should use +0.05 min as a retention time window.

9.4.3 The laboratory must calculate retention time windows for each standard on each GC column and whenever a new GC column is installed. The data must be retained by the laboratory.

9.4.4 The analyst should generate a value for both gasoline range organics and gasoline. Identification of gasoline is performed by comparing the retention times and patterns of the peaks in the sample chromatogram to those of the peaks in the standard chromatogram. The experience of the analyst weighs heavily in the interpretation of the chromatogram. Quantitation of the gasoline range organics is based on summation of all peaks eluting between 2-methylpentane and 1,2,4-trimethylbenzene. Quantitation of gasoline may include all the same peaks or the analyst may eliminate some peaks. Peaks could be deleted due to unusual peak shape or if they did not match the gasoline pattern. References 7, 8, 9, and 10 contain some background information on hydrocarbon pattern recognition. Environmental samples may contain more than one type of product, and loss of light end components may mean the product has been in the subsurface a longer period of time.

9.4.4.1 Other organic compounds, including chlorinated solvents, ketones, and ethers are measurable by this method and will be reported as gasoline range organics. These interference compounds should not be included in the gasoline total area.

9.4.4.2 Note: Although the retention time window definition (2-methylpentane to 1,2,4-trimethylbenzene) introduces a bias (55% to 75% for gasoline in Ottawa Sand), it improves precision and reduces interferences from non-gasoline components.

9.5 Gas Chromatograph Analysis:

9.5.1 Water Samples: Introduce volatile compounds into the gas chromatograph using the purge-and-trap method. Add 5.0-μL of surrogate standard to the sample prior to purging.

9.5.1.1 Adjust the purge gas flow rate (nitrogen or helium) to 25–40 mL/min on the purge-and-trap device.

9.5.1.2 Remove the plunger from a 5-mL syringe and attach a closed syringe valve. Open the sample or standard bottle, which has been allowed to come to ambient temperature, and carefully pour the sample into the syringe plunger and compress the sample. Open the syringe valve and vent any residual air while adjusting the sample volume to 5.0 mL. This process of taking an aliquot destroys the validity of the liquid sample for future analysis; therefore, if there is only one 40-mL vial, the analyst should fill a second syringe at this time to protect against possible loss of sample integrity. This second sample is maintained only until such time when the analyst has determined that the first sample has been analyzed properly. Filling one 5-mL syringe would allow the use of only one syringe. If a second analysis is needed from a syringe, it must be analyzed within 24 hours. Care must be taken to prevent air from leaking into the syringe.

9.5.1.3 The following procedure is appropriate for diluting purgeable samples. All steps must be performed without delays until the diluted sample is in a gas-tight syringe.

9.5.1.4 Dilutions may be made in volumetric flasks (10-mL to 100-mL). Select the volumetric flask that will allow for the necessary dilution. Intermediate dilutions may be necessary for highly concentrated samples.

9.5.1.5 Calculate the approximate volume of reagent water to be added to the volumetric flask selected and add slightly less than this volume of reagent water to the flask.

9.5.1.6 Inject the proper aliquot of samples from the syringe prepared in Paragraph 9.5.1.2 into the flask. Aliquots of less than 1 mL are not recommended. Dilute the sample to the mark with reagent water. Cap the flask, invert, and shake three times. Repeat the above procedure for additional dilutions. Alternatively the dilutions can be made directly in the glass syringe to avoid further loss of volatiles.

9.5.1.7 Fill a 5-mL syringe with diluted sample as in Paragraph 9.5.1.2.

9.5.1.8 Add 5.0 μL of surrogate spiking solution through the valve bore of the syringe; then close the valve.

9.5.1.9 Attach the syringe-syringe valve assembly to syringe valve on the purging device. Open the syringe valves and inject sample into the purging chamber.

9.5.1.10 Close both valves and purge the sample for 12 min.

9.5.1.11 At the conclusion of the purge time, attach the trap to the chromatograph, adjust the device to the desorb mode, and begin the gas chromatographic temperature program and GC data acquisition. Concurrently, introduce the trapped materials to the gas chromatographic column by rapidly heating the trap to 180°C and backflushing the trap with inert gas between 20 and 60 mL/min for 4 minutes.

9.5.1.12 While the trap is desorbing into the gas chromatograph, empty the purging chamber. Wash the chamber with a minimum of two 5-mL flushes of reagent water (or methanol followed by reagent water) to avoid carryover of pollutant compounds into subsequent analyses.

9.5.1.13 After desorbing the sample, recondition the trap by returning the purge-and-trap device to the purge mode. Wait 15 sec; then close the syringe valve on the purging device to begin gas flow through the trap. The trap temperature should be maintained at 180°C. Trap temperatures up to 220°C may be employed; however, the higher temperature will shorten the useful life of the trap. After approximately 7 to 35 min, turn off the trap heater and open the syringe valve to stop the gas flow through the trap. When cool, the trap is ready for the next sample.

9.5.1.14 If the initial analysis of a sample or a dilution of the sample has a concentration of analytes that exceeds the initial calibration range, the sample must be reanalyzed at a higher dilution. When a sample is analyzed that has a saturated response from a compound, this analysis must be followed by a blank reagent water analysis. If the blank analysis is not free of interferences, the system must be decontaminated. Sample analysis may not resume until a blank can be analyzed that is free of interferences.

9.5.1.15 All dilutions should keep the response of the major constituents (previously saturated peaks) in the upper half of the linear range of the curve.

9.5.2 Methanol Extraction for Soil/Sediment: Method is based on extracting the sediment/soil with methanol. The soil sample is either extracted or diluted, depending on solubility in methanol. An aliquot of the extract is added to reagent water. This is purged at the temperatures indicated in Table 8.11. A screening analysis is recommended (see 9.1).

Table 8.11. Purge and Trap Operating Parameters

	Analysis Method
	8020
Purge gas	Nitrogen or Helium
Purge gas flow rate (mL/min)	40
Purge time (min)	12.0 ± 0.1
Purge temperature (°C)	Ambient
Desorb temperature (°C)	180
Backflush inert gas flow (mL/min)	20–60
Desorb time	4

9.5.2.1 If available, obtain the field sample collected in methanol (Section 8.2). Weigh the sample vial to determine the actual weight. Shake for 2 min. Proceed to 9.5.2.4. If the field sample is not available, proceed to 9.5.2.2.

9.5.2.2 The sample (for volatile organics) consists of the entire contents of the sample container. Do not discard any supernatant liquids. In order to obtain representative analytical results, gently mix the contents of the sample container with a narrow metal spatula. For sediment/soil and waste that are insoluble in methanol, weigh 10 g (wet weight) of sample into a tared 20-mL vial, using a top-loading balance. Note and record the actual weight to 0.1 gram. For waste that is soluble in methanol, weigh 1 g (wet weight) into a tared scintillation vial or culture tube or a 10-mL volumetric flask. (If a vial or tube is used, it must be calibrated prior to use. Calibrate by pipeting 10.0 mL of methanol into the vial and marking the bottom of the meniscus. Discard this solvent.)

9.5.2.3 Quickly add 9.5 mL of methanol; then add 0.5 mL of the surrogate spiking solution (50 μg/mL) to the vial. Cap and shake for 2 min.

Note: Steps 9.5.2.2 and 9.5.2.3 must be performed rapidly and without interruption to avoid loss of volatile organics. These steps must be performed in a laboratory free from solvent fumes.

9.5.2.4 Allow sediment to settle, centrifuge if necessary. Pipet approximately 1 mL of the extract to a GC vial for storage, using a disposable pipet. The remainder may be disposed. If not analyzed immediately, these extracts must be stored at 4°C in the dark.

9.5.2.5 The GC system should be set up as in Section 9.0. This should be performed prior to the addition of the methanol extract to reagent water.

9.5.2.6 Table 8.9 can be used to determine the volume of methanol extract to add to the 5 mL of reagent water for analysis. If a screening procedure was followed, use the estimated concentration to determine the appropriate volume. The maximum volume of methanol is 100 μL. All dilutions must keep the response of the major constituents (previously saturated peaks) in the upper half of the linear range of the curve.

9.5.2.7 Remove the plunger from a 5.0-mL Luerlock type syringe equipped with a syringe valve and fill until overflowing with reagent water. Replace the plunger and compress the water to vent trapped air. Adjust the volume to 4.9 mL. Pull the plunger to 5.0 mL to allow volume for the addition of the sample extract and of surrogate standard. Add the volume of methanol extract determined from screening and a volume of methanol solvent to total 100 μL (excluding methanol in standards).

9.5.2.8 Attach syringe valve assembly to syringe valve on the purging device. Open the syringe valves and inject the sample into the purging chamber.

9.5.2.9 Proceed with the analysis as in 9.5.1.10-9.5.1.15. Analyze all reagent blanks on the same instrument as that used for the samples. The reagent blank should contain 100 μL of the methanol used to extract the samples.

9.5.3 Samples are analyzed in a set referred to as an analysis sequence. The sequence begins with instrument calibration followed by sample extracts interspersed with continuing calibration standards. The sequence ends when the set of samples has been injected or when qualitative and/or quantitative QC criteria are exceeded.

9.5.4 If the responses exceed the linear range of the systems, use a smaller amount of sample.

9.5.5 The calibration factor for each analyte to be quantitated must not exceed + 25% when compared to the initial standard of the analysis sequence. When this criteria is exceeded, inspect the GC system to determine the cause and perform whatever maintenance is necessary prior to recalibration and proceeding with sample analysis. All samples that were injected following the sample exceeding QC criteria must be reanalyzed.

9.6 Calculations:

9.6.1 External Standard Calibration: The concentration of Gasoline Range Organics in the sample is determined by calculating the absolute weight of analyte purged, from a summation of peak response for all chromatographic peaks eluting between 2- methylpentane and 1,2,4-trimethylbenzene, using the calibration curve or the calibration factor determined in paragraph 9.3.3. Refer to Section 9.4 (Retention Time Windows and Pattern Matching). The concentration of Gasoline Range Organics is calculated as follows:

Aqueous samples:

Concentration (μg/L) = [(Ax)(A)(D)]/[(As)(Vs)]
of Gasoline Range Organics

Where:

Ax = response for the Gasoline Range Organics in the sample, units in area.
A = Absolute weight of standard purged, μg.
As = Response for the external standard, units same as for Ax.
D = Dilution factor, if dilution was performed on the sample prior to analysis. If no dilution was made, D = 1, dimensionless.
Vs = Volume of sample extracted or purged, L.

Nonaqueous Samples (methanol extraction):

Concentration (μg/g) = [(Ax)(A)(D)Vt]/[(As)(W)Vi]
of Gasoline Range Organics

Where:

Vt = Volume of total extract (μL) (use 10000 μL or a factor of this when dilutions are made).
Vi = Volume of extract added for purging (μL)
W = Weight of sample extracted, g. The wet weight is used.

Ax, As, A, and D have the same definition as for aqueous samples.

9.6.2 The concentration of gasoline is determined using the techniques in 9.6.1. The peak area used may be the same as used in 9.6.1, or the analyst may eliminate peaks which are not representative of gasoline (see 9.4.4).

10. Quality Control

10.1 The laboratory must, on an ongoing basis, demonstrate through the analysis of quality control check standards that the operation of the measurement system is in control.

10.2 After successful calibration (Section 9.3), analyze a Surrogate Control Sample. This standard is also the reagent blank sample and is analyzed with every analytical batch or sequence. The surrogate recovery should be within established limits (Table 8.12) and the sample should not have Gasoline Range Organics above the practical quantitation limit.

Table 8.12. Acceptance Criteria for LCS and SCS

	Spike Concentration		Control Limits	
Analyte	Water μg/L	Soil μg/g[a]	% Recovery	Relative % Difference
LCS				
Gasoline Range Organics	500	25	50–100	20
SCS				
Trifluorotoluene	50	2.5	50–150	

[a]Methanol extraction method.

10.3 Every batch or 20 samples, duplicate Laboratory Control Samples must be analyzed. The accuracy and precision of the duplicate standards must be within established limits (Table 8.12).

10.4 If any of the criteria in 10.2 and 10.3 are not met, the problem must be corrected before samples are analyzed.

10.5 Calculate the surrogate standard recovery in each sample. If recoveries are outside established limits, verify calculations, dilutions and standard solutions. Verify instrument performance.

10.5.1 High recoveries may be due to a coeluting matrix interference—examine the sample chromatogram.

10.5.2 Low recoveries may be due to the sample matrix.

10.5.3 Low recoveries may be due to a poor purge (clogged purge tube). If this is suspected, reanalyze the sample while observing the purge tube.

10.6 Field Blanks, duplicates, and matrix spikes are recommended for specific sampling programs.

11. Method Performance

11.1 Single-lab method performance data for the methanol extraction method in Ottawa Sand and other soil types is presented below. Chromatograms for the method standard and PS-6 gasoline are in Figures 8.1 and 8.2.

11.2 Results for PS-6 spikes (Methanol extraction purge and trap) are shown in Table 8.13.

Table 8.13. Results for PS-6 Spikes (Methanol Extraction Purge and Trap)

Spike	PS-6 Spike Amount μg/g	Percent Recovery
Ottawa Sand	50	70
Ottawa Sand	500	78
Houston Black Clay	50	68
Houston Black Clay	50	66
Norwood Loam	50	60
Norwood Loam	50	57

11.3 The method detection limit calculated according to 40 CFR, Part 136, Appendix B was 0.5 μg/g gasoline for the methanol extraction of soils. The recommended Practical Quantitation Limit (PQL) is 5 μg/g.

REFERENCES

1. ASTM. ''Standard Methods for Comparison of Waterborne Petroleum Oils by Gas Chromatography,'' 3328–78.
2. USEPA. ''SW-846 Test Methods for Evaluating Solid Waste,'' 3rd Edition; Methods 5030, 8000, 8015, and 8020.
3. ASTM. ''Standard Practice For Sampling Waste and Soils for Volatile Organics'' Draft #1, 2/16/87.
4. American Petroleum Institute. ''Sampling and Analysis of Gasoline Range Organics in Soils,'' in preparation.
5. Parr, J. ''Sampling and Analysis of Soils for Gasoline Range Organics'' presented at First Annual West Coast Conference Hydrocarbon Contaminated Soils and Groundwater, 2/21/90.
6. American Petroleum Institute. ''Laboratory Study on Solubilities of Petroleum Hydrocarbons in Groundwater,'' API Publ. 4395, August, 1985.
7. ''Leaking Underground Fuel Tank (LUFT) Field Manual,'' State Water Resources Control Board, State of California, Sacramento, CA, May, 1988.
8. Fitzgerald, J. ''Onsite Analytical Screening of Gasoline Contaminated Media Using a Jar Headspace Procedure,'' in *Petroleum Contaminated Soils,* Vol. 2, (Chelsea, MI: Lewis Publishers, Inc., 1989).
9. Senn, R. B., and M. S. Johnson. ''Interpretation of Gas Chromatographic Data in Subsurface Hydrocarbon Investigations,'' *Ground Water Monitoring Review,* 1987.
10. Hughes, B. M., D. E. McKenzie, C. K. Trang, L. S. R. Minor. ''Examples of the Use of an Advanced Mass Spectrometric Data Processing Environment for the Determination of Sources of Wastes,'' presented at 5th Annual Waste Testing and Quality Assurance Symposium, July 24-28, 1989.

CHAPTER 9

A New Method for the Detection and Measurement of Aromatic Compounds in Water

John D. Hanby, Hanby Analytical Laboratories, Inc., Houston, Texas

INTRODUCTION

In the introduction to his encyclopedic treatise, *Friedel-Crafts Alkylation Chemistry: A Century of Discovery,* Royston M. Roberts makes the statement, "Probably no other reaction has been of more practical value." Professor Roberts goes on to say, "Major processes for the production of high octane gasoline, synthetic rubber, plastics and synthetic detergents are applications of Friedel-Crafts chemistry."[1] It is fitting that, over a century after this monumental discovery, a technique for the analysis of environmental contamination caused by-products of this reaction has been developed which employs the same chemistry.

The analysis of organic compounds in aqueous solution has long been recognized as problematical for many reasons. Primary among them, of course, is the limited solubility of nonpolar compounds in such an extremely polar solvent. In a recent laboratory study undertaken for the American Petroleum Institute (API) on the solubilities of petroleum hydrocarbons in groundwater, it was pointed out that it was not possible to obtain linear response when trying to directly inject water standards of various aromatic hydrocarbons into a gas chromatograph.[2] This irreproducibility in analysis of water samples has been a source of consternation to proponents of gas chromatography for a long time. A fairly comprehensive review of the sort of problems associated with the gas chromatography of water samples is presented by Grob in Chapter 5 of *Identification and Analysis of Organic Pollutants in Water* in their argument for the use of capillary versus packed column GC. A statement from that reference is particularly appropriate, "Environmental chemistry includes probably the most extreme branch of analytical chemistry . . . environmental samples should be analyzed with means and methods to provide maximum separation efficiency and resolutions."[3]

Certainly chromatography of all types has proved to be a technique of "maximum separation efficiency and resolution." With the advent of capillary columns of thousands of theoretical plates of separation efficiency, the ability to resolve picogram quantities of substances is available. However, the problem of obtaining representative samples and their subsequent quantitative as well as qualitative analysis remains as perhaps the dominant problem in environmental assessment. Among the criteria involved in sampler design discussed by Johnson et al. in a recent article in *Ground Water* are those which would "prevent changes in the analyte concentration due to: (1) sorption or degradation in the well; (2) changes in temperature or pressure; (3) cross-contamination between monitoring wells due to the sampling equipment."[4]

Each of these criteria might also be applied not only to the collection of samples but to their analyses as well. In the subsequent laboratory analysis of a sample which may have been very well collected, preserved, and transported to the laboratory, each of the above factors plays an analogous role: (1) sorption or degradation in the sampling container and analytical transference device, e.g., syringe, pipet, or beaker; (2) changes in temperature or pressure (particularly applicable to the extreme pressure/temperature changes occurring in the syringe and then the GC itself); (3) cross-contamination of syringes, purge and trap devices, sample lines, injectors, columns, and detectors.

The problem of sorption of organics in sampling devices and in the passage of samples through analytical tubing was addressed in an article by Barcelona et al. in *Analytical Chemistry.*[5] In that discussion, the sorption of various organic liquids in different organic materials is well documented. This problem is seemingly one of a particularly Sisyphean nature; i.e., the containment of a substance within a like substance is akin to rolling a stone up a hill only to have it immediately fall down the other side. Certainly the problems encountered by the industry in attempting to contain petroleum products in unlined fiberglass tanks attest to this dilemma. The development of permeation tube calibration systems is based on the phenomenon.[6]

The present method addresses all of the problems mentioned in that, put most succinctly, it combines immediacy and simplicity of analysis. That is, it is easily transportable to the field, which eliminates problems of sample transfer and storage, and it provides an immediate analysis of a large volume of sample that speaks generally to the problem of representativeness.

THE EXTRACTION/COLORIMETRIC TECHNIQUE FOR AROMATICS

The Hanby Field Test Method for aromatics in water, described here, comes in the form of a kit complete with necessary reagents and apparatus to perform immediate analyses at the groundwater well site. It is contained in a rugged plastic case with enough reagents to perform 30 field analyses. Within the case are contained: a 500 mL separatory funnel, a tripod ring stand, a 10 mL graduated cylinder, 2 reagent (liquid) bottles, one desiccant jar with 30 reagent (powder) vials, a color chart depicting test results for 11 typical aromatics, plastic safety glasses, and 12 pairs of gloves. Upon arrival at the site, the kit is opened and the tripod ring stand is assembled. A 500 mL water sample is introduced into the separatory funnel, which is placed in the ring stand. Next, 5 mL of the extraction reagent is poured into the separatory funnel using the 10 mL graduated cylinder. The sample is vigorously extracted for two minutes with occasional release of the slight pressure buildup which occurs. The funnel is placed back in the ring stand and the extraction phase is allowed to separate to the bottom for five minutes. After phase separation is complete, the lower extraction layer is drained into a test tube, allowing a small amount of the extraction solvent to remain in the separatory funnel. Then one of the reagent vials is opened and the contents immediately poured into the test tube. The tube is shaken for two minutes, allowing the catalyst to be dispersed well throughout the extraction reagent so that color development, which is concentrated in the powder, will be uniform. Hue and intensity of the color of the catalyst which has settled in the tube is now compared to the standard aromatics pictured in the color chart.

The wide range of intense colors produced in Friedel-Crafts reactions has been observed since the discovery of this reaction. A brief description of the chemistry of the reaction, as well as the color involved, is given by Shriner et al. in their widely used book.[7] In this novel adaptation of Friedel-Crafts alkylation chemistry, one of the reactants, the alkyl halide, is used as the extractant. The alkyl halide extractant plus the aromatic compound present in the water sample are caused to form electrophilic aromatic substitution products by the Lewis acid catalyst which is added in great enough amount to also act as the necessary dehydrant to allow the Friedel-Crafts reaction to proceed. These products are generally very large molecules; i.e., phenyl groups clustered around the alkyl moiety, which have a high degree of electron delocalization. These two factors are the principal reasons for the extreme sensitivity of this procedure; that is, large molecules are produced which are very intensely colored.

In the field conditions where this procedure is by and large carried out, the reaction is exposed to sunlight. This means that there will be a "window" in which to observe the color that is produced. This is due to the general instability of the reaction products to photochemical oxidation. Strong sunlight will cause most of the colors produced to fade to various shades of brown within just a minute or two; therefore, it is advisable to perform the test in a shaded area.

PURGE AND TRAP GC COMPARISON STUDIES

Comparison of the Field Test Kit method versus analyses performed with a purge and trap GC were made using standard solutions of benzene, toluene, ethyl benzene, and o-xylene (BTEX). The purge and trap/GC used for this study was a Tekmar LSC-3 and a Hewlett-Packard 5890, Supelcowax widebore capillary 60 meter column, programmed from 45°C (for 3 min.) to 120°C at 8°C/min. A 5 mL aliquot of the standard concentrations: 0.2, 1.0, and 10.0 ppm was injected into the purge vessel of the Tekmar. The sample was purged with helium for 10 minutes and desorbed for 4 minutes at 180°C. Purge flow was 20 mL/min helium. Two separate preparations of the BTEX standards were analyzed by the Field Test Kit and by purge and trap GC. Peak areas were compared for all chromatograms, and these results were normalized with regard to the theoretical response ratio of 0.2, 1.0, and 10.0. This simple average error calculation showed a typical purge and trap GC variation in analysis of ±9.1%.

The results of the Field Test Kit method, judging by comparison to the color chart are, of course, somewhat subjective. Rigorous evaluation of color intensities are being conducted using exact quantities of reagent and further UV/VIS spectrophotometric reflectance readings. These studies are being continued on a Varian DMS 300 as well as on a Cary 2200 UV/VIS/NIR spectrophotometer.

The slide of ethyl benzene analyses performed with the Field Test Kit method at 0.02, 0.1, 0.2, 0.3, 0.4, and 1.0 ppm, and scanned with the DMS 300, are indicative of the visual accuracy obtainable with the kit. The largest contributing factor in the variation of these results is the imprecision in amount of catalyst addition (±20%).

APPLICATIONS OF THE METHOD

Obviously, this method will have a wide variety of applications in field investigations. In fact, utilization of the Hanby Field Test by an environmental testing company has been going on since August 1987. Site investigations of hazardous waste-containing landfills and underground storage tank leaks have been conducted in several states thus far, and use of the kit has greatly facilitated sampling site locations. The first field use of the kit was in the establishment of groundwater monitoring well locations at an organic chemical processing unit. An article describing this first field use of the method is in preparation.

Recent regulations for the monitoring of underground storage tanks require that soil/groundwater investigations be carried out regularly to ensure that no leakage has occurred. It is clear that the use of this technique, which is easily learned and can be performed at an extremely low cost, will provide an immediate and definitive answer to these requirements.

OHIO RIVER STUDY

In the evening of January 2nd, 1988, the collapse of a tank containing approximately 3.5 million gallons of diesel fuel precipitated one of the worst inland oil spills in the country's history. Approximately one million gallons of the oil washed in a huge wave over the containing dikes around the tanks at the Ashland Oil plant at West Elizabeth, Pennsylvania and into the Monongahela River.

Monday morning, two days after the spill, I contacted Mike Burns of the Western Pennsylvania Water Company in regard to using the Hanby Field Test Kit at the company's water treatment facility on the Monongahela, south of Pittsburgh. Mike asked me to bring one of the kits to the plant. The next day I arrived in Pittsburgh and was met by Mike at the West Penn Water Works Treatment facility, where I demonstrated the use of the kit for the personnel at the plant. Mike suggested I call John Potter, the chief chemist at the Water Treatment Plant in Wheeling, West Virginia, which was the next major facility taking water from the Ohio river. John said the kit sounded like it would fill a real need for a rapid analysis of the river water at the plant's intakes. The next day, Wednesday, I was demonstrating use of the kit to the personnel at Wheeling. It was immediately put into use on an around-the-clock basis when they realized that in just a few minutes they could get visual indication down to 100 ppb of the diesel aromatic components.

The next morning I met with the West Virginia Department of Natural Resources personnel who were in Wheeling to monitor the oil spill. That afternoon I was invited by the office of the Environmental Protection Agency (EPA) in Wheeling to join EPA chemist Bob Donaghy, West Virginia Department of Natural Resources Inspectors Sam Perris and Brad Swiger, and the Ohio River Valley Water Sanitation Commission Coordinator of Field Operations, Jerry Schulte, on the river tugboat Debbie Sue to make a run up the Ohio River from Wheeling to try to locate the front of the spill.

The investigators met at the Debbie Sue at noon where it was tied up at the docks on the south side of Wheeling. A light snowfall had begun and the temperature was around 10°F as the boat pulled out into the Ohio, headed up stream.

Due to the fluctuating voltage from the tug's generator, the fluorometer readings exhibited a fairly wide swing during the ensuing measurements. As for the measurements performed with the Field Test Kit, I was primarily concerned with the sensitivity of the test in relationship to the near-zero temperature of the water. Reference to the API study of solubilities of petroleum hydrocarbon components in groundwater had indicated rather large decreases in partitioning of these components into water at lower temperatures.

There was no time, however, to spend worrying about these matters of close quantitation. The boat was soon into the channel and Jerry Schulte was bringing aboard the first bailer of water. On the first sample taken, just minutes after leaving the dock, an obviously detectable coloration was seen in the catalyst material of the Field Test. Reference to the color chart indicated presence of aromatic constituents at something less than 0.5 ppm diesel. Having no reference colors or data at these temperatures I arbitrarily chose this intensity to represent 0.1 ppm. The fluorometer was bouncing between zero and four on its movable dial indicator. (It was an old Turner model arbitrarily numbered from 0 to 100.)

We continued approximately 18 miles up the Ohio, taking samples from the surface and the bottom on the West Virginia side, mid-channel, and the Ohio side. As Table 9.1 indicates, the results from the EPA fluorometer and the Hanby Field Test Kit tracked each other fairly consistently at each point.

Table 9.1. Ohio River Sampling for Diesel Oil (January 7, 1988)

Ohio River Mile Point	Fluorometer Reading	Hanby Field Test (ppm)
89.0	4	0.10
85.5	8	0.15
85.5	9	0.15
85.5	6	0.20
85.0	8	0.20
85.0	11	0.20
84.5	6	0.20
84.5	10–15	0.50
84.5	10	0.20
82.0	20	1.00
81.0	25	1.00
80.0	30	1.50
79.0	33	2.50
77.0	57	10.00
76.0	43	8.00
75.0	35	7.00
74.0	48	5.00
70.0	29	3.00

VALDEZ OIL SPILL

At 12:04 A.M., March 24, 1989, the oil tanker *Exxon Valdez* ran onto Bligh Reef in Prince William Sound. Of the 1.26 million barrels of Prudhoe Bay crude oil the ship was carrying, approximately 10.1 million gallons immediately poured out of the ruptured tanks into the clear blue water of the sound. Of the series of unfortunate circumstances involved in this event, such as the unpreparedness of crew, port, and pipeline officials to immediately begin containment efforts, perhaps none was more critical than the imminent release of hundreds of millions of salmon fry which had been hatched and raised in the half dozen fish hatcheries in and around Prince William Sound.

On Monday, March 27, I contacted the Alaska Department of Environmental Conservation in regard to use of the test kit for onsite monitoring of aromatic contamination of the water caused by the crude oil. Prior analyses of Prudhoe Bay crude had revealed that in comparison to other sources of oil it was particularly high in aromatic content (25%), with the majority components of this fraction being naphthalenes (9.9%), phenanthrenes (3.1%), and pyrenes (1.5%). [Information supplied by the Alaska Department of Environmental Conservation (DEC).] David Kanuth of the Alaska DEC had obtained one of the test kits in January and had subsequently used it at a smaller oil release several weeks later. It was decided that the kit would be helpful at the Valdez incident and commissioner Dennis Kelso requested that I take a kit to three of the hatcheries which would probably be most impacted by the spill. On Thursday, March 30, at 8:30 A.M. Dick Fanell of the DEC, Mark Kuwada of the Alaska Fish and Game Department, and I took off in a pontoon-equipped Cessna 206 piloted by Ken Lobe, and flew SSW from the Valdez harbor to the three fish hatcheries we were scheduled to visit. Within minutes of being airborne we were able to see the huge fingers of oil spreading southerly from the area where the tanker was still impaled on the reef. Thirty minutes later our plane landed in the bay at the first hatchery, Esther.

After demonstrating the use of the test kit to the hatchery manager, two samples were obtained from the water near the fish pens. These samples were analyzed with the test kit, and the indications using the color chart results for gasoline as a guide were 0.2 ppm and 1.5 ppm. These results were unexpected, as no visible oil was apparent from the air as our plane flew in to the hatchery.

The next hatchery visited was the Main Bay Hatchery. A test kit was also left there following demonstration of its usage. One sample was taken near the fish pens, which indicated a concentration of 0.5 ppm aromatics.

The third landing was made in the water of the Port San Juan hatchery. Two samples were taken, indicating concentrations of 0.5 and 0.6 ppm.

The following day a return to Houston was made in order to prepare a comprehensive series of standards from 125 ppb to 20 ppm of the Prudhoe Bay crude in seawater. It was necessary for this work to be performed at the Houston laboratory because a large assortment of glassware and appropriate reagents were required to simultaneously prepare the 11 different standards that were utilized in the study.

The procedure for the 11 concentrations employed called for first preparing a solution of the Prudhoe Bay crude in hexane (5:100), then making a 100:1 dilution of the hexane stock in acetone. This resulted in a 500 ppm stock, which was then used to prepare 500 mL standards of: Blk, 125, 250, 500, 750, 1000, 2000, 3000, 4000, 5000, 10000, and 20000 ppb. When the 500 mL standards were prepared (in Galveston Bay saltwater), each was extracted according to the test kit method with 5 mL extractant reagent for two minutes. After all extractions had been performed and 3 mL amounts of the extractant had been collected in 16×100 mL test tubes, the color development catalyst was added and the tubes were shaken for two minutes in a darkened room. Upon completion, each tube was placed in a rack previously prepared and labeled, then the rack containing the 12 tubes was taken into a sunlit room for the photographs (f16 at 125th sec). This whole procedure was repeated twice more, as a quality control measure.

A remarkable difference in hue is observed as the concentration of the crude oil in the water increases. At approximately 1 ppm the blue colors of the polynuclear aromatics begin to neutralize the more orange colors of the single ring aromatics. This effect is particularly noticeable in the extractant (liquid) phase of the test tube, so that the liquid actually appears colorless above the gray color of the catalyst. In the photographs taken with the liquid phase masked, an evenly progressive increase in intensity of color or saturation is evidenced.

Enlarged copies of these photographs were sent to the various entities utilizing the Field Test Kit, including the Alaska Department of Environmental Conservation, the Department of Fish and Wildlife, and eight of the fish hatcheries themselves. Due to the extent of the spread of the oil and its contamination of the surrounding shoreline it must be assumed that the partitioning of these aromatic components into this prolific marine area will continue for quite a long period.

UV/VIS SPECTROPHOTOMETER STUDIES

Determinations of principal wavelengths and reflectance data were made in correspondence with aromatic compounds depicted on the color chart. These investigations were conducted by preparing a range of concentrations of selected aromatic compounds, performing the Hanby extraction/colorimetric procedure and then immediately measuring the reflectance of the catalyst.

METHOD

Ten parts-per-million (vol/vol) solutions of benzene, toluene, o-xylene, special unleaded gasoline, naphthalene, and diesel were prepared by injecting 20 mL amounts of each compound into 2.0 L of deionized water at 20°C to 21°C and stirring for one hour. Dilutions from the stock solutions were prepared to 0.01, 0.02, 1.0, and 5.0 ppm. The extraction/colorimetric procedure employed with the Field Test Kit was modified to fit the requirements of the UV/VIS reflectance apparatus. Four microliters of the extraction reagent were used to extract the water samples for two minutes. The extraction solvent was then drained into a cuvette. Two grams of the catalyst material was added to the cuvette, which was covered with its Teflon® cap, and the mixture was shaken vigorously for three minutes. The cuvette was placed in the spectrophotometer and scanned over a range of 350 nm to 600 nm.

INSTRUMENTAL PARAMETERS

For this study a Varian DMS 300 UV/VIS spectrophotometer was utilized. Instrument settings were: slit width 2 nm, tungsten source, scan rate 50 nm/min. All of the scans were corrected to 100% transmittance baseline using a blank sample which was scanned in reference to a barium sulphate reflectance disk. The sample compartment was fitted with a diffuse reflectance accessory which was modified by blocking out the top portion of the light path so that only the catalyst in the bottom half of the cuvette would be scanned.

DISCUSSION

Figures 9.1 through 9.3 show the spectrograms for each of the three substances scanned. The concentration for each of the plots is as follows (ppm by volume): A=0.1, B=0.2, C=1.0, D=5.0, E=10.0. These concentrations exhibit well defined differences in the traces of their reflectance curves for each of the substances.

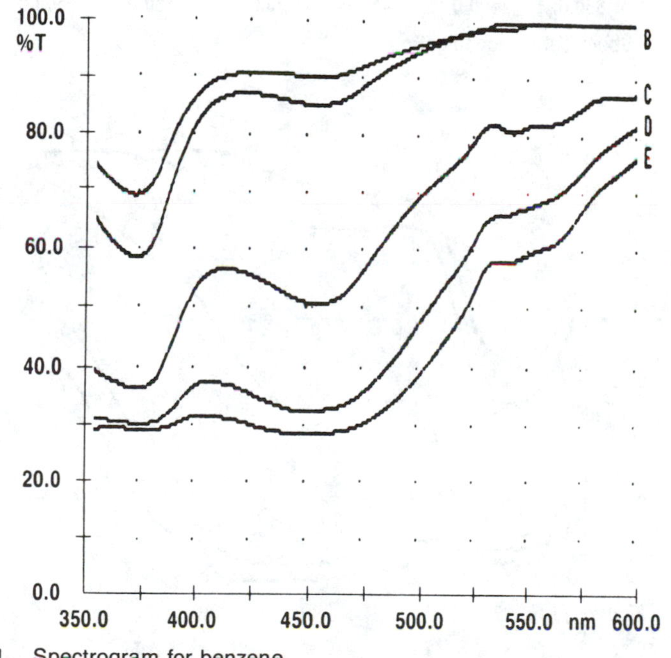

Figure 9.1. Spectrogram for benzene.

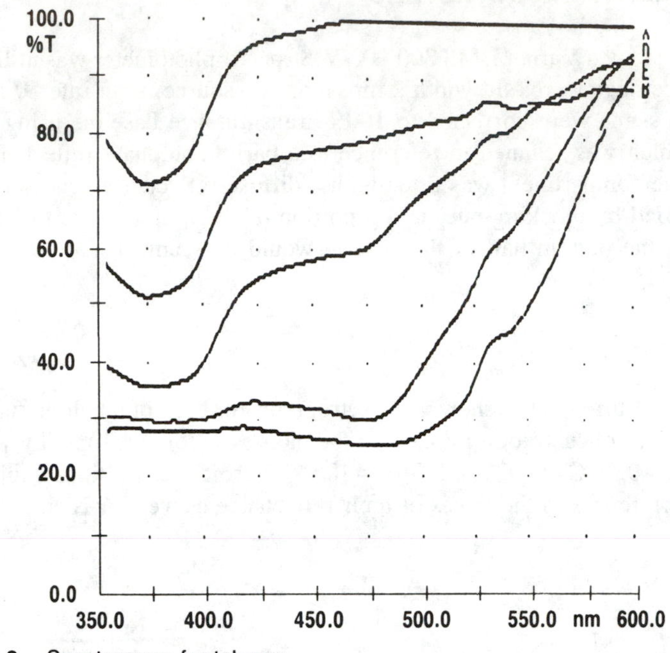

Figure 9.2. Spectrogram for toluene.

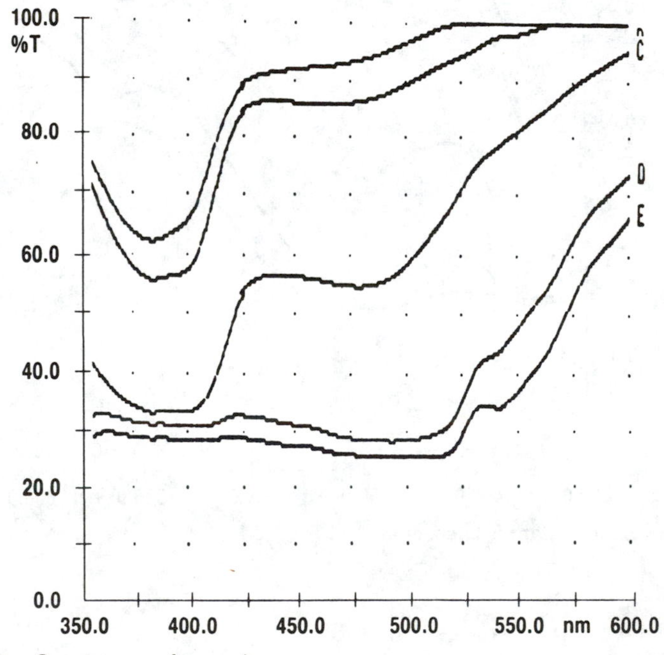

Figure 9.3. Spectrogram for o-xylene.

Figure 9.4 shows the reflectance trace for different concentrations of benzene. These concentrations are (ppm by volume): A=0.01, B=0.05, C=0.25, D=1.0. These runs were scanned on the spectrophotometer from 250 nm to 700 nm at different instrument settings: slit width=1.0 nm, scan rate=20 nm/min, smoothing constant=5 (sec). The different instrument parameters, plus the fact that a special cuvette was constructed to cover a larger area of the reflectance opening, contributed to the greatly enhanced differences in the traces at these, even lower, concentrations.

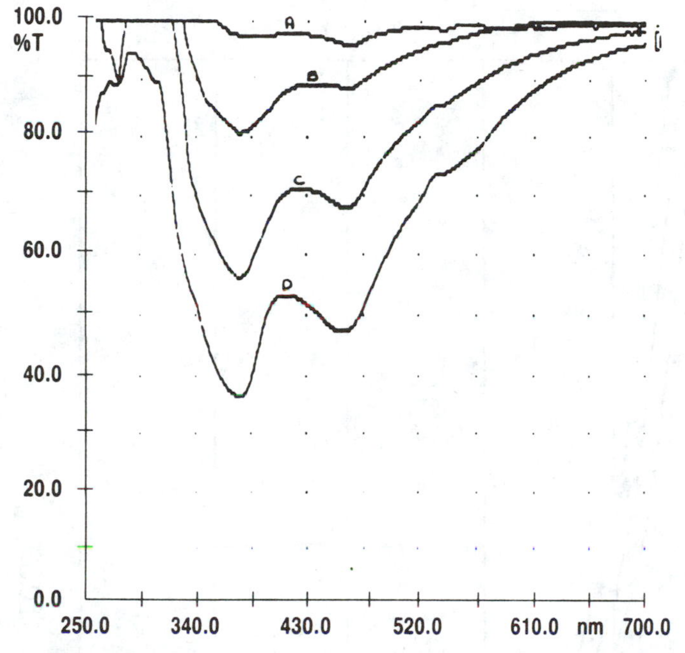

Figure 9.4. Reflectance trace for different concentrations of benzene.

Additional studies have been undertaken to determine the relationship of color intensity with concentration. These studies were performed using a hand-held colorimeter that measured the reflectance of the Field Test Kit color development reagent at 10 nanometer increments from 400 to 600 nm. The plot of these measurements (Figure 9.5) demonstrates the expected log relationship of concentration to reflectance—which can be closely correlated with transmittance. The regular test procedure employing 100 grams of soil was utilized. Gasoline (Exxon Plus Unleaded) was added to the soil to give concentrations as follows (mg/kg): 0, 10, 25, 50, 75, 100, 150, 200, and 400. The three plots in this figure are the reflectances measured at 430, 450 and 470 nm (1, 2, 3).

In conjunction with these analyses using the Field Test Kit, spiked soil samples were prepared for purge and trap GC/FID analyses. In this procedure approximately 10 g aliquots of clean soil were weighed into 40 mL VOC vials. Appropriate

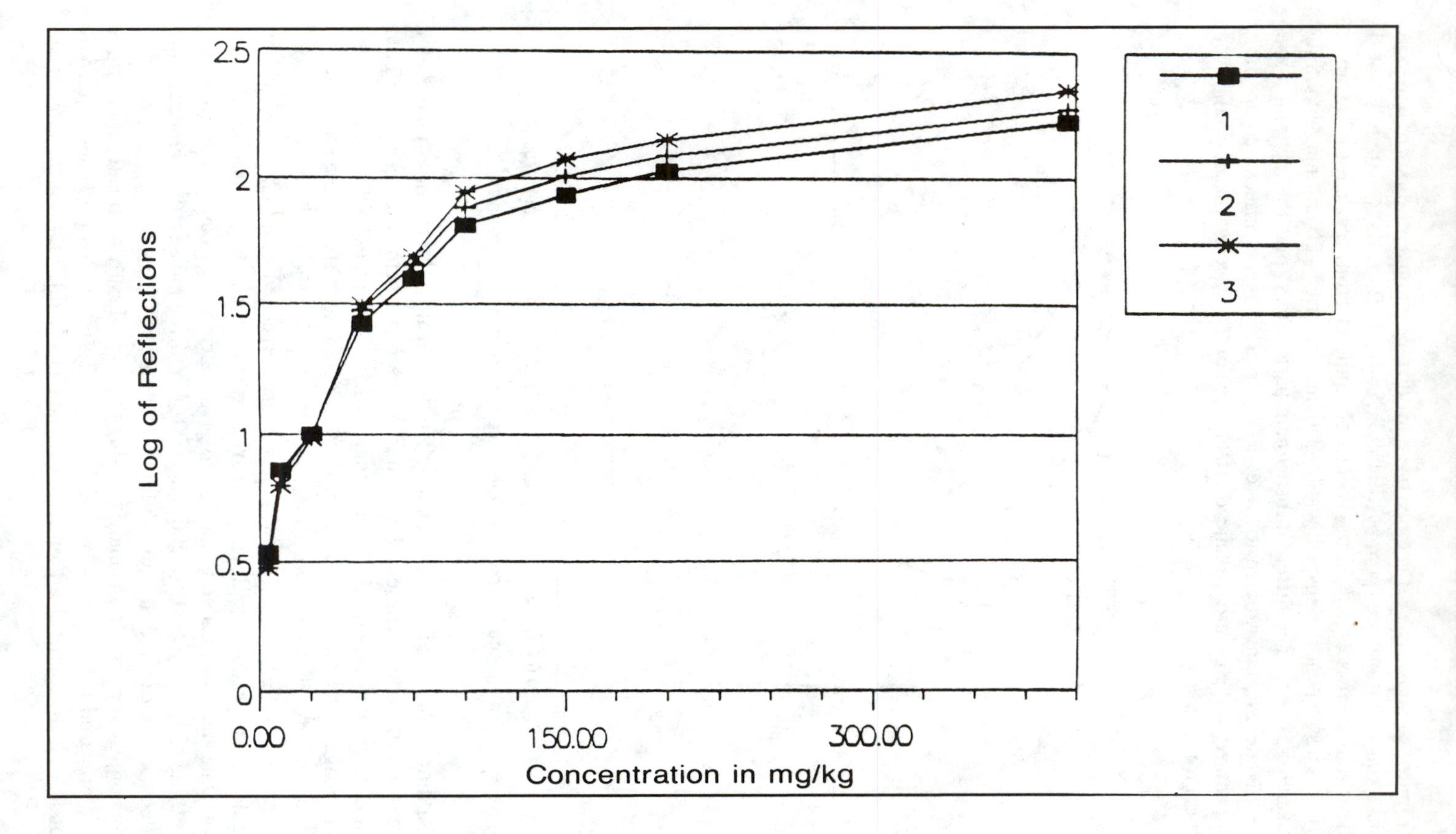

Figure 9.5. Reflectance of the Field Test Kit color development reagent.

additions of gasoline-spiked water was added to the vials to give the same concentrations as used in the Field Test Kit study. Figure 9.6 shows the results of the plot of spiked concentrations vs the recovered concentrations with the P&T/GC/FID analyses. In this instance, again, we see an exponential decline; however, here we are seeing a phenomenon of detector saturation as opposed to the $\log_{10}P_o/P$ (P_o=incident radiant energy, P=transmitted energy) plot of the test kit method.

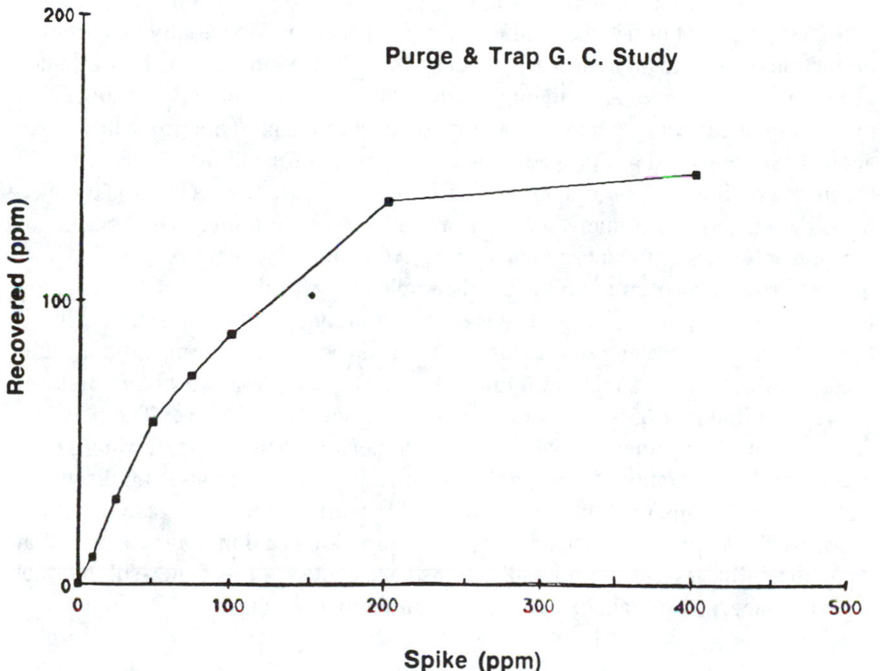

Figure 9.6. Plot of spiked concentrations vs recovered concentrations.

APPLICATIONS OF THE METHOD

In the two years since the introduction of the kit it has seen a wide variety of applications. Most prominently, perhaps, has been the utilization of this method in testing soil at underground storage tank removal and remediation sites. The ability to perform accurate, onsite evaluations of the level of fuel contamination in soil has provided contractors a low-cost and reliable means of exhaustively checking a large number of samples to ensure that soil removal operations have proceeded to a point well below maximum allowable levels.

Prior to the advent of this extraction/colorimetric technique, onsite soil analyses had been almost exclusively vapor or headspace methods. Several factors intrinsic to these methods can be seen as playing a fairly large role in their accuracy.

Probably most important among them are: the age of contamination, fuel type, and soil type. All three of these factors relate to the volatility of the analyte. Old gasoline spills, for instance, generally show a loss of the more volatile components. The range of volatility from gasolines through kerosene, diesel, and the heavier fuel oils; e.g., #5, #6, is very wide and the different soil types from loose, dry sands to moist, compact clays have, of course, an extremely deterministic effect on analyses that depend on vapor permeation.

Two approaches to soil analyses have been developed for the test kit. The first is a direct extraction method in which, typically, a 10 g soil sample is washed in a 20 mL amount of the alkyl halide extractant solvent. Essentially quantitative extraction of petroleum hydrocarbons in soil is achieved in three or four minutes of washing the sample. An aliquot (4.2 mL) of this extractant is then transferred to a test tube for catalyst addition and color development. This procedure takes about 10 minutes and has proved to be very effective for old fuel spills in which the more volatile components have essentially disappeared. This method has also been shown to be extremely sensitive in heavier fuel oil situations because of the presence of significant amounts of PNAs in these fuel types.

The second approach to soil analysis employed with the test kit is an analog to the EP toxicity or TCLP procedures in that an aqueous wash is first utilized to partition the petroleum hydrocarbon from the soil, simulating groundwater seepage effects. In this method, typically, a 100 g sample is washed in 500 mL of water containing 65 g of an inorganic salt flocculant mixture. The washing is carried out intermittently over a 30 minute period, then the wash water is extracted in the separatory funnel in the same fashion as the previously discussed water method. This method is appropriate for fresh solvent and gasoline situations, particularly where groundwater contamination is a dominant concern, but its applicability lessens considerably in instances where a less water-soluble fuel type is concerned or, again, with aged contamination sites.

CONCLUSIONS

The development of a field method for the analysis of organic contaminants at sub-part-per-million levels in water has proved to be a valuable tool in the establishment and the sampling of groundwater monitoring wells. The accuracy of the method has proved to far exceed that of direct injection gas chromatography. A rapid soil-wash method has also been developed employing the Hanby Field Test Kit technique which has proved to be effective on top and deep soils over a range of 5 mg/kg to 10,000 mg/kg gasoline in soil. Development of instrumental spectrophotometric techniques will allow even greater sensitivity and qualitative analysis of aromatic contaminants in soil and groundwater.

A variation of the procedure involving the extraction of a sample with an aromatic solvent and then addition of the Lewis acid catalyst allows for the determination of the presence of alkyl halides; e.g., trichloroethylene. In this version of the test, a reflectance adapter for the spectrophotometer is not necessary since the color is not concentrated in the catalyst but is developed in the extractant solvent.

REFERENCES

1. Roberts, R. M., and A. A. Khalaf. *Friedel-Crafts Alkylation Chemistry: A Century of Discovery.* (New York, NY: Marcel Dekker, Inc., 1984), p. 790.
2. TRC Environmental Consultants, Inc. "Laboratory Study on Solubilities of Petroleum Hydrocarbons in Groundwater," American Petroleum Institute, Washington, DC, 1985.
3. Keith, L. H. *Identification and Analysis of Organic Pollutants in Water* (Ann Arbor, MI: Ann Arbor Science Publishers, Inc., 1981).
4. Johnson, R. L., J. F. Pankow, and J. A. Cherry. "Design of a Ground-Water Sampler for Collecting Volatile Organics and Dissolved Gases in Small-Diameter Wells, *Ground Water,* 25:448–454 (1987).
5. Barcelona, J. J. "Sample Tubing Effects on Ground-Water Samples," *Anal. Chem.*, 57:460–465 (1985).
6. O'Keefe, A. E. "Primary Standards for Trace Gas Analysis," *Anal. Chem.*, 30:760–768 (1966).
7. Shriner, R. L., R. C. Fuson, D. Y. Curtin, and T. C. Morrill. *The Systematic Identification of Organic Compounds.* (New York, NY: John Wiley & Sons, 1980).

CHAPTER 10

A Critical Review of Site Assessment Methodologies

Douglas A. Selby, Las Vegas Valley Water District, Las Vegas, Nevada

INTRODUCTION

The ability to rapidly and accurately assess the nature and extent of petroleum hydrocarbon contamination in soil and groundwater is of significant importance to property owners, regulatory agencies, consultants, and others involved in site investigations. As state and federal underground storage tank regulations have been enacted, the need for assessment has increased dramatically. This demand has resulted in development of some accepted "standard" site assessment techniques and a number of innovative assessment methods which have yet to be proven. The site assessment normally forms the basis upon which subsequent actions are taken that may result in long-term liability and/or major expenditures for site remediation. As a result, it is extremely important that site assessment methods provide the most accurate and reliable information attainable. Many methods presently in use have assessment limitations that are either inherent to the technique or vary with site characteristics. These limitations must be recognized by both those who generate the assessment data and those who use that data as a basis for future actions. This chapter focuses on the methods available to assess the extent, magnitude, and nature of petroleum hydrocarbon contamination in soil and groundwater. It includes an examination of factors that influence the accuracy of the techniques to assist investigators in the determination of appropriate methods for specific sites. In doing so it also furnishes guidance for evaluating the strength of a site database in support of subsequent investigation or remediation activities.

INVESTIGATION METHODS

Current methods of assessment of hydrocarbon contaminated sites are the result of a gradual evolution of techniques which have been adapted from applications that were primarily exploratory in their scope. None of the techniques presently in use resulted from major technological breakthroughs, but rather are the products of innovation and adaptation to the specific demands of the growing field of hydrocarbon contamination assessment. In this chapter the focus is on the commonly practiced assessment techniques. It is recognized that there are other techniques, as well as many variations within each technique, that may serve to address some of the limitations identified here. It is also recognized that regional practices may be somewhat different due to regulatory demands and local environmental conditions.

Hydrocarbon site assessment methods can be categorized broadly into those that merely attempt to detect the presence or absence of contamination and those that attempt to quantify the magnitude and extent of contamination. Pedestrian surveys and soil gas surveys fall into the qualitative category, while soil and groundwater sampling have a quantitative goal. A brief description of principal hydrocarbon assessment methods follows:

Pedestrian Surveys

Certainly the most common of all assessment techniques is the pedestrian survey in which the investigator walks a site looking for visual signs of contamination or indications of hydrocarbon presence. This type of assessment does not claim to be a rigorous scientific technique, but proves to actually be a quick and inexpensive approach. While "nasal appraisals" cannot be recommended due to health concerns associated with inhalation of volatile organic compounds, the incidental detection of such odors is a clear indication of contamination. Visually, hydrocarbon stained soils and pavements are often readily apparent even to the untrained observer. The appearance of an oily film on water surfaces is also an obvious sign of hydrocarbon contamination. Seldom will a pedestrian survey be sufficient for site assessment, but it offers valuable guidance for planning of subsequent work.

Soil Gas Surveys

The soil gas survey has been adapted from its early use in petroleum resource exploration to its use today in the evaluation of contamination based upon the measurement of volatile compounds present in soils containing spilled petroleum products.[1] Soil gas surveys take advantage of the presence of readily detectable volatile organic compounds in fuels that migrate into the soil vapor as a function of vapor pressure and concentration gradients. Theoretically, these compounds respond according to Henry's law.

This technique has many variations, but basically there are two types of soil gas surveys: the active survey and the passive survey. Both techniques share the need to collect and analyze a sample of soil atmosphere. The passive technique accomplishes this by the placement of an adsorptive media, typically activated carbon, in soil excavations for periods of days to weeks, following which the media is collected and compounds desorbed in the laboratory for analysis. In the active technique, soil gas samples are extracted in a grab fashion for analysis onsite. The active technique is the most commonly applied to hydrocarbon investigations, due to its ability to be used in a search sampling mode to find hot spots and achieve plume boundary definition. The technique for collection of the samples varies, but generally involves some form of soil penetration or excavation, followed by withdrawal of a soil gas sample. The sample is then directly injected into a portable gas chromatograph for analysis, or measured in a semiquantitative manner using an organic vapor analyzer.

The type of instrument used to analyze the gas sample also plays a significant role in the technique. Instruments in general use are of the Flame Ionization Detector (FID) or Photoionization Detector (PID) types and may either detect gross total contamination, or through use of capillary chromotography columns identify and quantify concentrations of selected volatile compounds.

Soil Sampling

The collection and analysis of soil samples has become a standard practice in site assessment and along with groundwater sampling, one which has gained the wide acceptance of regulatory agencies. There are many variations in the method of collecting soil samples, but they all basically result in collection of a sample of soil in a fashion that minimizes changes in its hydrocarbon content. Physical access to the soil may be provided in a variety of ways, from hand tools to truck-mounted drill rigs. The method depends almost entirely upon site conditions and investigation goals. The sampling plan may be equally varied. Due to the cost involved in analysis of soil samples, it is also common to screen samples in the field for evidence of contamination, using organic vapor analyzers. Samples showing positive indications of contamination are then submitted for laboratory analysis.

Groundwater Sampling

Groundwater sampling is perhaps one of the most widely practiced methods of assessment because of the immediate threat of contamination of drinking water and the resulting public health concerns. Groundwater may be sampled through existing water supply wells, or most commonly from specially constructed monitoring wells. Monitoring wells for hydrocarbon assessment require screening above the water table to be able to adequately detect free-phase product, but in construction are otherwise similar to small, shallow domestic water wells. Groundwater may also be sampled on a grab basis using driven well points of various designs. These have the advantage of being able to sample the groundwater at more sampling points in a shorter period of time, but sacrifice the ability to make the repeat measurements possible with a monitoring well. Analysis of groundwater may be made using field test kits, mobile analytical laboratories, or fixed base analytical laboratories.

Geophysical Techniques

Several geophysical site assessment techniques have been attempted on an experimental basis. One of these is ground penetrating radar.[2] In theory, this technique can identify free product plumes based upon the characteristics of return radar signals. The hydrocarbon signal is differentiated from other subsurface reflections by calibration on known areas of free-product contamination based on monitoring well data.

Another geophysical method that has been explored is induced polarization.[3] When a current is applied to a conductive media it alters the distribution of negative and positive ions, creating a gradient effect. When the current is terminated the gradient relaxes, resulting in a current flow and measurable voltage termed the induced polarization effect. In theory and experimentally there is evidence that the presence of relatively large nonpolar organic molecules will reduce the induced polarization effect. However, the response occurs in a nonpredictable manner that is greatly influenced by soil characteristics.

Electrical resistivity surveys are commonly used to identify contamination by inorganic materials that cause an increase in the electrical conductivity of soils. This same technique has been applied in attempts to define the extent of hydrocarbon contamination by searching for areas of reduced conductivity that might be characteristic of organic contamination.

In general, geophysical methods are, at present, very limited in application to the assessment of hydrocarbon contamination, due to the difficulty in separating the subtle responses of dissolved or sorbed hydrocarbon material from the natural variable background responses of native soils. They should be considered as experimental in their development and may ultimately show their greatest promise in assessment of large, severely contaminated sites.

EVALUATION OF METHODS

An unbiased evaluation of assessment techniques must consider multiple factors ranging from site limitations to amount of money and time available for the survey. Seldom, if ever, does the investigator have an ideal site or infinite resources to commit to a site assessment, so ultimately every assessment is somewhat of a compromise that hopefully yields a reliable and accurate result. The experience and wisdom of the investigator is by far the most significant factor in a site assessment. It is the investigator who decides which technique is applicable, how that technique is applied, the level of effort appropriate, and who interprets the results. Virtually any of the techniques discussed below will yield reliable results if applied intensively enough to a given site and if those results are interpreted in a meaningful manner by an experienced scientist or engineer. The following evaluation is presented with the recognition that there are many varied approaches to each method, some of which very effectively overcome limitations; however, most methods have the following characteristics:

Pedestrian Survey

Positive Aspects

Inexpensive, rapid, good gross integration of many variables, no permits, unobtrusive, provides good guidance for future work.

Limitations

Depends upon experience of investigator, subjective, limited to surface assessment, personal protective equipment may limit sensitivity.

The pedestrian survey is of limited value in making absolute measurements of concentration and extent of hydrocarbon contamination. As a result, it will seldom satisfy the requirements of regulators or potential purchases of property if there is any site history of petroleum use or storage. The value of a pedestrian survey is that in a short period of time and with little expense, an experienced investigator can develop an opinion regarding potential areas of contamination which should be targets for future investigation. During the pedestrian survey the investigator can assess hydrocarbon storage and housekeeping practices, identify utility trench locations which may be possible conduits for contamination migration, assess proximity of any sensitive neighbors, and examine any open excavations, manholes, or utility vaults, on- or offsite, to gain insight into subsurface conditions. Because this is a qualitative approach that relies on human sensory abilities and the judgment of the individual in the field, there is no way to critically assess this method other than to recognize that it cannot provide the type of data typically necessary to reach conclusions about the magnitude and extent of hydrocarbon contamination.

Soil Gas Sampling

Positive Aspects

Inexpensive, rapid, quantitative, immediate results, limited regulatory permitting, minimal disruption to operating facilities, low public visibility, does not generate waste, excellent guidance for further investigation.

Limitations

Does not work well in wet or clayey soils, limited depth penetration, not accepted by regulatory agencies, requires experienced field personnel, results can be overinterpreted, usually requires soil or water sampling to confirm results, quality control often lacking.

The accuracy and value of soil gas sampling as a site assessment technique for hydrocarbon contamination has been the subject of recent reviews.[5,6] The technique has proven to be reliable in determining the presence or absence of contamination in groundwater on the basis of a whole site survey with an accuracy of approximately 80%. That is, when groundwater is contaminated with hydrocarbons, eight out of ten times it will be detected in a soil gas survey regardless of the site variables. When individual soil gas sampling points are correlated with nearby groundwater samples, the influence of site variables is revealed. Correlations of soil type and the distance between the vapor sample and groundwater show that the best results are achieved when soil gas samples are taken as close to the potential source of hydrocarbon vapors as possible and when soils are porous enough to result in good gas transport. Analysis of field data and observations by investigators indicate that water-saturated and clay soils are not adequately sampled with this method, primarily due to the lack of good gas transport. An advantage to soil gas sampling is that the sample density is high, resulting in better site coverage compared to soil and groundwater sampling alone. During a typical field day 20 or more vapor points can be measured. It has not proved to be a technique which can reliably predict the concentrations of hydrocarbons in soil or groundwater based upon theoretical relationships between liquid and vapor phase hydrocarbons, and for this reason is normally limited to a site screening role.

Soil Sampling

Positive Aspects

Quantitative, direct measurement of contaminants of concern, generally accepted by regulatory agencies.

Limitations

Sample collection and analysis can be expensive; sample numbers typically low, due to cost per sample; spatial distribution of contaminants in soils causes high sample variability; may require special regulatory permits; generates waste soil necessitating special handling and disposal.

Soil sampling may be accomplished using a variety of methods ranging from simple hand augers to drill rigs. It is commonly viewed as a definitive method of characterizing magnitude and extent of contamination; however, it has some significant and often overlooked limitations. Most notable is the fact that only a very small fraction of the total volume of soil at a site is sampled and an even smaller portion is actually analyzed. This, coupled with the patchy distribution of hydrocarbons in soils results in a high degree of variability in results. Soil hydrocarbon data collected at a fuel contaminated site with a uniform and relatively homogeneous sandy soil demonstrates this variability.[7] Twenty-four triplicate sample sets were collected and analyzed for total petroleum hydrocarbons. Each triplicate sample set was composed of samples taken at the same depth within a three-foot radius of one another. Within all sample sets the variation between samples was high, in one case ranging from 390 to 10,300 mg/kg. An analysis of the data to determine the optimum number of samples needed to achieve a 95% confidence of being within 250 mg/kg of the true value showed that over 1,000 samples would have to be collected and analyzed.[8] Clearly this is impractical from many perspectives. This type of result emphasizes the limitations of soil sampling for purposes of accurately assessing the contamination status of a site, especially as it relates to meeting a regulatory cleanup criteria. Whether a site meets a cleanup criteria or not may be as much a function of this variability as it is of true site conditions.

Groundwater Sampling

Positive Aspects

Quantitative, direct measurement of contaminants of concern; provides essential data for risk assessment; can also incorporate soil sampling; accepted by regulatory agencies; monitoring may be continued over time to evaluate trends.

Limitations

Expensive, so a limited number of wells usually installed; analytical costs high; special permits required; generates waste for disposal; time required for permits, drilling, analyses; long-term monitoring commitment may be necessary; highly visible to public; may be disruptive to operating facilities; access for drill equipment may limit use.

Groundwater sampling is considered the backbone of site assessment. Over many years it has proved to be a reliable and direct indication of the magnitude and extent of groundwater contamination when performed using appropriate procedures. It also has the advantage of permitting long-term monitoring of a site to assess temporal variability which may result from remediation and other activities. Groundwater samples tend to produce reproducible results, due to the uniform solubility of hydrocarbons in water as opposed to soil. This increases the level of confidence in the data, which is why it requires relatively few samples.

From an environmental risk perspective, contamination of groundwater is far more important due to the human exposure potential.

The installation procedure and sampling methods for groundwater monitoring wells are of critical importance in the reliability and accuracy of water quality data. Different well drilling techniques have been shown to greatly affect the conclusions reached regarding the extent and level of contamination. Mud rotary drilling although rapid, is of limited use for site assessment work because of the potential for mud residue to clog the formation and to affect the contaminant chemistry, thereby distorting the true subsurface conditions. Air rotary drilling removes the concerns over formation clogging and drilling fluid effects by using compressed air to remove cuttings. It lacks the ability to maintain a stable borehole in unconsolidated formations although this may be somewhat mitigated by the addition of drilling fluids. The major limitations from a site assessment standpoint come from the potential for introduction of oils and contaminants into the well from poorly filtered air, the difficulty in collecting accurate samples, potential for vertical contaminant migration during drilling, and the risk of exposure for the drill crew to toxic volatile vapors when pockets of soil or water contamination are encountered during drilling. The cable tool method is man's oldest drilling technique, dating back to the early Chinese about 4000 years ago.[9] Cable tool drilling has been promoted as perhaps the best drilling technique for site assessment due to its limited disturbance of the formation being penetrated, the stability of the borehole as a result of the driven casing, and the limited potential for cross contamination which is also the result of using a driven casing. The drawbacks to cable tool drilling are that it is a very slow well installation technique which increases the cost of each well and requires special care in the placement of the well casing and filter pack to avoid bridging and pulling of the well casing during withdrawal of the driven casing. Recommendations which have been made for limiting the problems with cable tool drilling include the following:

1. Use a driven casing of a size sufficient to leave at least a 3-in. annulus between the driven casing and the well casing;
2. Use fine uniform gravel for the filter pack to avoid bridging by poorly sorted large materials;
3. Allow for slumping of emplaced materials by a factor of 30% more than the calculated annular space volume;
4. Place filter pack and withdrawal driven casing in no more than 5-ft increments;
5. When drilling dry clays use a tricore bit or under reamer to prevent the driven casing from becoming entrapped in swelling clays.

Hollow stem augering is a common method for soil sampling and well installation at contaminated sites. It is the standard method for hydrocarbon site investigations in many areas. Hollow stem augering is relatively fast and inexpensive, allows for discrete sample collection during drilling, and introduces no drilling fluids or other contaminants. Its limitations include: the possible smearing of clays into permeable formations which may result in reduced estimates of aquifer transmissivity; its potential for cross contamination either by bringing contaminated soil in contact with shallower uncontaminated soils via the auger flights or by downward movement of groundwater from shallow to deeper aquifers; its relatively limited depth of penetration; the possibility of sand heaving up the auger; and its inability to effectively penetrate rock and large cobbles. From a site assessment perspective, the possible cross contamination aspects of this method are severe, as may be the occluding of permeable zones by clays and silts during the drilling process. Both may lead investigators to reach erroneous conclusions regarding distribution and potential for movement of contaminants in the aquifer. A modification of the hollow stem technique has been proposed that eliminates these two problems.[10] The modification involves the installation of a driven casing in conjunction with the drilling process, similar to the cable tool method. The casing helps preserve the integrity of the formations being penetrated by limiting the smearing and the potential for vertical movement of contaminants in the borehole.

Well development can affect the ability to accurately assess groundwater. The goal of well development is to eliminate, to the extent possible, the sediments, silts, drilling fluids, and other residues that may adversely affect the well yield and its long-term ability to provide representative water samples and aquifer measurements. Because most monitoring wells have some form of high porosity filter pack, the goal of development is to restore permeability of the disturbed natural formation of the borehole. Among the more effective well development techniques as measured by specific capacity of developed wells is jetting, but jetting has limitations for contaminant assessment work. Water jetting, wherein high pressure water is forced through the well screen, will introduce water into the formation that may distort subsequent sample results. Air jetting does not introduce water and can be quite effective; however, it carries the risk of introducing oil from the air compressor into the formation. Either jetting process must be used with care on PVC well screens, due to the potential for abrasion from high velocity sediments associated with these types of operations. They may also force fines into well screen openings.

Mechanical and air lift surging offer two development options that depend primarily upon the ambient well water for flushing sediments from the gravel pack. In mechanical surging, a surge block or plunger with a diameter approximating that of the inside diameter of the well is moved up and down in a plunger fashion to alternately force and draw water through the well screen. In air lift surging, compressed air is used to raise the water in the well to near the surface, at which point the air is cut off and the downward velocity of the falling water surges the well. This technique is usually accompanied by pumping to remove loosed sediments. After jetting, surging is the next most effective well development technique. It can create problems in monitoring well installation by forcing fines into screen openings and disrupting the filter pack, creating voids and channels. As in air jetting, the compressed air used in air lift surging can be a source of hydrocarbon contamination that will affect subsequent sampling results.

Well development by overpumping is a final technique for use in monitoring wells.[10] It is not the best method for developing a water supply well, because it is not generally vigorous enough to dislodge all sediment that may ultimately be freed from the formation; however, it is less likely to damage the well, disturb the filter pack, or introduce contaminants into the well than other methods. In overpumping, water is pumped or bailed from the well at a rate in excess of that expected during subsequent use. For monitoring wells this is relatively easy to achieve since all subsequent activity involves relatively brief and passive purging and sampling. Because it is not as effective a technique, overpumping developed wells will generally produce a water with greater potential for turbidity in samples. If the well drilling technique used has not introduced foreign material (drill fluids, water, oil) into the aquifer formation, the sample results obtained from wells developed by overpumping, although potentially containing sediment, should provide reliable indications of ambient groundwater quality. It has been recommended that the efficiency of overpumping can be improved by a second overpumping approximately 24 hours after the first.[10]

Purging of monitoring wells prior to sampling is critical to obtaining groundwater samples which are representative of the aquifer. The amount of water to purge continues to be the subject of debate within the industry. Purging of three to five casing volumes is often recommended as a rule of thumb; however, the definition of casing volume can vary. In some contexts it is represented as the calculated volume within the casing; in other cases it may be intended to include the casing plus the filter pack. The difference in the two may be insignificant in shallow wells where filter pack pore volume is low. Another common criteria by which to judge adequate purging is evidence of stability in pH, specific conductance, and temperature. Unfortunately, these parameters may not relate well to stability in organic contaminant concentrations, due to the vastly different behavior of dissolved organic compounds in comparison to these more or less gross physical measurements. The conclusion of a reviewer, that the inherent variability in hydrocarbon contamination in groundwater may outweigh the variability due to incomplete purging, still seems appropriate.[10] Perhaps the best guidance in this regard is to use a consistent purging protocol to limit variability attributable to procedure. This approach may impose a bias, but it should be one which is consistent enough to allow comparison between sampling events.

Assuming all issues of well installation, development, and purging have been addressed, the method for collection of groundwater samples becomes the next key area to hydrocarbon contaminant assessment. The principal concerns are (a) the sample is not contaminated by the equipment or procedure, and (b) volatile hydrocarbon constituents are not lost during the collection process. The contamination concern is addressed with a rigorous equipment decontamination program and strict adherence to quality control sample procedures, including collection of equipment rinse blanks. Decontamination in the field is often difficult, especially where free phase product may be encountered. It is often better to use individual laboratory cleaned samplers for each well to avoid possible cross contamination due to inadequate field decontamination. Another, although more expensive option, is the use of dedicated well samplers. The loss of volatile constituents during sampling is often underestimated or ignored in common groundwater sampling methods. A number of reviews have identified concerns and limitations of sampling devices.[11-13] A not uncommon practice is the use of bottom-filling open-top bailers to collect groundwater samples. Samples are collected by lowering the bailer into the well, allowing it to fill via a bottom check valve, retrieving the bailer, and decanting the sample into a VOA vial from the top of the bailer. Obviously, this results in atmospheric exposure and agitation of the sample, no matter how carefully the procedure is carried out. Special valve assemblies available for withdrawal of bailer samples from the bottom of the bailer reduce the agitation of the sample and its exposure to atmospheric losses. Syringe samplers overcome most of the concerns over volatile losses due to agitation and degassing. Their principal limitations are in the relatively small volumes that can be collected and their susceptibility to clogging in wells with elevated suspended sediment levels. A variety of pumps have been evaluated for collection of samples. The bladder type of gas displacement pump, although relatively expensive, has application to hydrocarbon sampling. Most are designed specifically for sampling low levels of contaminants and have a wide degree of flexibility in pumping rates and depths, providing them with the ability to both purge and sample a well. Since the gas used for driving the pump only serves to inflate a displacement type bladder, it does not come into contact with the water being sampled so the loss of volatiles through aeration and degassing is minimized. This is not the case for the gas-lift or gas-drive type of pumps, which may cause significant volatile losses due to turbulence and direct exposure of the sample to the drive gas. Pumps of the suction lift type expose samples to negative pressure, which may result in significant loss of volatile compounds through degassing.

GUIDELINES FOR APPLICATION OF ASSESSMENT TECHNIQUES

1. Start with comprehensive pedestrian and literature surveys and use site characteristics obtained to guide method selection.
2. For rapid screening investigations use an active soil gas survey, but interpret the results with caution if soils show evidence of low gas permeability.
3. Do not rely on theoretical relationships between gas and liquid hydrocarbon phases to predict groundwater or soil contaminant concentrations.
4. Always use calibrated equipment and enforce a rigid quality control program.
5. Do not base site evaluations exclusively on a limited number of soil samples, due to the highly variable nature of soil samples. If remediation is anticipated, conduct more focused and intensive soil sampling to define horizontal and vertical extent of contamination.
6. Do not base any groundwater remediation or risk assessments on one set of samples.

7. Accept the fact that you will seldom be permitted the luxury of enough money or time to perform a sampling program that yields results of high statistical confidence. Professional judgment and experience are essential in extrapolating results to reach conclusions about concentration and extent of contamination.
8. Be suspicious of failure to find evidence of contamination with any method at sites which have a fuel storage or use history.

REFERENCES

1. Reisinger, H. J., D. R. Burris, L. R. Cessar, and G. D. McCleary, ''Factors Affecting the Utility of Soil Vapor Assessment Data,'' in Proc. First National Outdoor Action Conference on Aquifer Restoration, Groundwater Monitoring and Geophysical Methods, Las Vegas, NV, 1987.
2. Ricketts, B. M., and B. D. O'Flanagan. ''Field Study: A Petroleum Contamination Investigation in DeLeon Springs, Florida,'' in Proc. Petroleum Hydrocarbons and Organic Chemicals in Groundwater, Houston, TX, 1988.
3. Krumenacher, M. J., and R. W. Taylor. ''Innovative Application of Induced Polarization for Detecting Organic Groundwater Contamination,'' in Proc. Petroleum Hydrocarbons and Organic Chemicals in Groundwater, Houston, TX, 1988.
4. Andres, K. G., and R. Canace. ''Use of the Electrical Resistivity Technique to Delineate a Hydrocarbon Spill in the Coastal Plain Deposits of New Jersey,'' in Proc. Petroleum Hydrocarbons and Organic Chemicals in Groundwater, Houston, TX, 1984.
5. Greensfelder, R. W., M. Singh, and G. Davitt, ''Sampling Soil Gas in Tight Soils,'' in Proc. Petroleum Hydrocarbons and Organic Chemicals in Groundwater, Houston, TX, 1989.
6. Marks, B. J., D. A. Selby, R. E. Hinchee, and G. Davitt, ''Soil Gas and Groundwater Levels of Benzene and Toluene—Qualitative and Quantitative Relationships,'' in Proc. Petroleum Hydrocarbons and Organic Chemicals in Groundwater, Houston, TX, 1989.
7. Selby, D. A., Unpublished results, 1989.
8. Gilbert, R. O., Statistical Methods for Environmental Pollution Monitoring (New York: Van Nostrand Reinhold Co., 1987).
9. Driscoll, F. G. Groundwater and Wells (St. Paul, MN: Johnson Division, 1987), p. 268.
10. Keeley, J. F., and K. Boateng, ''Monitoring Well Installation, Purging, and Sampling Techniques—Part 1: Conceptualization,'' Groundwater 25(3):300–313 (1987).
11. Barcelona, M. J., J. A. Helfrich, E. E. Garske, and J. P. Gibb, ''A Laboratory Evaluation of Ground Water Sampling Mechanisms,'' Groundwater Monitoring Review 4(2):32–41 (1984).
12. Nielsen, D. M., and G. L. Yeates, ''A Comparison of Sampling Mechanisms Available for Small-Diameter Ground Water Monitoring Wells,'' Groundwater Monitoring Review 5(2):83–99 (1985).
13. Johnson, R. L., J. F. Pankow, and J. A. Cherry, ''Design of a Ground Water Sampler for Collecting Volatile Organics and Dissolved Gases in Small-Diameter Wells,'' Groundwater 25(4):448–454 (1987).

CHAPTER 11

Where Do Organic Chemicals Found in Soil Systems Come From?

James Dragun, Sharon A. Mason, and **John H. Barkach,**
The Dragun Corporation, Berkley, Michigan

INTRODUCTION

Today's regulatory climate encourages the private sector to assess the environmental condition of their facilities. An environmental assessment often includes the collection of soil samples. In fact, today's general trend is to obtain data on the concentration of dozens of organic chemicals that could be present in soil. Because many state regulatory agencies require soil to be remediated to background concentrations, the determination of background concentrations of naturally-occurring organic chemicals in soil is of utmost importance to the regulated community.

Despite the trend to obtain reams of numbers to show the presence of chemicals, many misconceptions exist among environmental scientists and engineers regarding the interpretation of those numbers. For example, many environmental scientists and engineers mistakenly believe that the presence of benzene, toluene, xylene, or ethylbenzene at any concentration in soil is *prima facie* evidence that a release of a petroleum product has occurred.

The presence of organic chemicals in soil may or may not be problematic. This depends primarily upon the source. If an industrial point source is responsible for the spill or bulk release, then remedial activity usually ensues. However, if the source is *not* an industrial release, then remedial activity may not be required.

This chapter will briefly discuss the sources, other than industrial point sources, responsible for the presence of organic chemicals in soil systems.

MAN-MADE SOURCES OF ORGANIC CHEMICALS FOUND IN SOIL

Many organic chemicals that are found in soil are the result of human activities. These include agricultural amendments, atmospheric deposition, oil field brines, ''inert'' fill material, and septic systems.

Common agricultural amendments, such as fertilizer, sewage sludge, and pesticides, are comprised of hundreds of organic and inorganic chemicals. In particular, pesticides are often diluted with petroleum distillates. Petroleum distillates contain benzene, toluene, xylene, and ethylbenzene (BTXE). The detection of BTXE from soils where pesticides have been applied would not be considered exceptional.

In many regions, sewage sludge has been added to agricultural soils to increase the organic matter content of the soil. Because sewage sludge has a high adsorption capacity for volatile organic chemicals (VOCs), semivolatile organic chemicals (SVOCs), polychlorinated biphenyls (PCBs), and pesticides, the application of sewage sludge obtained from an industrial area may result in the detection of a variety of organic chemicals in soil.

In some areas, oil field brines are applied to roads to control the formation of ice during winter months. Laboratory data have revealed the presence of BTXE in oil field brines. In the authors' experience, the application of oil field brines to a dirt road resulted in a state regulatory agency mistakenly accusing an adjacent property owner of contaminating subsurface soils with VOCs.

Atmospheric deposition can result in the detection of small concentrations of a number of organic chemicals in soil. Atmospheric deposition can result in the discharge of VOCs, SVOCs, particulates, and chlorinated solvents onto the soil surface. Common sources of organic chemicals attributed to atmospheric deposition include automobiles, trucks, power plants, municipal incinerators, and paint operations.

Historically, ''inert'' fill material has been utilized to fill low-lying areas (e.g., wetlands, construction sites, and shore lines). Many industrial by-products have been used as fill material and include foundry sand waste, municipal incinerator ash, power plant ash, and manufacturing and processing aggregates. In particular, foundry sand can contain a number of organic chemicals in concentrations up to 10,000 to 50,000 parts per million (ppm). Organic chemicals detected in foundry sand include phenol, formaldehyde, kerosene, fuel oil, chlorinated organic solvents, cresols, etc.

Septic systems discharge thousands of gallons of wastewater into soil. Because many household and commercial products contain organic solvents, the presence of BTXE and chlorinated organic chemicals from septic field soils is not surprising.

In general, man-made sources of organic chemicals represent a minor contribution with respect to the total mass of organic chemicals in soil. However, chemicals deposited by man-made sources are frequently mistaken for chemicals discharged by onsite industrial or commercial operations.

THE NATURALLY-OCCURRING SOURCE OF ORGANIC CHEMICALS FOUND IN SOIL: BIOTA

The second source of organic chemicals in soil is biota. Soil biota can be placed into two categories: (a) vegetation, and (b) soil fauna and soil flora.[1] These two categories are responsible for the bulk of the organics present in soil. Topsoil can typically contain from 1% to 8% organic matter (i.e., 10,000 to 80,000 ppm). In the vadose zone, subsoil can contain from 0.1% up to 1.0% organic matter (i.e., 1,000 to 10,000 ppm).

The first category, vegetation, is comprised of the residues, dead parts, and exudates of green plants (i.e., grasses, trees, and shrubs), that enter soil. The residues and dead parts of vegetation can enter soil when foliage falls onto the soil surface and decays. The residues of agricultural plants typically are plowed into soils after harvest. Quantities of vegetation added to soils range from 200 to 18,000 kg/ha/yr, with an intermediate value of about 4,000 to 5,000 kg/ha/yr.[1]

The second category, soil fauna and soil flora, is comprised of thousands of species of soil microorganisms and macroorganisms, some of which are present in tremendous numbers in soil. For example, the average agricultural soil contains 100,000,000 bacteria per gram of soil. The typical subsoil contains from 1,000 to 10,000,000 bacteria per gram of soil. Typical populations for bacteria in groundwater range from 100 to 200,000 individual organisms per milliliter of groundwater.

Soil fauna and flora exude various organic chemicals as a part of their metabolic functions. Also, when these soil organisms expire, cellular lysis occurs and causes the release of cell contents into the soil. In other words, hundreds of organic chemicals identified in the many metabolic pathway diagrams, one of which is illustrated in Figure 11.1, are released into the soil after cell lysis.

The organic chemicals within the two categories identified above generally include alkanes, alkanoic acids, alkanols, alkanoids, cyclic alkanes, methyl alkanones, organic cyanides, and numerous aromatic derivatives including polynuclear aromatic hydrocarbons.[2] In addition to the above chemicals, benzene, toluene, ethylbenzene, and xylenes have been known since 1966 to be naturally-occurring in soil systems at 1 to 5 ppm.[3] The presence of these aromatics in soil should come as no surprise, because naturally-occurring humic and fulvic acids, which comprise a significant portion of soil organic matter, contain benzene in their chemical structure. For example, fulvic acid typically contains from 15% to 45% benzene in its chemical structure.[4] Dutch reference levels for judging soil

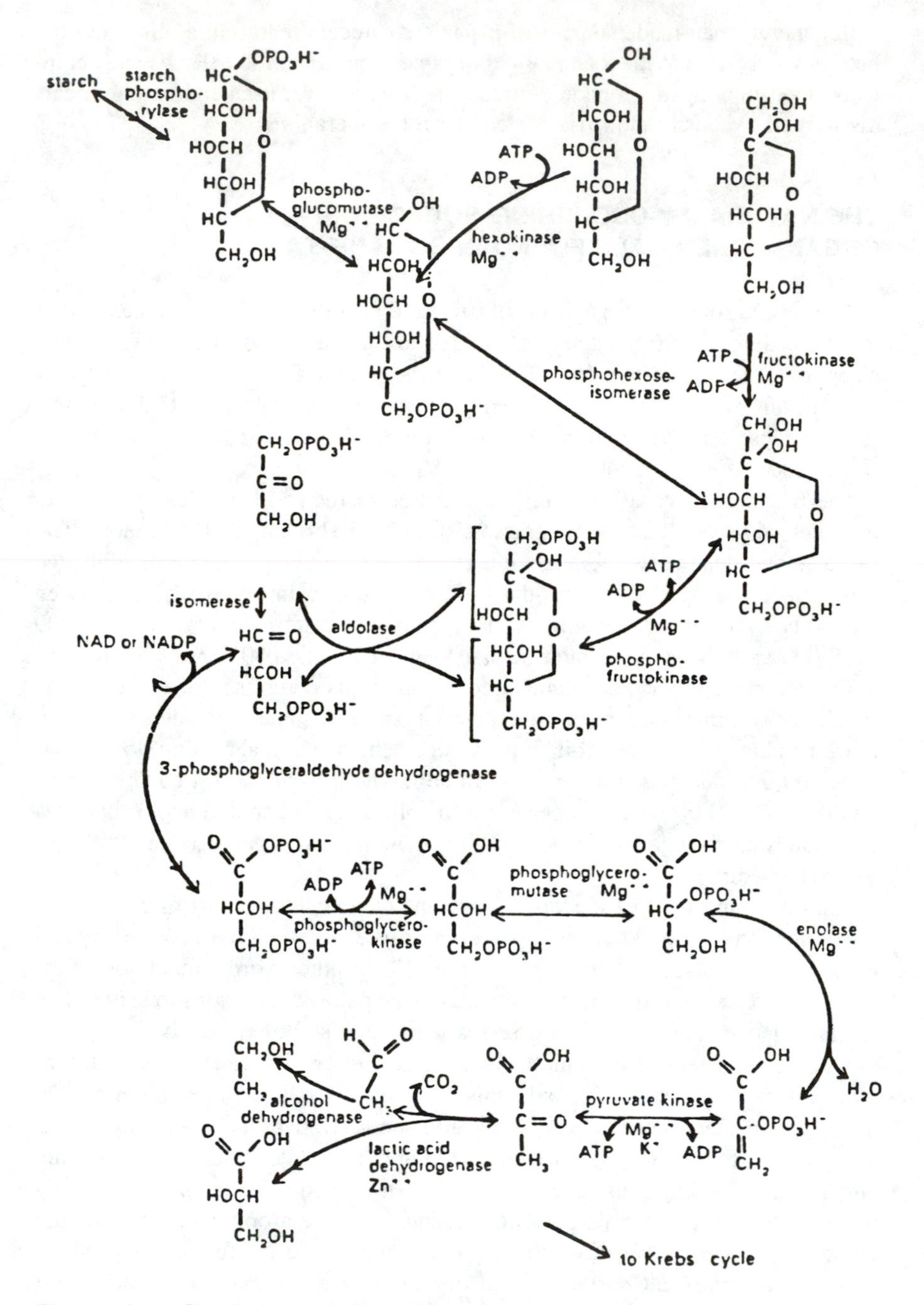

Figure 11.1. Glycolysis metabolic pathway.

contamination[5] account for the existence of this source; Dutch soils containing less than 7 ppm of these mononuclear aromatics can be utilized as fill material.

It is important to recognize that not all naturally-occurring soil organic chemicals have been identified; more are discovered as each year passes. Also, more industrially-derived organic chemicals will be identified as naturally-occurring as our knowledge of the reactivity of those naturally-occurring soil organic chemicals increases. For example, propane is a naturally-occurring organic chemical found in soil.[6,7] The well-documented hydrolysis of the beta carbon of straight-chain alkanes by the enzyme monooxygenase during the beta-oxidation metabolic pathway[2,8] could convert propane into acetone, which is a methyl alkanone. Therefore, it would not be surprising to find the presence of acetone in some soil samples.

Unlike metals which do not change their chemical structure, these naturally-occurring organic chemicals are reactive and, as a result, are transient in soil systems. For example, benzene is a reactive chemical that will react chemically (e.g., surface catalyzed oxidation) and microbiologically (e.g., hydroxylation, nitration, ring fission).[2,8] Therefore, the persistence of benzene in soil will depend upon the physical, chemical, and biological properties of the soil system.

CONCLUSIONS

The first source of organic chemicals present in soil is human activities which result in an ''across-the-board'' type of deposition into or onto soils. Many organic chemicals that are found in soil are the result of agriculture, atmospheric deposition, fill material, septic systems, and other sources.

The second source of organic chemicals in soil is biota, which fall into two categories: (a) vegetation, and (b) soil fauna and soil flora. Vegetation is comprised of the residues, dead parts, and exudates of green plants (i.e., grasses, trees, and shrubs). Soil fauna and soil flora are comprised of thousands of species of soil microorganisms and macroorganisms, which exude various organic chemicals as a part of their metabolic functions. When these soil organisms expire, cellular lysis occurs and causes the release of cell contents into the soil.

Unlike metals which do not change their chemical structure, these naturally-occurring organic chemicals are reactive and, as a result, transient in soil systems.

Not all naturally-occurring soil organic chemicals have been identified; more are discovered as each year passes, and more industrially-derived organic chemicals will be identified as naturally-occurring, based on our knowledge of the reactivity of those naturally-occurring soil organic chemicals that have been identified.

REFERENCES

1. Fairbridge, R. W., and C. W. Finkl, Jr. *The Encyclopedia of Soil Science Part* 1. *Physics, Chemistry, Biology, Fertility, and Technology* (Stroudsburg, PA: Dowden, Hutchinson & Ross, 1979).
2. Dragun J. *The Soil Chemistry of Hazardous Materials*. (Silver Spring, MD: Hazardous Materials Control Research Institute, 1988).
3. Simonart, P., and L. Batistic. ''Aromatic Hydrocarbons in Soil,'' *Nature* 212:1461–1462 (1966).
4. Murray, K., and P. W. Linder. ''Fulvic Acid: Structure and Metal Binding. I. A Random Molecular Model,'' *J. Soil Sci.* 34:511–523 (1983).
5. Nunno, T. J., J. A. Hyman, and T. Pheiffer. ''Part 7: European Approaches to Site Remediation,'' *Hazardous Materials Control* 2(5): 38–46 (1989).
6. El-Sebaay, A. S., O. Van Cleemput, and L. Baert. ''Theoretical Considerations on the Oxidation of Some Gaseous Hydrocarbons in the Soil Atmosphere,'' *Soil Sci.* 147(2):103–106 (1989).
7. Van Cleemput, O., and A. S. El-Sebaay. ''Gaseous Hydrocarbons in Soil,'' *Advances in Agronomy* 38:159–181 (1985).
8. Brink, R. H. ''Biodegradation of Organic Chemicals in the Environment,'' in J. D. McKinney, Ed., *Environmental Health Chemistry* (Ann Arbor, MI: Ann Arbor Science, 1981).

CHAPTER 12

Fate of Hydrocarbons in Soils: Review of Modeling Practices

Marc Bonazountas, National Technical University, Athens, Greece

INTRODUCTION

The recent literature in soil pollution modeling related to hydrocarbon fate in soil systems is reported in Table 12.1. This chapter describes concepts, uses, and limitations of state-of-the-art mathematical environmental pollutant fate modeling for use in hydrocarbon fate studies and analyses of environmental quality.

Models are used in a variety of ways to assist in decisionmaking and will be used to a greater extent in the future; specific statutes or regulations require the use of models in certain situations. Additionally, provisions of the National Environmental Policy Act of 1969 (NEPA), as well as judicial decisions constituting NEPA and other environmental statutes, should facilitate the increased use of mathematical models.

Terrestrial chemicals fate modeling has traditionally been performed for three distinct sub-compartments: (1) the land surface (or watershed); (2) the unsaturated soil (or ''soil'') zone; and (3) the saturated (or ''groundwater'') zone of the region. In general, the mathematical simulation is structured around two major processes, the hydrologic cycle and the pollutant cycle, each of which is associated with a number of physicochemical processes. Land surface models also account for a third cycle, sedimentation.

An evaluation of the fate of inorganic compounds in soil and groundwater requires a detailed consideration of the physical, chemical, and biological processes

Table 12.1. References Related to the Fate of Petroleum Products in Soils

References of information in package
U.S. Geological Survey
Baehr, A., and M. Corapcioglu. "A Compositional Multiphase Model for Groundwater Contamination by Petroleum Products," *Water Resour. Res.,* 23(1):191–213 (1987).
Baehr, A. "Selective Transport of Hydrocarbons in the Unsaturated Zone Due to Aqueous and Vapor Phase Partitioning," *Water Resour. Res.,* 23(10)1926–1938 (1987).
Baehr, A., G. Hoag, and M. Marley. "Removing Volatile Contaminants from the Unsaturated Zone by Inducing Advective Air-Phase Transport," *Contam. Hydrol.,* 4:1–26 (1989).
PCB Onsite Spill Model (POSSM)
Brown, S., and S. Boutwell. "Chemical Spill Exposure Assessment Methodology," EPRI. Prepared by CH2M HILL, January 1988.
Brown, S., and A. Silvers. "Chemical Spill Exposure Assessment," *Risk Analysis,* 6(3) (1986).
Pesticide Root Zone Model (PRZM)
Carsel, R., and C. Smith, L. Mulkey, J. Dean, and P. Jowise. "User's Manual for the Pesticide Root Zone Model (PRZM)." USEPA Environmental Research Lab, Athens, GA. December, 1984.
Dean, J., and R. Carsel. "A Linked Modeling System for Evaluating Impacts of Agricultural Chemical Use," USEPA, Athens, GA and Woodward-Clyde, 1988.
Smith, C. N., R. S. Parrish, R. F. Carsel, A. S. Donigian, and J. M. Cheplick. "Validation Status of Pesticide Leaching and Groundwater Transport Model," Prepared for Office of Pesticide Programs by Environmental Research Laboratory, Athens, GA. August, 1988.
Pesticide Analytical Model (PESTAN)
Enfield, C., R. Carsel, S. Cohen, T. Phan, and D. Walters. "Approximating Pollutant Transport to Groundwater," *Ground Water,* 20(6) November-December 1982.
GEOTOX—A Multimedia Compartment Model
McKone, T. E., L. B. Gratt, M. J. Lyon, and B. W. Perry. "Geotox—User's Guide and Supplement," Lawrence Livermore National Laboratory/U.S. Army Medical Research and Development Command. Project Order 83PP3818. May 1987.
McKone, T. E., and D. W. Layton. "Screening the Potential Risks of Toxic Substances Using a Multimedia Compartment Model: Estimation of Human Exposure," *Regulatory Toxicology and Pharmacology.* 6:359–380 (1986).
McKone, T. E., and D. W. Layton. "Exposure and Risk Assessment of Toxic Waste in a Multimedia Context," Lawrence Livermore National Laboratory. Presented at Air Pollution Control Association. May 1986.
McKone, T. E. "Geotox—Simulating Contaminant Behavior and Human Exposure," *E&TR* pp. 14–20. May 1987.
Seasonal Soil Compartment Model (SESOIL)
Bonazountas, M., and J. Wagner. "Sesoil" A Seasonal Compartment Model. ADL and DIS/ADLPIPE. For USEPA Office of Toxic Substances, May 1984.

continued

Table 12.1. Continued

References of information in package
Seasonal Soil Compartment Model (SESOIL)—continued
Leu, D. "California Site Mitigation Decision Tree Manual," Prepared by the Dept. of Health Services, Toxic Substances Control Division, May 1986.
Wagner, J., and M. Bonazountas. "Potential Fate of Buried Halogenated Solvents via SESOIL," ADL for USEPA Office of Toxic Substances, January 1983.
Watson, D., and S. Brown. "Testing and Evaluation of the SESOIL Model," Anderson-Nichols & Co. for USEPA Environmental Research Lab, Athens, GA, August 1985.
Personal Computer-Graphical Exposure Modeling System/SESOIL (PCGEMS/SESOIL)
"Personal Computer Version of the Graphical Exposure Modeling System," User's Guide. Prepared for USEPA/OTS Contract # 68024281. Prepared by General Sciences Corp. September 1989.
Preliminary Pollutant Limit Value Approach
Small, Mitchell. "The Preliminary Pollutant Limit Value User's Manual," U.S. Army Biomedical Research & Development Laboratory, Ft. Detrick, Frederick, MD. Technical report 8918. July 1988.
Background
Enfield, C., and R. Carsel. "Mathematical Prediction of Toxicant Transport Through Soil," USEPA Environmental Research Lab, Ada, OK.
Hern, S., and S. Melancon. "Vadose Zone Modeling of Organic Pollutants." (Chelsea, MI: Lewis Publishers, Inc., 1986).
Langlois, C., E. Calabrese, and C. Gilbert. "A Review of Three Soil Fate Models," University of Massachusetts, 1989.
Melancon, S., J. Pollard, and S. Hern. "Evaluation of SESOIL, PRZM and PESTAN in a Laboratory Column Leaching Experiment," *Environ. Toxicol. Chem.*, 5:865–878 (1986).
USEPA. "Selection Criteria for Mathematical Models Used in Exposure Assessments: Groundwater and Surface Water Models," USEPA Office of Health and Environmental Assessment, 1986.
Other
Senes Consultants (Prepared for the Decommissioning Steering Committee). "The Development of Soils Cleanup Criteria in Canada. Volume 2—Interim Report on the Demonstration Version of the AERIS Model." 1988-12-15.
Senes Consultants (Prepared for the Decommissioning Steering Committee). "Contaminated Soil Cleanup in Canada. Volume 5—Development of the AERIS Model Final Report." September 1989.
Senes Consultants (Prepared for the Decommissioning Steering Committee). "Contaminated Soil Cleanup in Canada. Volume 6—User's Guide for the AERIS Model." September 1989.
LUFT Task Force, State of California. "Leaking Underground Storage Tank Manual: Guidelines for Site Assessment, Cleanup, and Underground Storage Tank Closure." May 1988.
RAFT—"User's Manual for Risk Assessment/Fate and Transport (RAFT) Modeling System." Prepared for the Penn. Bureau of Waste Management by the Scientific Services Section. 1989.

and reactions involved, such as complexation, absorption, precipitation, oxidation-reduction, chemical speciation, and biological reactions to determine the free metal concentration in soil solutions. These processes can affect such characteristics as species solubility, availability for biological uptake, physical transport, and corrosion potential.

SOURCES AND EMISSIONS

Through numerous human activities, pollutants are released to the soil compartment. The particular particle significantly influences the fate of pollutants in the soil and groundwater zones. Releases include both point source and area loading. They may be intentional, such as landfills and spray irrigation of sewage; unintentional, such as spills and leaks; or indirect, through pesticide drift or surface run-offs. The point of release may be at the soil surface, or from a source buried deep in the soil. Substances released are in liquid, semiliquid, solid, or particulate form. In some cases a waste material will be pretreated or deactivated prior to disposal to limit its mobility in soil. The rate of release may be continuous, such as at a municipal landfill; intermittent; on a "batch" basis such as practiced by some industries; or as a one-off episode such as the uncontrolled disposal of barrels of waste or a spill.

Landfills are principally disposal sites for municipal refuse and some industrial wastes. Municipal refuse is generally composed of 40% to 50% (by weight) of organic matter, with the remaining consisting of moisture and inorganic matter such as glass, cans, plastic, and pottery. Under aerobic decomposition, carbonic acid that is formed reacts with any metals present and calcareous materials in the rocks and soil, thus increasing the hardness and metal content of the leachate. Decomposition of the organic matter also produces gases, including CO_2, CH_4, H_2S, H_2, NH_3, and N_2, of which CO_2 and CH_4 are the most significant soil contaminants.

FATE OF CONTAMINANTS IN SOIL SYSTEMS

General

The mechanisms of contamination from various activities and the transport of organic contaminants in groundwater are described by Hall and Quam,[1] Mackay and Roberts,[2] Mercer,[3] Faust,[4] Abriola and Pinder,[5] Fried et al.[6] and many other researchers. Information presented below is mainly obtained from the work of Mackay and Roberts, and Hall and Quam.

Organic contaminants can reach the groundwater zone either dissolved in water or as organic liquid phases that may be immiscible in water. These contaminants travel with the soil moisture and are retarded in their migration by various factors.

The subsurface transport of immiscible (nonaqueous phase liquid, NAPL) organic liquids is governed by a set of factors different from those for dissolved contaminants. However, some components of organic liquids can dissolve into the groundwater; therefore, the process of dissolved pollutants has been of primary importance in the past. Metals are subject to different processes.

Nature of Contamination

Hall and Quam[1] report on the nature of contamination from petroleum products, for example, that one of the most important factors in contamination of groundwater is the extremely low concentration of the product which can give rise to objectionable tastes and odors. The specific aspects of contamination may be broadly classified into:

- the formation of surface films and emulsions, and
- the solubility in water of certain petroleum products.

The water-soluble components of petroleum products which give rise to taste and odor problems are the aromatic and aliphatic hydrocarbons. Phenols and cresols are examples of these compounds, and are known to generate taste and odors at concentrations as low as 0.01 mg/L. When chlorine is added to drinking water, as in most municipal water supplies, it reacts with the phenols to form chlorophenols which give rise to objectionable tastes and odors at concentration as low as 0.001 mg/L. It is thus apparent that very small quantities of hydrocarbons can give rise to widespread contamination of water resources.

Transport and Distribution of Contaminants

Major transport and transformation processes for organic dissolved contaminants are advection, dispersion, sorption and retardation, and chemical and biological transformation. The migration of an immiscible organic liquid phase is governed largely by its density, viscosity, and surface-wetting properties.

Density differences of about 1% are known to influence fluid movement significantly. With few exceptions, the densities of organic liquids differ from that of water by more than 1%. In most cases the difference is more than 10%. The specific gravities of hydrocarbons may be as low as 0–7%, and halogenated hydrocarbons are, almost without exception, significantly more dense than water.

Mackay and Roberts[2] designate organic liquids less dense than water as "floaters," which spread across the water table, and organic liquids more dense than water as "sinkers," which may plummet through sand and gravel aquifers to the underlying aquitard (relatively impermeable layer) where present.

ENVIRONMENTAL FACTORS AND CHEMISTRY

General

The chemical, physical, and biological properties of a substance, in conjunction with the environmental characteristics of an area, result in physical, chemical, and biological processes associated with the transport and transformation of the substance in soil and groundwater.

Soil models tend to be based on first-order kinetics and thus employ only first-order rate constants, with no ability to correct these constants for environmental conditions in the simulated environment which differs from the experimental conditions. The processes and corresponding physical parameters that are important in determining the behavior and fate of a chemical are different in the analysis of dissolved trace-level contaminants to analyses of contaminants from large-scale releases (e.g., spills). Models are available for organic dissolved compounds, immiscible fluids, and inorganic compounds. Physical, chemical, and biological processes important to the above modeling categories are given below.

Miscible/Organic Compounds

Physical processes affecting dissolved organic compound fate in soil systems are mainly: (1) advection, dispersion, volatilization; and (2) sorption, ion-cation exchange.

Chemical processes affecting dissolved contaminant migration are mainly ionization, hydrolysis, oxidation/reduction, and complexation. Biological processes affecting dissolved contaminant migration are mainly bioaccumulation and biodegradation.

In general, the processes of advection, dispersion, volatilization, and sorption are most important to both trace-level analyses and large-scale release analyses. Bulk properties (e.g., viscosity, solubility) are usually only important in simulations involving large amounts of contaminants.

Adsorption and desorption can drastically retard the migration of pollutants in soils; therefore, knowledge of this process is of importance when dealing with contaminant transport in soil and groundwater. The type of pollutant will determine to what kinds of material the pollutant will sorb. For organic compounds it appears that partitioning between water and the organic carbon content of soil is the most important sorption mechanism (Mackay 1982).

Ion exchange (an important sorption mechanism for inorganics) is viewed as an exchange with some other ion that initially occupies the adsorption site on the solid.

Cation exchange can be quite sensitive to other ions present in the environment.

Ionization is the process of separation or dissociation of a molecule into particles of opposite electrical charge (ions). The presence and extent of ionization has a large effect on the chemical behavior of a substance. An acid or base that is extensively ionized may have markedly different solubility, sorption, toxicity, and biological characteristics than the corresponding neutral compound.

Hydrolysis is one of a family of reactions which leads to the transformation of pollutants. Under environmental conditions, hydrolysis occurs mainly with organic compounds. Hydrolysis is a chemical transformation process in which an organic RX reacts with water, forming a new molecule.

For some organic compounds such as phenols, aromatic amines, and alkyl sulfides, chemical *oxidation* is an important degradation process under environmental conditions. Most of these reactions depend on reactions with free radicals already in solution and are usually modeled by pseudo-first-order kinetics.

Complexation, or chelation, is the process by which metal ions and organic or other nonmetallic molecules (ligands) can combine to form stable metal-ligand complexes. The complex that is found will generally prevent the metal from undergoing other reactions or interactions that the free metal cation would undergo. Complexation may be important in some situations; however, the current level of understanding of the process is not very advanced, and the available information has not been shown to be particularly useful to quantitative modeling.

Bioaccumulation is the process by which terrestrial organisms such as plants and soil invertebrates accumulate and concentrate pollutants from the soil. Bioaccumulation has not been examined in soil modeling, apart from some nutrient cycle (phosphorus, nitrogen) and carbon cycle bioaccumulation attempts.

Biodegradation refers to the process of transformation of a chemical by biological agents, usually by microorganisms, and it actually refers to the net result of a number of different processes, such as mineralization, detoxication, cometabolism, activation, and change in spectrum.

Immiscible Fluids

The transport of an organic liquid phase also is influenced by its viscosity and its surface-wetting properties, compared with those of water. According to Mackay and Roberts[2] and the literature reviewed, halogenated aliphatics tend to spread by capillary action into aquifer media and to be retained in amounts of about 0.3% to 5% by volume, following the passage of the organic liquid. For the above reasons, it also is difficult to completely remove hydrocarbons from granular aquifers as the capillary attraction between smaller particles may be sufficient to prevent bubbles of, for example, gasoline passing between the grains.

Miscible Inorganic Compounds

Important processes and mechanisms of environmental interactions of inorganic materials are volatility, solubility, fast aqueous reactions, slow aqueous reactions, speciation, soil interaction, and bioaccumulation. Reactions of inorganic dissolved species in aqueous media can be classified under the following general categories:

- reaction/dissociation (with solvent molecules)
- substitution reactions (with solvent or dissolved species)
- redox reactions (with dissolved gases or other ions).

Slower Reactions in Aqueous Media

In the category of slower reactions in aqueous media one may consider the following reactions:

- ligand substitution reactions of relatively kinetically inert ions
- electron transfer reactions involving inner sphere mechanisms for relatively kinetically inert ions and some outer sphere reactions
- reactions with dissolved gaseous species/bacterial catalyzed reduction (e.g., oxidation by dissolved oxygen)
- precipitation of solids by formation of insoluble species via substitution; and/or
- oxidation reactions (e.g., formation of metal hydroxides and metal sulfides).

The above slower reactions impact the speciation and concentration of the inorganic species in the aqueous phase and determine the type and extent of further interactions in the same manner as the fast reactions, although a different time frame will be required to achieve completion of some of these reactions. For the relatively kinetically inert metals (e.g., Cu^{3+}, Fe^{3+}, Cr^{3+}, and other such inorganic ions), immediate modifications of the speciation of the ion may not occur upon mixing with other aqueous phases (e.g., leachate with groundwater under oxidizing conditions), and the reaction may not proceed prior to the mixture reacting with soil strata.

Speciation

An evaluation of the fate of trace metals in surface and subsurface waters requires the consideration of speciation, adsorption, and precipitation. These processes can affect metal solubility, toxicity, availability, physical transport and corrosion potential. As a result of a need to describe the complex interactions involved in these situations, various models have been developed to address a number of specific situations. Steps in speciation calculations are not described in this publication. Speciation computerized packages (models) are described in parts of a later section.

Soil and Sediment Interactions

Important interactions of dissolved aqueous inorganic species occur with soils. The more important mechanisms of interaction include formation of precipitates, adsorption of components onto soil surfaces, modification of speciation by soil constituents (solid and liquid phase), and reactions induced by bacteria present in soil.

Bioaccumulation

A great deal of information exists concerning the effects of specific inorganic constituents on a large variety of organisms in specific environment. However, such information has generally been difficult to use in the development of broadly generalized techniques for estimation of effects. Estimation techniques for understanding the impact potential of inorganic compounds on organisms in aquatic or terrestrial environments have not at present been developed.

Processes Related to Terrestrial Plants

If there are exceptions to the general lack of estimation techniques and models relating to inorganic chemical concentrations, biological effects and fate, they are likely to be found in agricultural theory and modeling. Application of soil amendments such as fertilizers for macronutrients (e.g., phosphates, nitrates) or micronutrients (e.g., Mn), or for improving the soil chemical environment to release nutrients (liming), is a relatively well-developed science.

MATHEMATICAL MODELING—OVERVIEW

Mathematical models can greatly assist decisionmakers in determining the importance of pollutant pathways in the environment, as long as they are used properly and with an understanding of their limitation.

In general, soil/groundwater modeling concepts deal mainly with point source pollution and can be categorized into (1) unsaturated soil zone (or soil), (2) saturated soil zone (groundwater), (3) geochemical, and (4) ranking. The first two categories follow comparable patterns of mathematics and approach; the third enters into chemistry and speciation modeling; and the fourth follows a screening approach. The fist two modeling categories are designed to handle either organic or inorganic dissolved pollutants, or immiscible fluids for nonaqueous phase (NAPL) chemicals.

In developing a model, scientists attempt to reach an optimal compromise among the above approaches, given the level of detail justified by both the data availability and the study objectives. Deterministic model formulations can be further classified into simulation models which employ a well-accepted empirical equation that is forced, via calibration coefficients, to describe a system, and analytic models in which the derived equation describes the physics/chemistry of a system.

Four solution procedures are mainly followed: the analytical; the numerical (e.g. finite different, finite element); the statistical; and the iterative. Numerical techniques have been standard practice in soil quality modeling. Analytical techniques are usually employed for simplified and idealized situations. Statistical techniques have academic respect, and iterative solutions are developed for specialized cases. Both the simulation and the analytic models can employ numerical solution procedures for their equations. Although the above terminology is not standard in the literature, it has been used here as a means of outlining some of the concepts of modeling.

Generally speaking, a deterministic or stochastic soil quality model consists of two major parts or modules:

- the flow module or moisture module, or hydrologic cycle module—aiming to predict flow or moisture behavior (i.e., velocity, content in the soil), and
- the solute module—aiming to predict pollutant transport, transformation, and soil quality in the soil zone.

DISSOLVED POLLUTANT MODELING

In this section more emphasis is placed on the unsaturated soil zone than on groundwater modeling. This emphasis can be justified by the fact that similar modeling concepts govern both environments.

Unsaturated Soil Zone (Soil) Modeling

Soil modeling follows three different mathematical formulation patterns: (1) Traditional Differential Equation (TDE) modeling, (2) compartmental modeling, and (3) stochastic modeling. Some researchers may categorize models differently as, for example, into numerical or analytic, but this categorization applies more to the techniques employed to solve the formulated model rather than to the formulation per se. A model has a flow (moisture) module and a quality module.

TDE Modeling

The TDE moisture module (of the model) is formulated from three equations: (1) the water mass balance equation, (2) the water momentum, and (3) the Darcy equation, and also other equations such as the surface tension of a potential energy equation. The resulting differential equation system describes moisture movement in the soil and is written in a one-dimensional, vertical, unsteady, isotropic formulation as

$$d[K(\psi)(d\psi/dz+1)]/dz = C(\psi)d\psi/d\tau+S \tag{1}$$

$$\upsilon_z = -K(z,\psi)d\varphi/dz \tag{2}$$

where:

z = elevation (cm)
ψ = pressure head, often called soil moisture tension head in the unsaturated zone (cm);
$K(\psi)$ = hydraulic conductivity (cm/min);
$C(\psi) = d\theta/d\psi$ = slope of the moisture (φ) versus pressure head (ψ) (cm^{-1});
t = time (min);
S = water source or sink term (min^{-1});
φ = $z+\psi$; and
v_z = vertical moisture flow velocity (cm/s). The moisture module output provides the parameters v and θ as input to the solute module.

The TDE solute module is formulated with one equation describing the pollutant mass balance of the species in a representative soil volume, $d\,V = dxdydz$. The solute module is frequently known as the dispersive, convective differential mass transport equation, in porous media, because of the wide employment of this equation, that may also contain an adsorptive, a decay, and a source or sink term. The one-dimensional formulation of the module is

$$d(\theta c)/dt = [d(\theta\chi\ K_o\ dc/dz)] - [d(vc)/dz] - [\rho \times d_s/dt] \pm \Sigma P \qquad (3)$$

where:

θ = soil moisture content;
c = dissolved pollutant concentration in soil moisture;
K_o = apparent diffusion coefficient of compound in soil-air;
v = Darcy velocity of soil moisture;
ρ = soil density;
s = adsorbed concentration of compound on soil particles;
ΣP = sum of sources or sinks of the pollutant within the soil volume;
z = depth.

At this point it is important to note that the flow model (a hydrologic cycle mode) can be absent from the overall model. In this case the user has to input to the solute module the temporal (t) and spatial (x,y,z) resolution of both the flow (i.e., soil moisture) velocity (ν) and the soil moisture content (θ) of the soil matrix. This approach is employed by Enfield et al.[7] and other researchers. If the flow (moisture) module is not absent from the model formulation (e.g., Huff[8]), then users are concerned with input parameters that may be frequently difficult to obtain. The approach to be undertaken depends on site specificity and available monitoring data.

Some principal modeling-specific deficiencies when modeling solute transport via the TDE approach are reported by Bonazountas.[9]

Compartmental Modeling

Compartmental soil modeling is a concept available since 1984 (SESOIL Model, Bonazountas and Wagner[10]) and can apply to both modules. The solute fate module, for example, consists of the application of the law of pollutant mass conservation to a representative user-specified soil element. The mass conservation principle is applied over a specific time step, either to the entire soil matrix or to the subelements of the matrix such as the soil-solids, the soil-moisture, and the soil-air. These phases are assumed to be in equilibrium at all times; thus, once the concentration in one phase is known, the concentration in the other phases can be considered, whereas phases and subcompartments can be interrelated with transport, transformation, and interactive equations.

Compartmental models may bypass the deficiencies of the TDE modeling because they may handle geochemical issues in a more sophisticated way if required, but this does not imply that compartmental models are ''better'' than TDE models. They are simply different.

Stochastic, Probabilistic, and Other Modeling Concepts

Stochastic or probabilistic techniques can be applied to either the moisture module or the solution of other equations. Stochastic or probabilistic modeling is mainly aimed at describing ''breakthrough'' times of overall concentration threshold levels, rather than individual processes or concentrations in individual soil compartments. Coefficients or response functions for these models have to be calibrated to field data, since major processes are studied via a black-box or response function approach and not individually. Other modeling concepts may be related to soil models for solid waste sites and specialized pollutant leachate issues (Schultz[11]).

Physical, Chemical, and Biological Processes Modeled

Modelers should be fully aware of the range of applicability and processes considered by a computerized package. There exists some disagreement among soil modelers as to whether there is a need for increased model sophistication, since almost all soil modeling predictions have to be validated with monitoring data, given the physical, chemical, and biological processes that affect pollutant fate in soil systems. Because of the latter consideration, many simplified models may provide excellent results, assuming accurate site-specific calibration is achieved. Nevertheless, model sophistication is reflected in the process modeled, but model selection is mandated by the project needs and data availability. The important physical processes of a typical soil model are reported by Bonazountas.[9]

Saturated Soil Zone (Groundwater) Modeling

Saturated soil zone (or groundwater) modeling is formulated almost exclusively via a TDE system consisting of two modules, the flow, and the solute module. The two modules are written as[12]

$$\nabla\,(\rho k/\mu)\,(\nabla p - \rho g \nabla Z) - q = d(\varphi)/dt$$

$$\nabla[\rho C(k/\mu)(\nabla p - \rho g \nabla Z)] + \nabla(\rho E)\nabla C = d(\rho\varphi C)dt \tag{4}$$

where:

C = concentration, mass fraction;
E = dispersion coefficient;
g = acceleration due to gravity;
k = permeability;
p = pressure;
q = mass rate of production or injection of liquid per unit volume;
t = time;
Z = elevation above a reference plane;
φ = porosity;
ρ = density;
μ = viscosity.

Mathematical groundwater modeling has been the least problematic in its scientific formulation, but has been the most problematic model category when dealing with applications, since these models have to be calibrated and validated as described later. Actually we have only TDE and some other (e.g., stochastic) formulations. The proliferation of literature models is mainly due to different model dimensionalities (zero, one, two, three); model features (e.g., with adsorption, without absorption terms); solution procedures employed (e.g., analytic, finite difference, finite element, random walk, stochastic); sources and sinks described; and the variability of the boundary conditions imposed. Some of the principal modeling deficiencies discussed in the previous section apply to groundwater models also. In general, (1) there exists no "best" groundwater model and (2) for site-specific applications, groundwater models have to be calibrated.

Selected Models

Bonazountas[9] lists selected soil and groundwater models and their main features. Models listed are documented, operational, and very representative of the various structures, features, and capabilities.

Ranking Modeling

Ranking models are aimed at assessing environmental impacts of waste disposal sites. The first ranking models focused on groundwater contamination; later models had a wider scope (e.g., health consideration). These models rank or rate contaminant migration at different sites, as it is affected by hydrogeologic, soil, waste type, density, and site design parameters.

Modeling Issues

Model selection, application, and validation are issues of major concern in mathematical soil and groundwater dissolved pollutant modeling. For the model selection, issues of importance are: the features (physics, chemistry) of the model; its temporal (steady state, dynamic) and spatial (e.g., compartmental approach) resolution; the model input data requirements; the mathematical techniques employed (finite difference, analytic); monitoring data availability; and cost (professional time, computer time). For the model application, issues of importance are: the availability of realistic input data (e.g., field hydraulic conductivity, adsorption coefficient); and the existence of monitoring data to verify model predictions. Some of these issues are briefly discussed by Bonazountas.[9]

Input data have to be compiled and input to the model from site-specific investigations and analyses (e.g., leaching rates of pollutants, soil permeability); national databases (e.g., climatic data from the National Oceanographic Atmospheric Administration); and other sources (e.g., diffusion rates of pollutants from handbooks). Some data categories are pollutant source data, climatic data, geographic data, particulate transport data, and biological data.

Exact knowledge of the physics of the soil system—although essential—is impossible prior to employing any modeling package. Numerical (e.g., finite difference) TDE soil models require the net infiltration rainfall rate after each storm, even as an input parameter to their moisture module. The rate can be either a user input or can be generated by another model. The same models require the soil conductivity as a function of the soil moisture content as an input parameter. Its value can be obtained from field investigations, from laboratory data, or from references, but much uncertainty exists in this area of input data gathering.

Numerical soil models (time, space) provide a general tool for quantitative and qualitative analyses of soil quality, but require time-consuming applications that may result in high study costs. In addition, input data have to be given for each mode or element of the model: which model has to be run twice, and the number of rainfall events. On the other hand, analytic models obtained from analytic solutions of Equation 2 are easier to use, but can simulate only average temporal and spatial conditions, which may not always reflect real world situations. Statistical models may provide a compromise between the above situations.

Model output "validation" is essential to any soil modeling effort, although this term has a broad meaning in the literature. For the purpose of this section, we can define validation as "the process which analyses the validity of final model output,'' namely, the validity of the predicted pollutant concentrations or mass in the soil column (or in groundwater), to groundwater and to the air, as compared to available knowledge of measured pollutant concentrations from monitoring data (field sampling).

A disagreement of course in absolute levels of concentration (predicted versus measured) does not necessarily indicate that either method of obtaining data (modeling, field sampling) is incorrect or that either data set needs revision. Field sampling approaches and modeling approaches rely on two different perspectives of the same situation.

Important issues in groundwater model validation are the estimation of the aquifer physical properties, the estimation of the pollutant diffusion, and the decay coefficient. The aquifer properties are obtained via flow model calibration (i.e., parameter estimation) and by employing various mathematical techniques such as kriging. The other parameters are obtained by comparing model output (i.e., predicted concentrations) to field measurements—a quite difficult task, because clear contaminant plume shapes do not always exist in real life.

Model Applications

Numerous applications are available in the literature. Case studies are presented in earlier publications.[9]

IMMISCIBLE CONTAMINANT MODELING

General

In an effort to quantify nonaqueous phase liquid (NAPL), or immiscible contaminant migration, from waste sites and oil spills noted with increasing frequency, various mathematical models were developed in the 1980s, and considerable research is continuously directed toward improving our understanding of these immiscible fluid processes in the soil and the groundwater zones of an area.

As reported by Mercer,[3] many of the recently developed theoretical concepts and modeling approaches pertaining to this problem had originated in the petroleum industry. However, because of the different physical environment of deep petroleum reservoirs compared to the shallow aquifers, as well as different incentives and areas of concern in the petroleum industry, there is a great need to adopt and extend this work.

According to Abriola and Pinder,[5] Van Dam presented the first detailed analysis of hydrocarbon pollution of groundwater as a two-phase problem. He examined the stages of contaminant infiltration and incorporated a capillary pressure term in his expression for fluid potential. Many researchers followed his work, among them Faust[4] and others reported in the reference section of this chapter.

Schwille[13] provides a review on physical, chemical, and biological parameters affecting NAPL migration in porous media, with an emphasis on mineral oil products and chlorohydrocarbons.

Modeling Concepts

Four phases separated by distinct interfaces exist in a soil matrix: solid (soil), water, gas, and contaminant. In the real physical system, each phase could possibly be formed by a number of chemical components or species, and mass transfer could occur across phase boundaries. In that respect, a contaminant may be available in four phases: adsorbed, dissolved, gaseous, and NAPL. Migration of such a contaminant can be modeled as a three-phase flow process.

Basically there exist two concepts (ways) in modeling three-phase flow:

- by employing the three-phase flow (air, moisture, NAPL) equations in porous media jointly with a set of relationships; and
- by employing the governing equations of a two-phase flow in porous media (e.g., water-NAPL, air-NAPL, or air-NAPL/water) and by adjusting the coefficients (e.g., relative permeability of fluid) of the equations to reflect the specific problem.

The second approach eliminates the need for an equation governing the third phase (e.g., Faust[4]), but introduces the need for determining the "relative" permeability of the "second" phase (e.g., NAPL-water) to the "first" phase (e.g., air). The model presented by Abriola and Pinder[5] follows the first concept. The model developed by Faust[4] follows the second concept. Most of the computer codes consider three phases: air, water, and NAPL, but the necessary input data to drive models do not exist. A series of applications is reported in earlier publications.

Selected Models

The U.S. Geological Survey in Reston, Virginia, has developed a series of computer codes for NAPL fluids. These codes are available in the public domain and can be easily obtained. Many researchers have developed proprietary codes, information on which is obtained from the publications of their research, as reported by Bonazountas.[9]

Modeling Applications

As an example only, we report the work of Faust.[4] He demonstrates the impact of an undetected leak of an NAPL on a surficial unit considered to be an aquitard. Two base cases were simulated:

- a NAPL more dense than water, where we expect gravity effects to be dominant and create a downward migration; and
- a NAPL less dense than water, where we expect the contaminant to pool near the water table.

The above two cases and expected conditions are observed in the results of the simulation cited by Bonazountas.[9]

AQUATIC EQUILIBRIUM MODELING

An evaluation of the fate of inorganic compounds in soil and groundwater requires more detailed consideration of the chemical and biological processes involved, such as complexation, adsorption, coagulation, oxidation-reduction, chemical speciation, and biological activity. As a result of a need to describe the complex interactions involved in these situations, various models have been developed to address specific needs. We have, for example, adsorption models, surface complexation models, constant capacitance models, cation-exchange models, and overall fate modeling packages which account for one or more geochemical processes.

Excellent state-of-knowledge reviews on chemical equilibria codes (models) of inorganic pollutants in soils are presented by Sposito,[14] Cederberg et al.,[15] Kincaid et al.,[16] Miller and Benson,[17] Jennings et al.,[18] Theis et al.[19] Mattigod and Sposito,[20] Jenne et al.,[21] Nordstrom et al.,[22] and many others. The chapter has benefited from the reviews of the above researches, and particularly from the work of Sposito.[14] Readers interested in details should refer to the original publications and to the publication of Calabrese et al.[23]

PARAMETERS AFFECTING MIGRATION

A quantitative evaluation of critical parameters affecting dissolved organic contaminant migration through soils when conducting mathematical modeling is presented by Tucker et al.[24] Critical parameters affecting inorganic species and NAPL migration are discussed in the corresponding chapters.

REFERENCES

1. Hall, P. L., and H. Quam. "Countermeasures to Control Oil Spills in Western Canada," *Ground Water* 14(3):163 (1976).
2. Mackay, D. M., and P. V. Roberts. "Transport of Organic Contaminants in Groundwater," *Environ. Sci. Technol.* 19(5):384 (1985).

3. Mercer, J. W. "Miscible and Immiscible Transport in Groundwater," EOA, U.S. Geological Survey, 1984, p. 691.
4. Faust, C. R. "Transport of Immiscible Fluids Within and Below the Unsaturated Zone—A Numerical Model," Geotrans Report No. 84-01, Geotrans, Herndon, VA, 1984.
5. Abriola, L. M., and G. F. Pinder. "A Multiphase Approach to Modeling of Porous Media Contamination by Organic Compounds: 1. Equation Development; 2. Numerical Solution," *Water Resour. Res.* 21(1):11–32 (1985).
6. Fried, J. J., P. Muntzer, and L. Zilliox. "Ground-Water Pollution by Transfer of Oil Hydrocarbons," *Ground Water* 17(6):586 (1979).
7. Enfield, C. G., R. F. Carsel, S. Z. Cohen, T. Phan, and D. M. Walters. "Approximating Pollutant Transport to Groundwater," U.S. Environmental Protection Agency, RSKERL, Ada, OK (unpublished paper), 1980.
8. Huff, D. D. "TEHM: A Terrestrial Ecosystem Hydrology Model," Oak Ridge National Laboratory, Oak Ridge, TN, 1977.
9. Bonazountas, M. "Mathematical Pollutant Fate Modeling of Petroleum Products in Soil Systems," in *Soils Contaminated by Petroleum: Environmental & Public Health Effects,* E. J. Calabrese and P. T. Kostecki, Eds. (New York: John Wiley & Sons, 1988), pp. 31–111.
10. Bonazountas, M., and J. Wagner. "Modeling Mobilization and Fate of Leachates Below Uncontrolled Hazardous Waste Sites," in *Proceedings of the 5th National Conference on Management of Uncontrolled Hazardous Waste Sites,* November 7–9, Hazardous Materials Control Research Institute, Silver Spring, MD, 1984.
11. Schultz, D. "Land Disposal of Hazardous Waste," *Proceedings of the 8th Annual Research Symposium,* H. J. Hutchell, Kentucky, March 8–10, U. S. Environmental Protection Agency, Cincinnati, OH, 1982.
12. Bachmat, Y., J. Bredehoeft, B. Andrews, D. Holz, and S. Sebastian. *Groundwater Management: The Use of Numerical Models,* Water Resources Monograph, American Geophysical Union, Washington, DC, 1980.
13. Schwille, F. "Groundwater Pollution in Porous Media by Fluids Immiscible with Water," *The Science of the Total Environment* 21:173–185 (1981).
14. Sposito, G. "Chemical Models of Inorganic Pollutants in Soils," *CRC Critical Reviews in Environmental Control* 15(1): (1985).
15. Cederberg, G. A., R. L. Street, and J. O. Leckie. "A Groundwater Mass Transport and Equilibrium Chemistry Model for Multicomponent Systems," *Water Resour. Res.* 21(8):1095–1104 (1985).
16. Kincaid, C. T., J. R. Morrey, and J. E. Rogers. "Geochemical Models for Solute Migration, Vol. 1: Process Description and Computer Code Selection," Report No. EA-3417, Electric Power Research Institute, Palo Alto, CA, 1984.
17. Miller, C. W., and L. V. Benson. "Simulation of Solute Transport in a Chemically Reactive Heterogeneous System: Model Development and Application," *Water Resour. Res.* 19(2):381–391 (1983).
18. Jennings, A. A., D. J. Kirkner, and T. L. Theis. "Multicomponent Equilibrium Chemistry in Groundwater Quality Models," *Water Resour. Res.* 18(4):1089–1096 (1982).
19. Theis, T. L., D. J. Kirkner, and A. A. Jennings. "Multi-Solute Subsurface Transport Modeling for Energy Solid Wastes," Technical Progress Report, Department of Civil Engineering, University of Notre Dame, Notre Dame, Indiana, 1982.

20. Mattigod, S. V., and G. Sposito. "Chemical Modeling of Trace Metal Equilibria in Contaminated Soil Solutions Using the Computer Program GEOCHEM," in *Chemical Modeling in Aqueous Systems,* E. A. Jenne, Ed. American Chemical Society Symposium, Series No. 93, American Chemical Society, Washington, DC, 1979, pp. 837–856.
21. Jenne, E. A. "Chemical Modeling—Goals, Problems, Approaches and Priorities," in *Chemical Modeling in Aqueous Systems,* E. A. Jenne, Ed. American Chemical Society Symposium, Series No. 93, American Chemical Society, Washington, DC, 1979.
22. Nordstrom, D. K., L. N. Plummer, T. M. L. Wigley, T. T. Wolery, J. W. Ball, E. A. Jenne, R. L. Bassett, D. A. Crerar, T. M. Florence, B. Fritz, M. Hoffman, G. R. Holdren, G. M. Lafon, S. V. Mattigod, R. E. McDuff, F. Morel, M. M. Reddy, G. Sposito, and J. Throil Kill. "A Comparison of Computerized Chemical Models for Equilibrium Calculations in Aqueous Systems," in *Chemical Modeling in Aqueous Systems,* E. A. Jenne, Ed. American Chemical Society Symposium, Series No. 93, American Chemical Society, Washington, DC, 1979.
23. Calabrese, E. J., and P. T. Kostecki. *Soils Contaminated by Petroleum: Environmental & Public Health Effects* (New York: John Wiley & Sons, 1988).
24. Tucker, W. A., E. V. Dose, G. J. Gensheimer, R. E. Hall, D. N. Koltuniak, C. D. Pollman, and D. H. Powell. "Evaluation of Critical Parameters Affecting Contaminant Migration Through Soils," *National Conference on Environmental Engineering,* July 1–3, 1985, Northeastern University, Boston, Massachusetts, 1985. Unpublished: available from Dr. W. A. Tucker, Environmental Science and Engineering, Inc., Gainesville, Florida.

CHAPTER 13

Soil Vapor Extraction Research Developments

George E. Hoag, Environmental Research Institute and School of Civil Engineering, The University of Connecticut, Storrs, Connecticut, and VAPEX Environmental Technologies, Inc., Canton, Massachusetts

Michael C. Marley, Bruce L. Cliff, and **Peter Nangeroni,** VAPEX Environmental Technologies, Inc., Canton, Massachusetts

INTRODUCTION

Recently, in situ subsurface remediation processes have been the focus of significant attention by the scientific community involved with the cleanup of volatile and semivolatile environmental contaminants. Of the in situ processes researched to date, vapor extraction holds perhaps the most widespread application to the remediation of these types of organic chemicals frequently found in the subsurface. The vapor extraction process has been successfully employed at many types of sites as a stand-alone technology, and may also be considered a synergistic technology to other types of in situ subsurface remediation technologies, such as bioremediation and groundwater pump, skim, and treat.

In the past five years, in situ vapor extraction has been applied at many sites by means of significantly different approaches. These range from ''black box'' DESIGN WHILE YOU DIG TECHNIQUES to those utilizing sophisticated numerical models interfaced with laboratory, pilot, and full-scale parameter determination for design purposes. The extent of success in field application of vapor extraction is varied, in many cases related to monitoring and interpretive limitations employed before, during, and after the remediation. Because application of the technology is quite recent, many remediations are still in progress; thus, interpretation and publishing of final results in refereed scientific journals is limited.

Research Milestones in Vapor Extraction

Thornton and Wootan[1] introduced the concept of vertical vapor extraction and injection wells for the removal gasoline product, as well as vapor probe monitoring for the quantitative and qualitative analysis of diffused hydrocarbon vapors. A further enhancement of this research was published by Wootan and Voynick,[2] in which various venting geometries and subsequent air flow paths were hypothesized and tested in a pilot-sized soil tank. In their first study, 50% gasoline removal was achieved, while in their second study up to 84% removal of gasoline was observed.

Local Equilibrium Concept

In controlled laboratory soil column vapor extraction experiments by Marley and Hoag[3] and Marley,[4] 100% removal of gasoline at residual saturation was achieved for various soil types (0.225 mm to 2.189 mm average diameters), bulk densities (1.44 g/cm to 2.00 g/cm), moisture contents (0% to 10% v/v) and air flow rates (16.1 $cm^3/(cm^2/min)$ to 112.5 $cm^3/(cm^2/min.)$ They also successfully developed an equilibrium solvent-vapor model using Raoult's Law to predict concentrations of 52 components of gasoline in the vapor extracted exhaust of the soil columns.

Baehr and Hoag[5] adapted a one-dimensional three-phase (immiscible solvent, aqueous, and vapor phases) local equilibrium transport model developed by Baehr[6] to include air flow as described by Darcy's Law for compressible flow. This first deterministic one-dimensional model effectively predicted the laboratory vapor extraction results of Marley and Hoag[3] and provided the basis for higher-dimensional coupled air flow contaminant models for unsaturated zone vapor extraction.

Porous Media Air Flow Modeling

Because local equilibria prevailed in the above studies, a higher-dimensional model, developed by Baehr, Hoag, and Marley[7] was used to model air flow fields under vapor extraction conditions. The three-dimensional radially symmetric compressible air flow model is used to design vapor extraction systems using limited lab and/or field air flow pump tests. A steady state in situ pump test determination of air phase permeability is preferred over laboratory tests because an accounting is possible of the presence of an immiscible liquid, anisotropy, soil surface, variations in soil water conditions, and heterogeneity in air phase permeabilities. The numerical solution developed can simulate flow to a partially screened well, and allow determinations of vertical and horizontal air phase permeabilities. Heterogeneous unsaturated zones can also be evaluated using the numerical simulation. Analytical solutions to radial flow equations, such as one developed early by Muskat and Botset[8] are generally restricted to determination of average horizontal air perambulated determination for impervious soil surfaces.

Removal of Capillary Zone Immiscible Contaminants

Hoag and Cliff[9] reported that an in situ vapor extraction system was effective in removing 1,330 L gasoline at residual saturation and in the capillary zone at a service station, and achieved cleanup levels to below 3 ppm (v/v) in soil vapor and below detection limits in soils. The entire remediation took less than 100 days. A groundwater elevation and product thickness log for the time period of before, during, and after the vapor extraction remediation is graphically shown in Figure 13.1. On day 250 only a skim of gasoline was present in this well, and on day 290 no skim was detected. One year after the vapor extraction remediation took place, groundwater samples were nondetected for gasoline range hydrocarbons, reflecting that at least advective dispersive transport and possible natural microbiological activity in the groundwater were mechanisms responsible for this effect.

Field Application of Porous Media Air Flow Models

Baehr, Hoag, and Marley[7] utilized the above site for a field air pump test to determine the horizontal air phase permeability and to simulate the sensitivity of the model to changes in air phase permeabilities utilizing site geometries and boundary conditions. Based upon a full-scale air flow pump test, the air phase permeability for the site was predicted to be $k = 7.0 \times 10^{-8}$ cm^2 for a mass air withdrawal rate of 11.1 g/sec and a normalized pressure of $P_s/P_{atm} = 0.9$. For reference 11.1 g/sec, assuming an air phase density of 1.2×10^{-3} g/cm^3, equals about 555 L/min.

To illustrate the sensitivity of the model to a range of air phase permeabilities, the above service station vapor extraction well geometry, depth to water table, and appropriate boundary conditions were used as input and air phase permeability and mass air withdrawal rates were varied. In Figure 13.2, the normalized air phase pressure in the well for various mass air withdrawal rates and air phase permeabilities are shown. Significant increase in the vacuum developed in the wells can be observed for order of magnitude decreases in air phase permeabilities and small increases in the mass air withdrawal rates. Review of Figure 13.2 indicates that if a mass air withdrawal rate of 40 g/sec was used at the service station site ($k = 7.0 \times 10^{-8}$), then the normalized air phase pressure at the well would be approximately $P_s/P_{atm} = 0.6$, within an acceptable range of operating conditions.

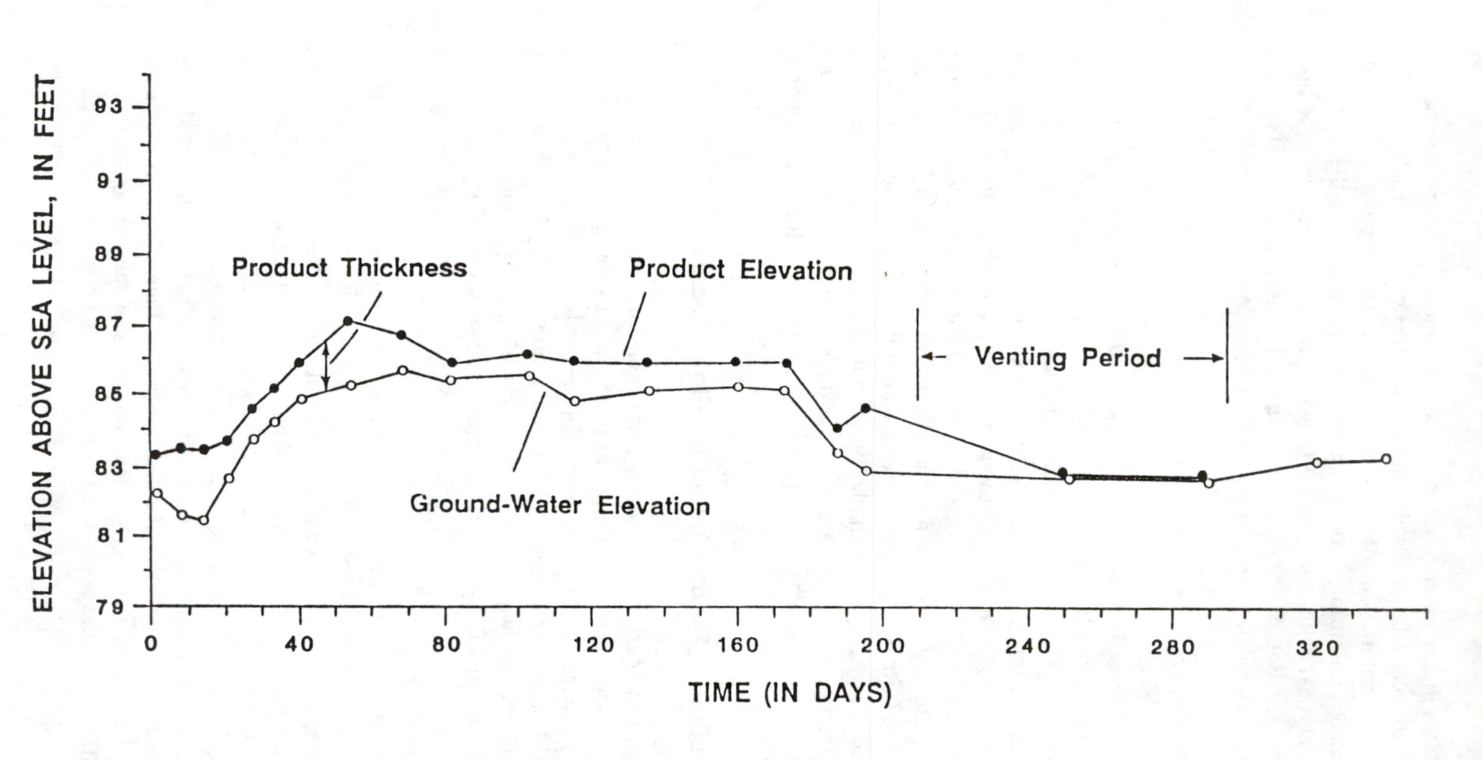

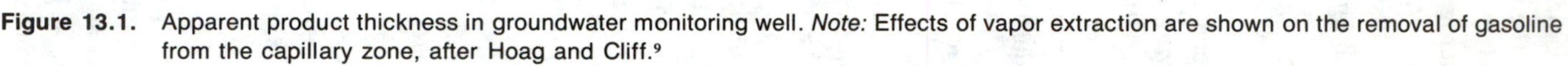

Figure 13.1. Apparent product thickness in groundwater monitoring well. *Note:* Effects of vapor extraction are shown on the removal of gasoline from the capillary zone, after Hoag and Cliff.[9]

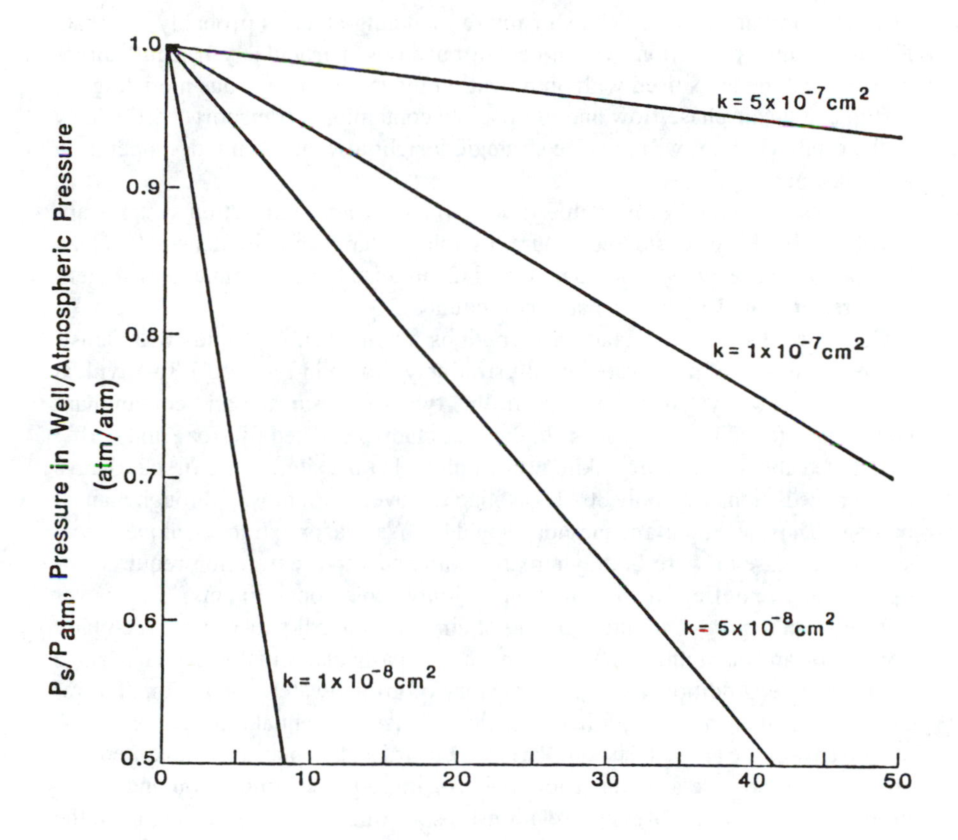

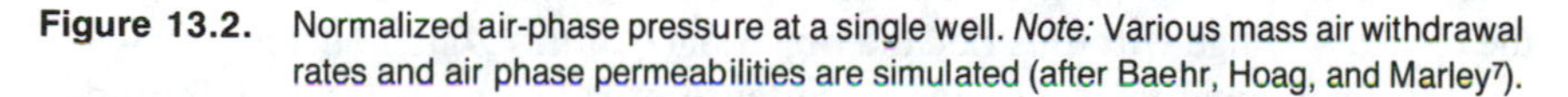

Figure 13.2. Normalized air-phase pressure at a single well. *Note:* Various mass air withdrawal rates and air phase permeabilities are simulated (after Baehr, Hoag, and Marley[7]).

A limitation of the model developed by Baehr, Hoag, and Marley[7] is that it is not coupled to contaminant transport. However, in the design of vapor extraction systems for volatile contaminants, this generally is not a fundamental need and can be accomplished by either laboratory venting tests, similar to those developed by Marley and Hoag,[3] or by utilization of the one-dimensional coupled model developed by Baehr and Hoag.[5]

RESEARCH NEEDS

A fundamental need of vapor extraction modeling occurs in the area of capillary zone/unsaturated zone interaction when immiscible phases are present on or in the capillary fringe. While air phase modeling alone is probably adequate for most vapor extraction system design purposes, particularly if a full three-dimensional model is used with optimization modeling, a rigorous modeling effort to couple air phase flow and immiscible contaminant transport, particularly in the capillary zone, will provide strategic insight to vapor extraction operation and planning.

To assess research needs in this area, two basic vapor extraction systems applications should be considered: (1) immiscible contaminant with density less than 1.0 (petroleum range hydrocarbons); and (2) immiscible contaminants with density greater than 1.0 (halogenated compounds).

Generalized subsurface phase distributions for immiscible liquids with densities less than that of pure water are illustratively shown in Figure 13.3. A typical vapor extraction system installation in this type of subsurface and contaminant condition is found in Figure 13.4. In the case study presented by Hoag and Cliff,[9] as detailed above, pump and skim was employed at the site for the first 210 days of the remediation, with only 300 L gasoline removed (i.e., mostly through manual bailing). Thus, an important question should be: Was vapor extraction alone necessary in this case or were both pump and skim and vapor extraction required for optimal or even effective remediation of immiscible contaminants? To answer this question requires an understanding of air-immiscible liquid-water three-phase conduction and distribution in the porous media, particularly in the capillary fringe areas at a site. Additionally, the site history of groundwater fluctuation and immiscible contaminant behavior in the capillary fringe is essential information necessary to answer the above question. Parker, Lenhard and Kuppusamy[10] and Lenhard and Parker[11] provide a parametric model for three-phase conduction and measurements of saturation-pressure relationships for immiscible contaminants in the unsaturated and capillary zones. However, to date, this author is not aware of the coupling of these types of models to air phase and contaminant transport models.

A more in-depth hypothetical examination of the possible relationships near the capillary fringe will illustrate the importance of this zone in determining the need for pump and treat, and the importance of solute mass transfer from the capillary zone into the saturated flow regime. In the case of a recent spill of an immiscible contaminant with density less than water, when relatively steady groundwater flow prevails, a zone may exist on the capillary fringe of floating product, as shown in Figure 13.5. Infiltrating water, under draining conditions, will reach an equilibrium with the immiscible liquid, resulting in a saturated solute condition. For hypothetical purposes only, if it is assumed that only vertical groundwater flow exists in the capillary zone, the rate of solute input to the saturated zone will be limited by the rate of infiltration and C_s. If it is assumed, again for illustrative purposes, that a horizontal flow boundary exists at the groundwater table, then mass transfer of solute from the capillary zone into the saturated zone will have only limited effects on the rate of solute input into the saturated zone. When considering the quantities of water infiltrating through the capillary zone per year in comparison to saturated flow rates, the above assumptions may

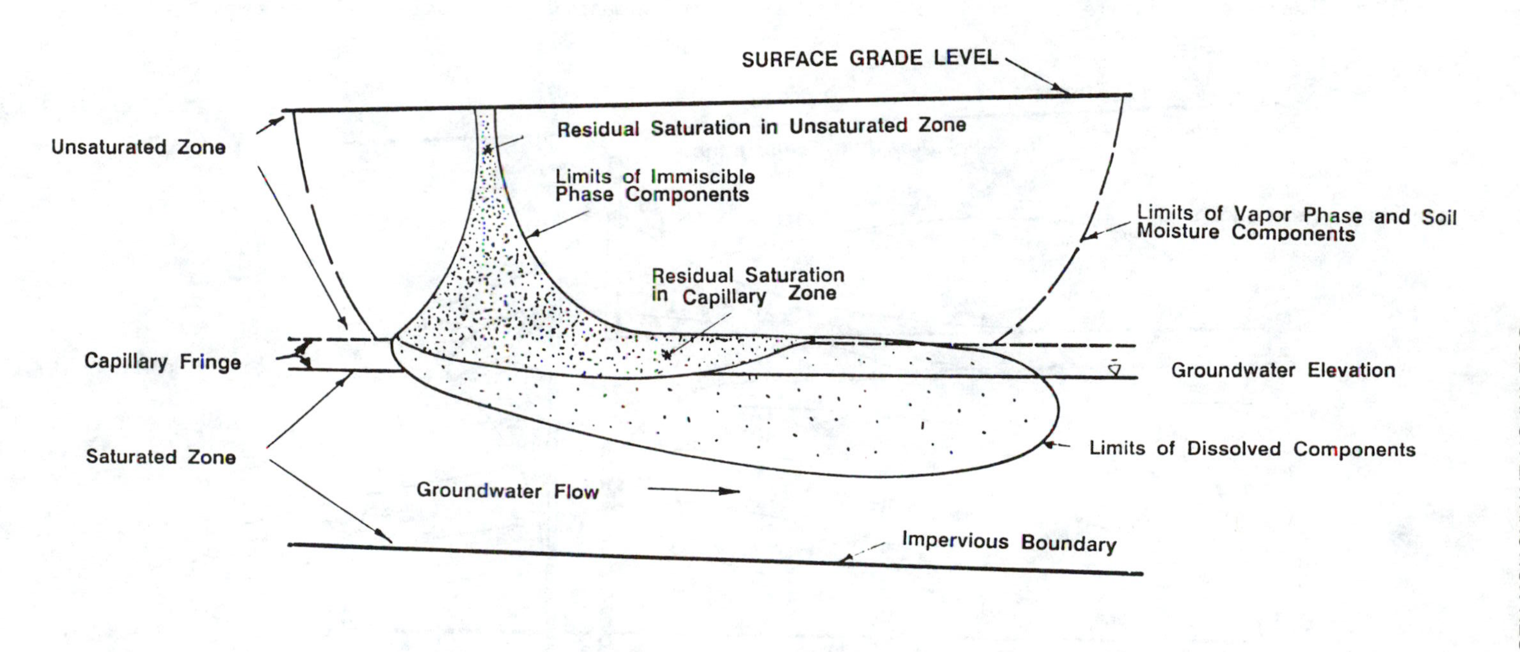

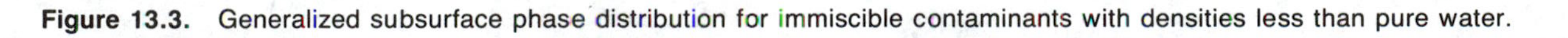

Figure 13.3. Generalized subsurface phase distribution for immiscible contaminants with densities less than pure water.

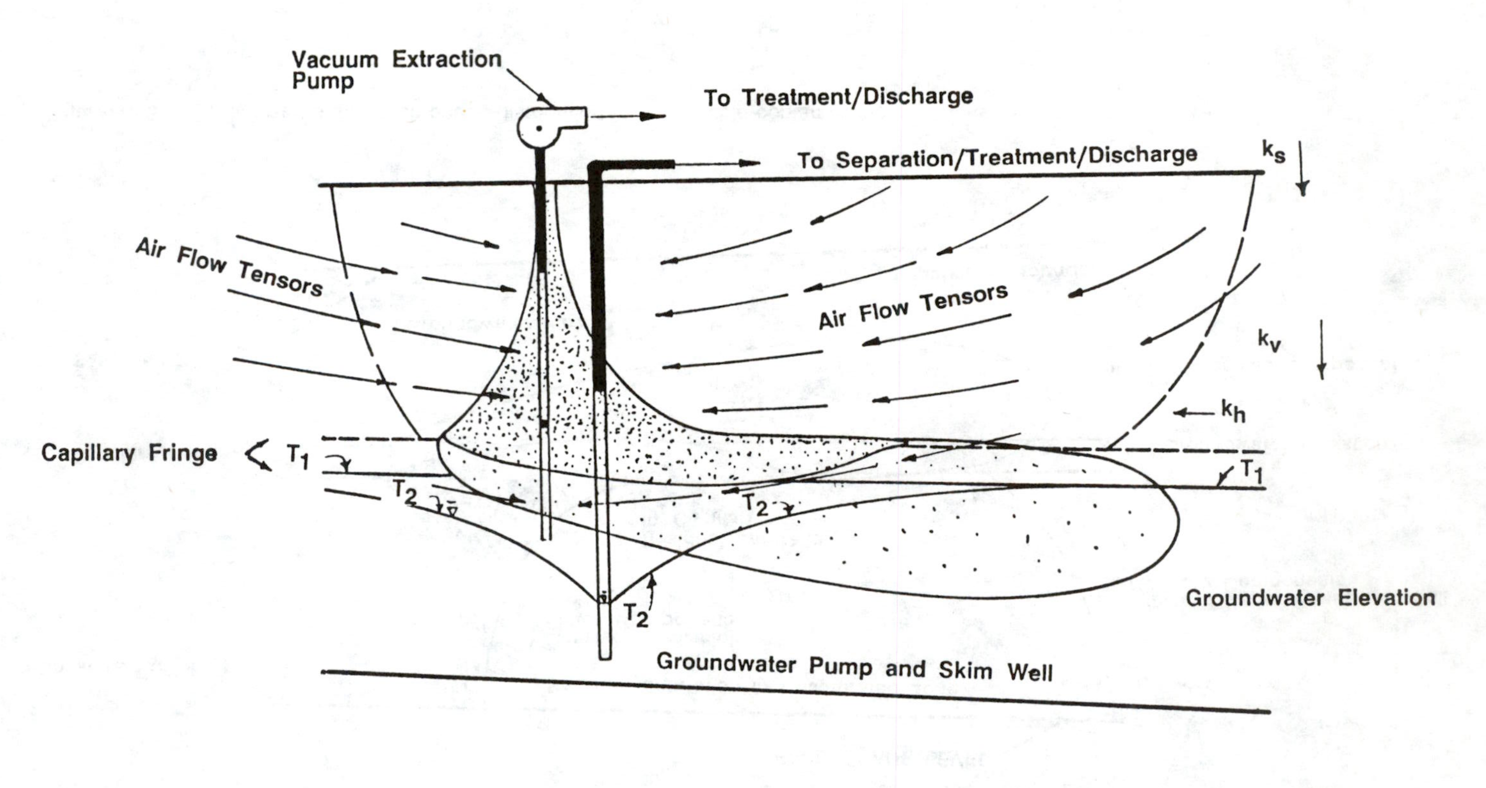

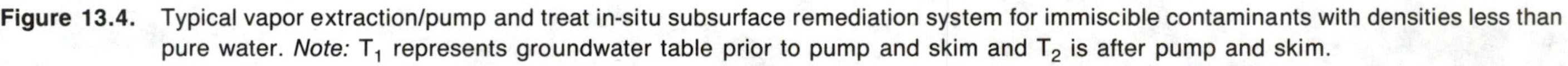

Figure 13.4. Typical vapor extraction/pump and treat in-situ subsurface remediation system for immiscible contaminants with densities less than pure water. *Note:* T_1 represents groundwater table prior to pump and skim and T_2 is after pump and skim.

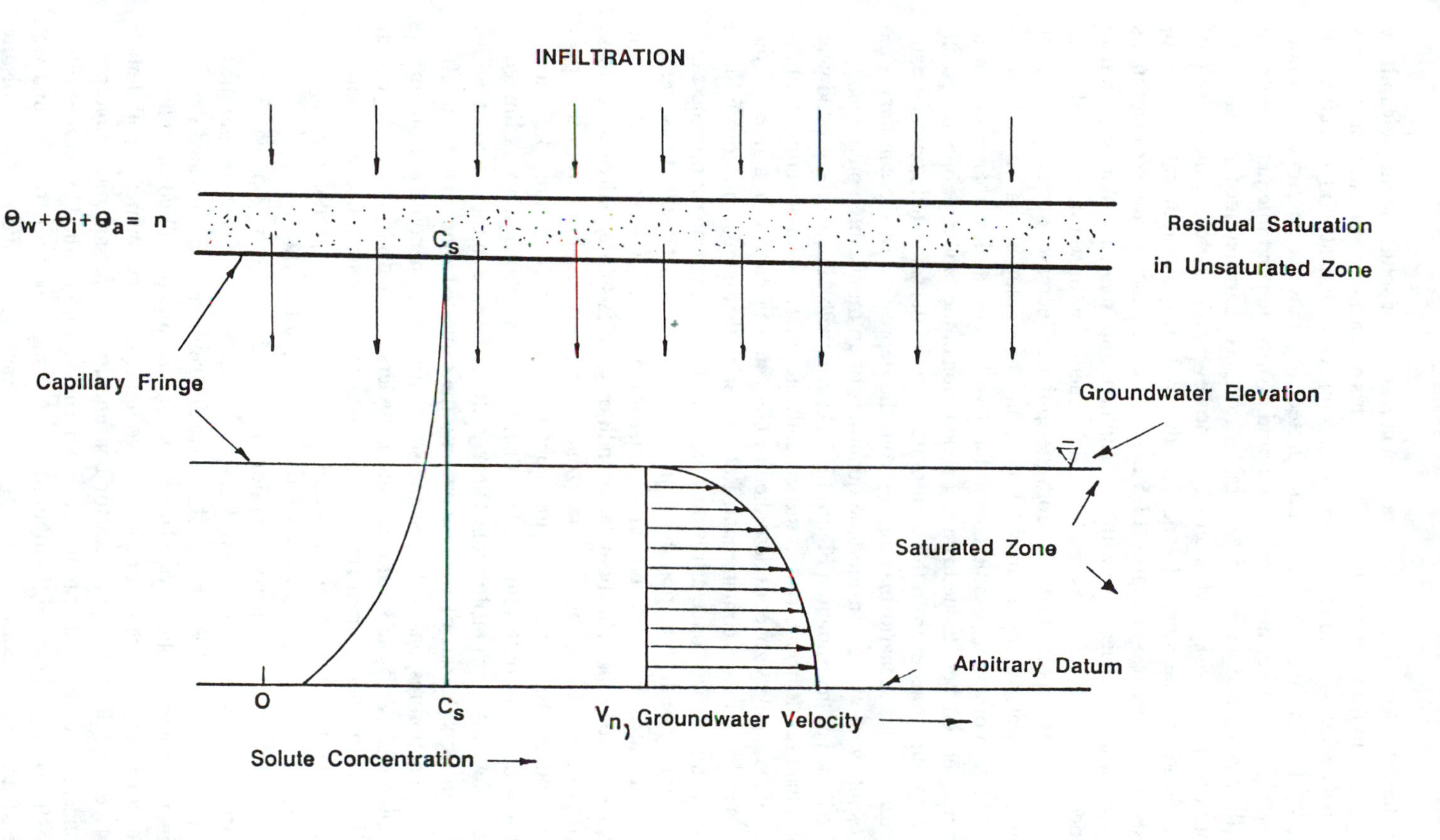

Figure 13.5. Immiscible contaminant on capillary zone, density less than pure water.

be valid. The result is relatively inefficient transfer of solutes from the capillary fringe zone to the saturated zone. Thus, in this scenario, pump and treat systems may not be necessary to remove the immiscible contaminants and advective dispersive dilution may be adequate to protect groundwater resources. Without knowledge a priori of the immiscible liquid distribution and interaction with the capillary zone, and advective-dispersive transport characteristics at a site, this approach may be risky. An alternative, however, may be close monitoring of groundwater in the saturated zone near the spill area, as vapor extraction proceeds. If the scenario in Figure 13.5 exists at a site, then solute concentrations in groundwater will decrease with time and no pump, treat, and skim system may be necessary to achieve desired levels of remediation in soil and groundwater. If near-field transport of solutes from the spill area increases steadily with time, then groundwater pumping may be necessary to employ at that time.

In the case of an immiscible contaminant with a density less than water, with impingement on the saturated zone by penetration of the capillary zone, the potential for solute transfer from the unsaturated zone to the saturated zone is greatly increased. This scenario may result from the depression of the capillary zone in a spill event where considerable quantities of an immiscible contaminant are spilled, such as that shown in Figure 13.3. Alternatively, fluctuating groundwater tables may result from a rise in the groundwater table through wetting (imbibition) of the capillary zone, as described by characteristic curves for a given porous media and immiscible contaminant. Remembering that immiscible contaminants become immobilized once at residual saturation, the result of wetting the capillary zone may be the condition shown in Figure 13.6. The net result of this scenario is that saturated solute concentrations exist at the top of the horizontal flow zone of the saturated zone. This boundary condition enables substantially greater mass transfer of solute into the saturated zone, principally resulting from the upper flow boundary being the immiscible contaminant itself. In Figure 13.5, the upper boundary was only solute at less than C_s, and solubilization was limited to that achieved through infiltration. Clearly, the difference in these two situations greatly affects the rate of solute input into the saturated zone and should affect remedial action responses. Unless the solute transport phenomena from an immiscible phase into the saturated zone is understood and physically defined at a site, then neither optimal remediation systems can be designed nor saturated zone solute transport models can be effectively utilized to predict the impact of immiscible liquid remediation on saturated zone solute transport.

Immiscible phase boundary conditions presented in Figures 13.5 and 13.6 also greatly affect the vapor diffusive flux rates from the capillary zone into the unsaturated zone. Bruell[12] and Bruell and Hoag[13] rigorously investigated the effect of immiscible liquid phase boundary conditions on subsequent hydrocarbon diffusive flux rates of benzene. For a given column geometry (diffusive path length of 47.6 cm), diffusive flux rates for benzene at 20°C for an immiscible phase boundary condition similar to that shown in Figure 13.5 resulted in benzene diffusive flux rates of 24.9 mg/(cm^2·min) and 6.1 mg/(cm^2·min), for dry and wet (i.e., field capacity moisture content) concrete sand, respectively. Thus, moisture

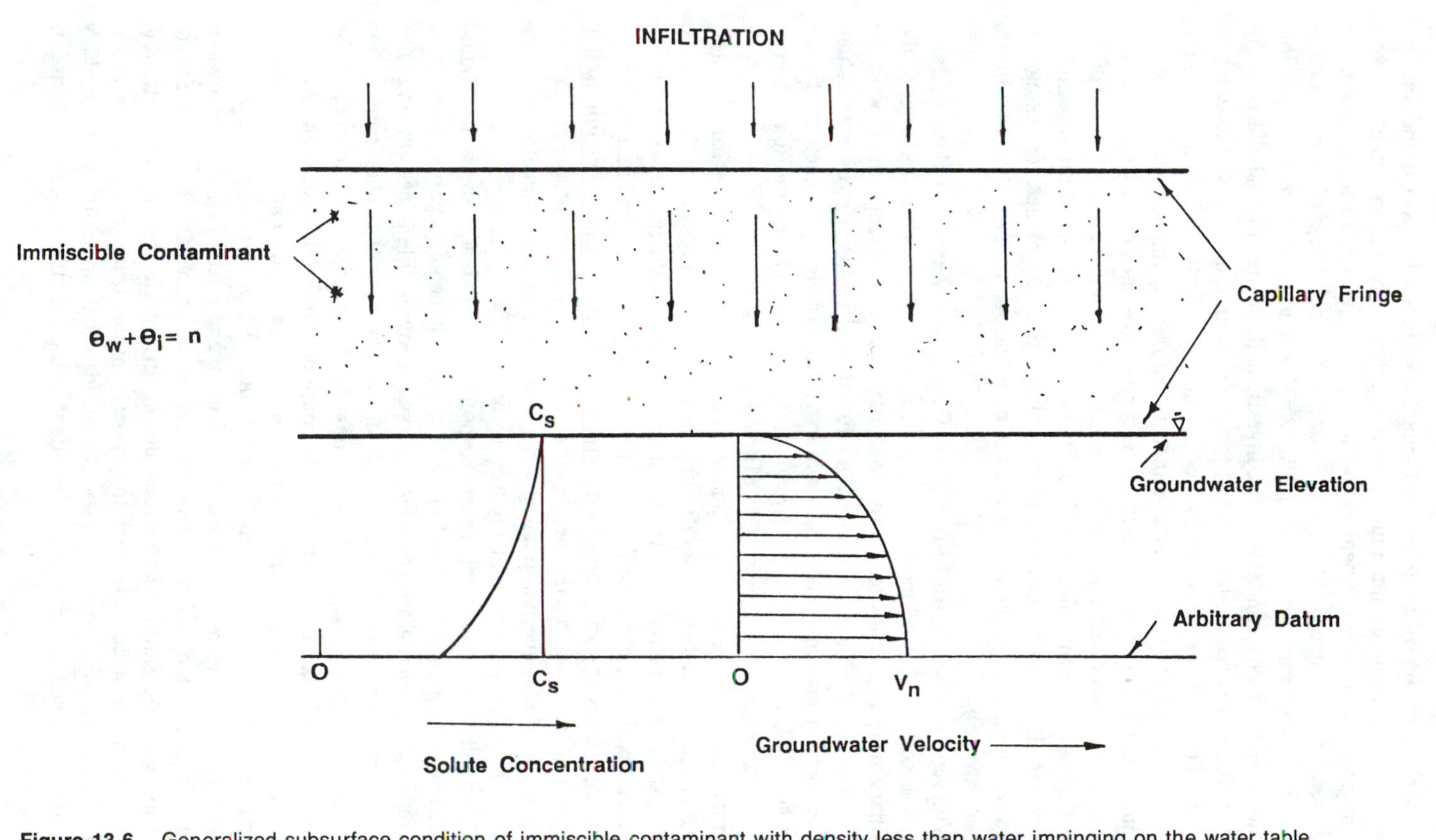

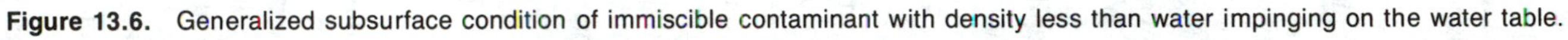
Figure 13.6. Generalized subsurface condition of immiscible contaminant with density less than water impinging on the water table.

content played a significant role in reducing the effective porosity of the concrete sand. When residual saturation immiscible liquid phase boundary conditions were investigated, the maximum benzene diffusive flux rate was 26.6 mg/(cm^2·min); however, the diffusive path length was only 22.4 cm. The moisture content in the residually saturated zone was 3.2% (v/v). When capillary zone immiscible liquid phase boundary conditions were investigated, the maximum benzene diffusive flux rate was reduced to 4.8 mg/(cm^2·min) with a diffusive path length of 22.4 cm. The moisture content in the capillary zone reflected saturated conditions (i.e., $\Theta_w = n$.) This research demonstrated that the immiscible liquid phase boundary condition greatly affects the diffusive flux rates of hydrocarbons that occur in the unsaturated zone. As the moisture content increases, then the diffusive flux rates of contaminants will decrease. The net result of these boundary conditions affects the concentrations of hydrocarbon vapors detected using soil gas assessment techniques and the rates of hydrocarbon recovery utilizing in situ vapor extraction.

With reference to Figure 13.6, and knowing that advective air flowrates also decrease with increasing moisture content, a condition exists in the area of the capillary zone where advective air flow may not be in direct contact with the immiscible phase. Thus, in this case, diffusion will be the controlling mechanism of contaminant removal during vapor extraction. In the case depicted in Figure 13.5, it is likely that some advective air flow will contact the immiscible phase, greatly increasing vapor extraction efficiency.

In the case of an immiscible liquid with a density greater than that of water, contaminant distribution is significantly different, given a hypothetical spill to the subsurface. Penetration of the capillary and saturated zones by the immiscible liquid is likely, given sufficient spill volumes as shown in Figure 13.7. Of great importance is the occurrence of groundwater flow through the immiscible liquid phase in the saturated zone, resulting in substantially greater solubilization rates of the immiscible phase and greater groundwater contamination potential than in the cases presented in Figures 13.5 and 13.6.

A typical in situ remedial action response to the dense immiscible liquid phase contamination is given in Figure 13.8. Simultaneous vapor extraction and ground-water pumping are necessary to expose immiscible phase contaminants to advective air flow and to increase diffusive flux rates of contaminants in the vicinity of the groundwater table at time=T_2. In this case, dewatering of the saturated zone in the area of immiscible phase contaminants is desirable. Long-term plume management interceptor pumping strategies, such as those developed by Ahlfeld, Mulvey, Pinder, and Wood[14] and Ahlfeld, Mulvey, and Pinder[15] should be implemented to optimally circumvent uncontrolled groundwater contamination and to maximize groundwater contaminant recovery rates. Strategies to maximize saturated zone dewatering in the vicinity of the immiscible phase liquids must be developed to properly implement this approach. Additionally, in situ bioremediation may be considered as an additional technology to further degrade the immiscible liquid, if complete subsurface dewatering is not possible.

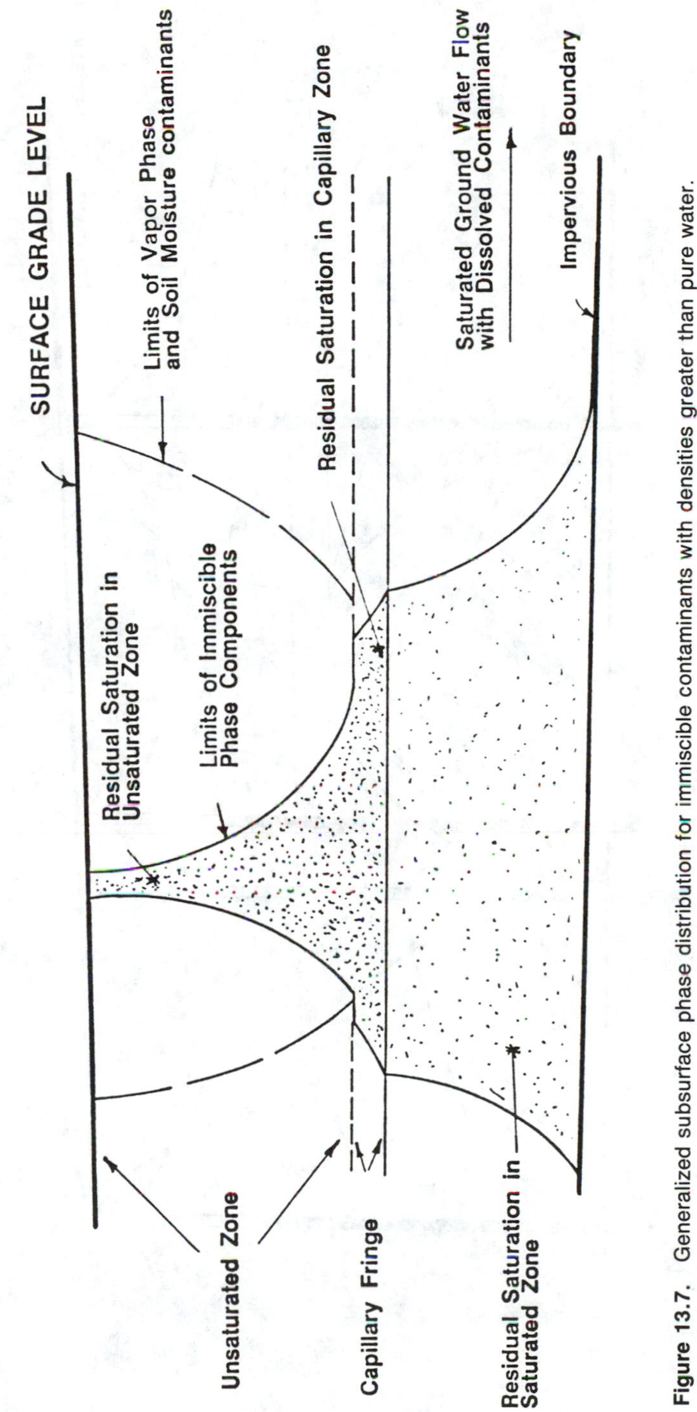

Figure 13.7. Generalized subsurface phase distribution for immiscible contaminants with densities greater than pure water.

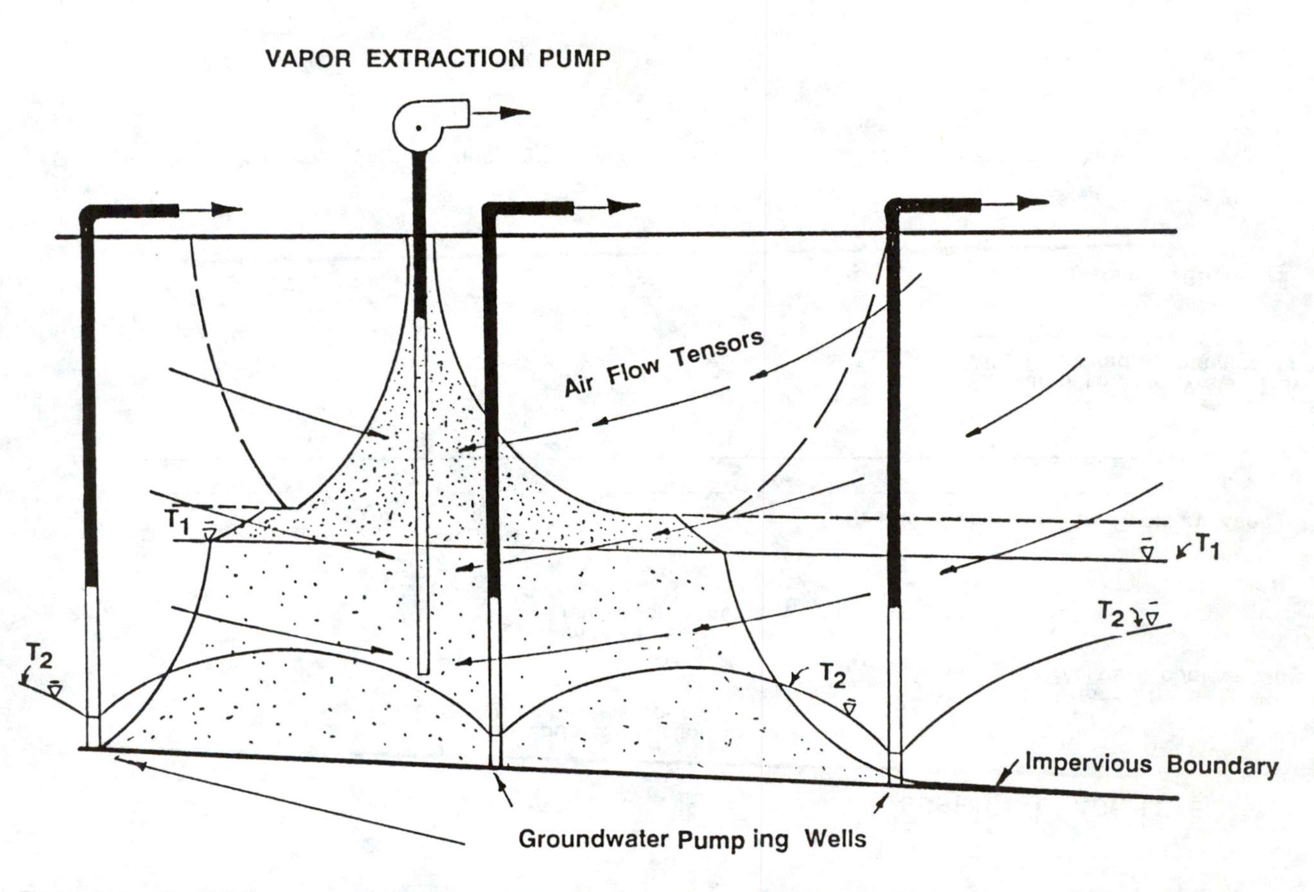

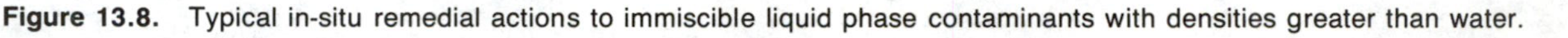
Figure 13.8. Typical in-situ remedial actions to immiscible liquid phase contaminants with densities greater than water.

SUMMARY

Significant advances have been made in the past five years in the understanding of volatile and semivolatile contaminant behavior as related to vapor extraction technologies. Coupled modeling of both contaminant behavior and advective air flow, however, remains limited to one-dimensional systems. Given the significant hydrogeological complexity of porous media and subsequent heterogeneous distributions of immiscible phase contaminants, the design utility of higher-dimensional coupled models is questionable. Higher-dimensional advective air flow models are being used to design vapor extraction systems. These models are generally dependent on site-specific parameters best determined in field air pumping tests, unless uniform hydrogeologic conditions prevail with quantifiable boundary conditions necessary for model design predictions. Three-dimensional models are being adapted to deal with nonradial symmetry and will be necessary to rigorously model multiple extraction well and extraction well/injection well applications.

Significant modeling and experimental research is needed to further understand immiscible contaminant behavior in the capillary zone and adjacent boundary conditions. The interaction of immiscible phase liquids in the capillary zone with unsaturated zone infiltration and saturated zone transport must be the focus of this research. The approach should include both hydrogeologic characteristics and testing procedures necessary to determine the influencing factors. Chemical fate and transport in the unsaturated zone under natural and advective air flow conditions must also be better understood to more effectively apply optimal in situ remediation processes.

Emphasis should be placed on basic research in the above areas, to be followed at the appropriate time by demonstration level projects. When demonstration level projects precede basic research needs, as has frequently been the case in the past five years, the results generally do not properly reflect necessary parameter control or monitoring, and either inconclusive or misleading results may be generated.

REFERENCES

1. Thornton, J. S., and W. L. Wootan. ''Venting for the Removal of Hydrocarbon Vapors from Gasoline Contaminated Soil,'' *J. Environ. Sci. Health* A17(1):31–44 (1982).
2. Wootan, W. L., and T. Voynick. ''Forced Venting to Remove Gasoline Vapor From a Large-Scale Model Aquifer,'' American Petroleum Institute, 82101-F:TAV, 1984.
3. Marley, M. C., and G. W. Hoag. ''Induced Soil Venting for Recovery and Restoration of Gasoline Hydrocarbons in the Vadose Zone,'' *Proc. of Pet. Hydro. and Org. Chem. in Ground Water: Prevention, Detection and Restoration.* National Water Well Association and the American Petroleum Institute, Houston, 1984, pp. 473–503.

4. Marley, M. C. ''Quantitative and Qualitative Analysis of Gasoline Fractions Stripped by Air From the Unsaturated Zone,'' Master's thesis, Department of Civil Engineering, The University of Connecticut, 1985.
5. Baehr, A. L., and G. E. Hoag. ''A Modeling and Experimental Investigation of Venting Gasoline Contaminated Soils,'' in E. J. Calabrese and P. T. Kostecki, eds. *Soils Contaminated by Petroleum: Environmental and Public Health Effects* (New York, NY: John Wiley & Sons, 1985).
6. Baehr, A. L. ''Immiscible Contaminant Transport in Soils with an Emphasis on Gasoline Hydrocarbons,'' PhD dissertation, Department of Civil Engineering, University of Delaware, 1984.
7. Baehr, A. L., G. E. Hoag, and M. C. Marley. ''Removing Volatile Contaminants From the Unsaturated Zone by Inducing Advective Air-Phase Transport,'' *Journal of Contaminant Hydrology,* 4:1–26 (1988).
8. Muskat, M., and H. G. Botset. ''Flow of Gas Through Porous Materials,'' *Physics* 1:27–47 (1931).
9. Hoag, G. E., and G. Cliff. ''The Use of the Soil Venting Technique for the Remediation of Petroleum Contaminated Soils,'' in E. J. Calabrese and P. T. Kostecki, eds. *Soils Contaminated by Petroleum: Environmental and Public Health Effects* (New York, NY: John Wiley & Sons, 1985).
10. Parker, J. C., R. J. Lenhard, and J. Kuppusamy. ''A Parametric Model for Constitutive Properties Governing Multiphase Fluid Conduction in Porous Media,'' Virginia Poly. Tech. and State University, 1986.
11. Lenhard, R. J., and J. C. Parker. ''Measurement and Prediction of Saturation-Pressure Relationships in Air-Organic Liquid-Water-Porous Media Systems,'' Virginia Poly. Tech. and State University, 1986.
12. Bruell, C. J. ''The Diffusion of Gasoline-Range Hydrocarbons in Porous Media,'' PhD dissertation, Environmental Engineering, University of Connecticut, Storrs, 1987.
13. Bruell, C. J., and G. E. Hoag. ''The Diffusion of Gasoline-Range Hydrocarbon Vapors in Porous Media, Experimental Methodologies,'' *Proc. of Pet. Hydro. and Org. Chem. in Ground Water: Prevention, Detection and Restoration.* National Water Well Association and the American Petroleum Institute, Houston, 1987, pp. 420–443.
14. Ahlfeld, D. P., J. M. Mulvey, G. F. Pinder, and E. F. Wood. ''Contaminated Groundwater Remediation Design Using, Simulation, Optimization and Sensitivity Theory: 1, Model Development,'' *Water Resources Research* 24(3):431–442 (1988).
15. Ahlfeld, D. P., J. M. Mulvey, and G. F. Pinder. ''Contaminated Groundwater Remediation Design Using, Simulation, Optimization and Sensitivity Theory: 2, Analysis of Field Site,'' *Water Resources Research* 24(3):443–452 (1988).

CHAPTER 14

Bioremediation of Hydrocarbon Contaminated Surface Water, Groundwater, and Soils: The Microbial Ecology Approach

Michael R. Piotrowski, Woodward-Clyde Consultants, San Diego, California

INTRODUCTION

Bioremediation is currently receiving considerable attention as a remediation option for sites contaminated with hazardous organic compounds. There is an enormous amount of interest in bioremediation, and numerous journals now publish research articles concerning some aspect of the remediation approach (Table 14.1). An example of the growth of knowledge in this field is provided by the National Technical Information Service (NTIS), which recently made available a database containing 259 citations concerning biodegradation of toxic wastes. Of those citations, 157 citations were added between the 1989 and 1990 database editions. The most recent high-profile attempt at using the technology is the U.S. Environmental Protection Agency's (EPA's) bioremediation efforts for alleviating the environmental impact of the Valdez oil spill, and initial indications are that it has been useful.

A review of the literature indicates that two basic forms of bioremediation are currently being practiced: the microbiological approach and the microbial ecology approach. Each form has its advocates and detractors, and the microbiological approach is generally advocated by most of the firms that practice bioremediation. In this chapter, the merits and disadvantages of these forms are reviewed and a conceptual approach is presented for assessing which form may be most useful for a particular contaminant situation. I conclude that the microbial ecology form of bioremediation may be the most useful for the majority of contaminant situations, and I will present two case histories in support of this hypothesis.

Table 14.1. Partial List of Journals Containing Articles Pertaining to Biodegradation

Annual Review of Microbiology	Ground Water
Applied and Environmental Microbiology	Ground Water Monitoring Review
Applied Microbiology and Biotechnology	Hazardous Waste and Hazardous Materials
Archives of Environmental Contamination and Toxicology	International Biodeterioration
Archives of Microbiology	Journal of Biotechnology
Bio-Cycle	Journal of Environmental Quality
Biogeochemistry	Microbiology
Biological Wastes	Journal of Industrial Microbiology
Biology and Fertility of Soils	Journal of Microbial Biotechnology
Bioprocess Engineering	Journal of Soil Science
Biotechnology Advances	Marine Environmental Research
Biotechnology and Applied Biochemistry	Marine Pollution Bulletin
Biotechnology and Bioengineering	Microbial Ecology
Biotechnology Letters	New Biotech
Biotechnology Progress	Research in Microbiology
Canadian Journal of Microbiology	Science
Canadian Journal of Soil Science	Soil Biology & Biochemistry
Chemosphere	Soil Science
Critical Reviews in Biotechnology	Soil Science Society of America Journal
Critical Reviews in Microbiology	Trends in Biotechnology
Environmental Pollution	Waste Management
Environmental Progress	Waste Management and Research
Environmental Science and Technology	Water, Air and Soil Pollution
Environmental Toxicology and Chemistry	Water Research
FEMS Microbiology Ecology	Water Resources Bulletin
FEMS Microbiology Reviews	Water Resources Research
Geomicrobiology Journal	Water Science and Technology

CURRENT BIOREMEDIATION APPROACHES

The Microbiological Approach

This approach involves augmentation of a contaminated site with one or more species of contaminant-specific microorganisms (the so-called "superbugs"). In this way, it is believed that the rate of contaminant biodegradation in the affected soil or water will be appreciably enhanced because the density of contaminant-specific degraders will have been artificially increased.

There are currently two basic means to achieve this end (Figure 14.1). The first form involves the use of prepackaged, contaminant-specific degraders that have been specifically selected due to their inherent or induced capability to degrade the contaminants of concern (e.g., microbial products available from Sybron Corporation or General Environmental Sciences). These microorganisms were generally derived from a contaminated site, stressed to degrade a specific contaminant at elevated concentrations, cultured in large numbers, and preserved in some manner to be stable and allow easy application to the contaminated soil or water. There are currently a number of firms that market a variety of prepackaged microorganisms possessing the capabilities to degrade a variety of organic contaminants.

Pre-Packaged "Superbugs"

1) Identify Contaminant(s) at Site
2) Select Pre-Packaged Microbes with Capability(ies) to Degrade Contaminant(s)
3) Apply Selected Microbes to Site at High Densities with Nutrients
4) Monitor Contaminant Reductions
5) Reapply as Needed

Site-Specific "Superbugs"

1) Identify Contaminant(s) at Site
2) Collect Site Samples and Return them to Laboratory
3) Isolate Series of Microbial Strains from Samples (Plating Techniques)
4) Expose Isolates to Elevated Concentrations of Contaminants
5) Isolate Contaminant-Specific Microbial Strains
6) Select Strains with Enhanced Degradative Capabilities or Growth Potential
7) Assess Nutrient Requirements for Selected Strain(s)
8) Mass-Culture High Biomass of Selected Strain(s)
9) Apply Microbial Culture(s) to Site with Nutrients
10) Monitor Contaminant Concentrations
11) Reapply Microbial Culture(s) and Nutrients as Needed

Figure 14.1. The microbiological approach to bioremediation.

Many of these strains of microorganisms were created by traditional selection techniques. Others have been created by forcing elevated rates of mutation using mutagenic chemicals or ionizing radiation and selecting for mutant strains with enhanced capabilities to degrade the organic contaminants.[1] Finally, attention has also been placed on directly creating desirable traits in microbial strains by using methods of genetic engineering to create genetically engineered microorganisms (GEMs).[2] In each case, microbial strains possessing potentially valuable biodegradative traits have been produced.

The second means currently practiced involves selection, culture, and application of site-specific strains that exhibit desirable biodegradative qualities.[3] In this form, the site is visited, the contaminated soil or water is sampled, and the samples are returned to the laboratory for microbiological analysis. The analysis generally includes isolation of microbial strains with desirable traits, selection of those strains exhibiting enhanced capabilities for the biodegradation of the contaminant(s) of concern, studies of nutritional requirements of the selected strains, mass-culture of the strain possessing the optimum qualities, and dispersion of the cultures at high densities in the affected soil or water along with the nutrients identified as being important for high activity rates of the microbial strains. This form has the advantage (relative to the previous form) of using site-specific strains which are presumed to be more capable of persisting (and biodegrading) at the site than prepackaged microbial strains that were originally isolated from other locations.

The Microbial Ecology Approach

In this approach, emphasis is placed on identifying and adjusting certain physical and chemical factors that may be impeding the rate of biodegradation of the contaminant(s) by the naturally occurring (indigenous) microbial community in the affected soil or water (Figure 14.2). Once the factors have been adjusted appropriately, contaminant biodegradation may proceed at appreciable and satisfactory rates.[4]

In deciding the feasibility of this approach, little attention is given to identification of the microbiological content of the affected soil or water; rather, the focus of the biological feasibility study is to identify the factor(s) anticipated to be limiting appreciable microbial metabolism of the contaminants of concern.[5] Once the factors have been identified, actions are taken to ameliorate the conditions suspected of impeding the indigenous microorganisms. The effect of changing the factors is then evaluated by monitoring contaminant concentrations over time. The implicit assumption of this approach is that the indigenous microbial community is capable of degrading the contamination once the proper site conditions have been achieved.

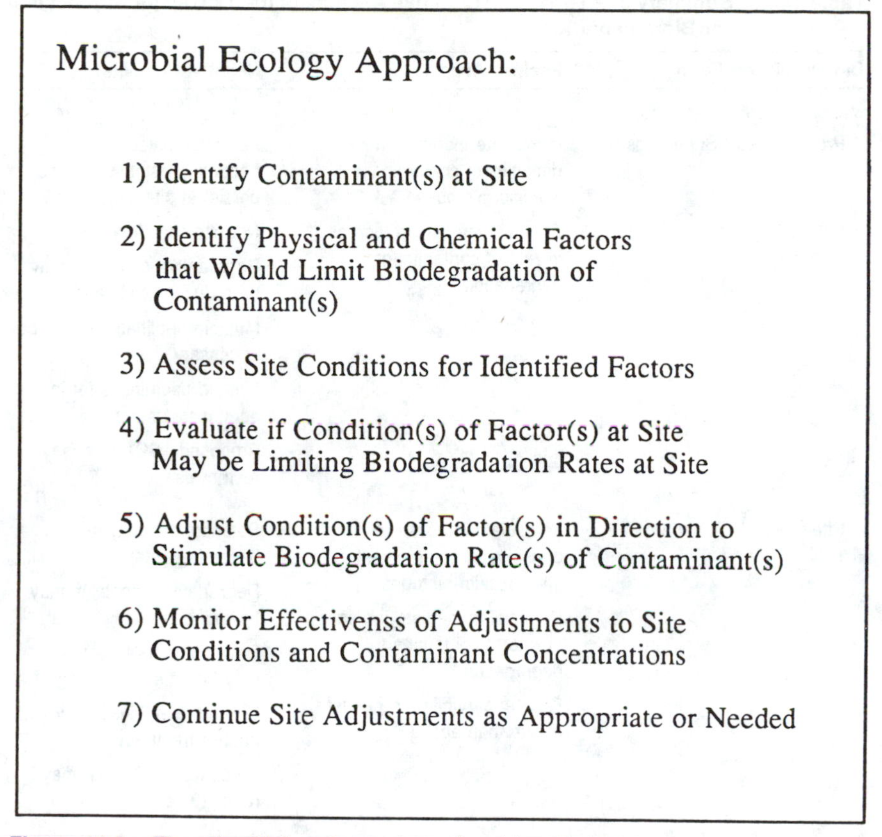

Figure 14.2. The microbial ecology approach to bioremediation.

ASSESSMENT OF THE TWO BIOREMEDIATION APPROACHES

Many studies have been conducted that have produced results in support of both bioremediation approaches. Studies have also been reported indicating that each approach has limitations. A summary of the positive and negative aspects of each approach is presented in Table 14.2 and discussed below.

The Microbiological Approach

This basic approach has the chief advantage of rapidly increasing the densities of contaminant-specific microorganisms in the affected medium. In this way, the

Table 14.2. Summary of Positive and Negative Aspects of the Two Major Approaches for Bioremediation

Bioremediation Form	Positive Aspects	Negative Aspects
Microbiological Approach		
Prepackaged "Superbugs"	Immediate increase in densities of contaminant-specific microbes Potential immediate increase in rate of contaminant degradation	Potential difficulty in becoming established and persist at site Not site-specific Degradative capability may be reduced relative to lab Multiple applications may be necessary Limited usefulness for in situ aquifer treatment Proposed TSCA rule may restrict use
Site-specific "Superbugs"	Immediate increase in densities of contaminant- and site-specific microbes Potential immediate increase in rate of contaminant degradation Greater capability to persist at site than above	Potential difficulty in becoming established at site Degradative capability may be reduced relative to lab Development of microbes may be time-consuming Limited usefulness for in situ aquifer treatment Proposed TSCA rule may restrict use
Microbial Ecology Approach	Less complicated to assess and apply More applicable for in situ aquifer treatment than above Less costly than above Proposed TSCA rule not applicable	Uncertainty over cleanup time Uncertainty over final contaminant concentrations achievable Not probably useful for fresh spills May not work with high concentrations of contaminants May not work with low concentrations of contaminants

rate of contaminant biodegradation is anticipated to be elevated in short order, and this has been reported in some cases.[1,6,7] However, each form of the microbiological approach (i.e., use of prepackaged or site-specific "superbugs") has drawbacks.

Prepackaged "Superbugs"

Prepackaged microbial strains have been observed in some cases to have difficulty becoming established in a contaminated medium that is already harboring indigenous microorganisms. A number of factors may be involved in this phenomenon, including low contaminant concentrations in nature relative to the concentrations that the added species are adapted to in the laboratory, the added microorganisms are removed by predation, and the presence of naturally occurring compounds that inhibit the growth of the added microorganisms.[8] It may also be that the indigenous microorganisms out-compete the added microorganisms for available organic substrates. As a result, it may be necessary to repeat the application of the packaged microbial strain several times during the course of treatment or increase the quantity released each time. Repetitive applications or increased quantities per application increase the cost associated with the bioremediation approach, in some cases, to unacceptable levels.[9]

In addition, it has been observed that biodegradation rates of the contaminants of concern in samples of soil or water that have been augmented with prepackaged microbial strains may not be significantly elevated over the rates in samples containing natural or nonselected microorganisms.[9-12] In other words, the indigenous or nonselected microbes produced essentially equivalent rates of contaminant biodegradation as the "superbugs," indicating that their application was not necessary.

Finally, it has been observed that the longer a microbial strain has been isolated from the natural environment, the less fit it becomes, and the less likely it will be to survive for any length of time once it is returned to the natural environment.[13] The analogy of the wild animal that is raised domestically and then released to the wild is appropriate. The released animal typically perishes in a matter of days. Therefore, it is important, when selecting a "superbug" for site application, to investigate how long the strain has been growing in a test tube and away from the natural environment.

Site-Specific "Superbugs"

The second form of the microbiological approach, development of a large biomass of one or more site-specific strains with degradative capabilities for reapplication to the contaminated site, is intended to overcome some of the limitations involved in using "superbugs" derived from a foreign site. However, it also has attendant problems, many of which are shared with the prepackaged strain approach.

First of all, the means of microbial isolation can result in undesirable artifacts. There is a fundamental bias in the way specific strains are isolated in the laboratory for purposes of bioremediation. The only available method for isolating these strains is to apply a dilute solution of the contaminated soil or water to an agar plate that contains macro- and micronutrients presumed to be required by the microorganisms. After sufficient incubation time, one or more colonies will grow on the plate. Individual colonies are then typically transferred to other plates that contain nutrient-amended agar solutions that have also been amended with various concentrations of one or more of the contaminants of concern. Specific colonies that grow on these plates generally possess the metabolic capability to degrade the contaminants at the concentrations tested. These colonies are then studied for contaminant degradation capabilities and strain-specific nutrient requirements, and strains possessing enhanced degradation capabilities are selected and developed.

This approach assumes that the microorganisms growing on the plates are the ones that are active and functionally important at the site, and this may not be the case. The agar plate and the incubation conditions impart a selective bias on which microbial strains will grow on the plates. Alexander[14] has pointed out that plating techniques typically recover less than 10% of the total microbial community that may be present in a soil. Therefore, the plating approach will not yield an accurate assessment of the composition of the entire microbial community and yields little information about the function of the natural community. Furthermore, the most opportunistic microbial strains will usually be the strains to grow most quickly on the plates, and there is a problem with using opportunistic species.

From an ecological point of view, opportunistic species rarely remain the major biotic components of an ecosystem as it matures. The nature of the opportunistic mode of life is to take advantage of the fact that a newly formed or denuded habitat does not contain an established community of organisms. Once the habitat becomes colonized by a variety of species, the opportunists typically decline in density and importance to the function of the ecosystem. Therefore, the presumption that the microbial species isolated on the plates are important and persistent components of the microbial community may not be valid.

The third limitation of the microbiological approach is that in the general practice of the approach, isolated strains are studied in pure culture. While the extensive degradative capabilities of numerous microbial species have been demonstrated in this manner, these capabilities may not occur to the same extent or in the same fashion when the species are active in a mixed microbial community such as occurs in nature.[14] Thus, it is commonly observed that many species of "superbugs" do not perform in the field as well as was predicted by the laboratory studies. This latter aspect is the major reason why the microbial ecology approach is an advance on the microbiological approach.

Microorganisms in nature coexist with numerous species as a community or consortium. Some species are competitors, some are predators, and in some cases the metabolic activities of a number of species are necessary for the complete degradation of an organic compound. The net interaction of these microbial components is the fundamental determinant for the overall biodegradative activity of the microbial community. Therefore, although studies of pure strains of microorganisms yield insight to microbial potential, they fail to address how the capabilities will be exhibited in a mixed culture, natural situation.

Another aspect of the microbiological approach which is a limitation is that it does not appear to be especially useful for in situ (in the ground) bioremediation of contaminated aquifers. Since pump-and-treat approaches have been recently declared by the EPA to have been largely ineffective for complete aquifer treatment,[15] in situ bioremediation remains essentially the only potentially feasible option available for aquifer treatment. However, augmentation of "superbugs" directly into a contaminated aquifer may not be effective because microorganisms tend to occupy particle surfaces in aquifers rather than exist as free-living or planktonic organisms.[16] Therefore, it can be anticipated that microorganisms injected into an aquifer would tend to colonize the surfaces of aquifer sediments in the immediate vicinity of the injection wells. It is also anticipated that microbial transport of the added microorganisms through an aquifer would be confined to the more porous regions of the aquifer. This reasoning suggests that the major beneficial effect of augmentation would be larely restricted to the vicinity of the well.

The fifth limitation to the microbiological approach is more of a political limitation than a scientific one. The EPA, as part of implementation of the Toxic Substances Control Act (TSCA), is in the process of developing a rule designed to regulate the practice of the microbiological approach for bioremediation. The initial form of the proposed rule was released in 1988 and included a stipulation that application of a naturally occurring microbial species to a contaminated site would be considered a microbial release (akin to the release of a genetically engineered organism) and therefore would require a permit. The original form of the proposed rule provoked extensive protest from those who advocate the microbiological approach, and led to the establishment of the Applied BioTreatment Association (ABTA) in 1989.

ABTA was formed by a coalition of firms that specialize in bioremediation and the microbiological approach, to lobby the EPA, Congress, and the public for acceptance and support of bioremediation for hazardous site treatment in general, but also to lobby specifically for the microbiological approach. Development of the rule was based on congressional and public concerns over the potential environmental impacts and hazards associated with environmental releases of GEMs. Because these impacts and hazards are difficult to predict, the EPA wanted to extend regulatory control to releases of naturally occurring microorganisms (i.e., the microbiological practice of bioremediation). The ABTA disagrees with the inclusion of naturally occurring microorganisms and wants the EPA to "leave natural microorganisms alone."[17] The proposed TSCA rule was to be released in the spring of 1990. However, it was not. The current status of the proposed rule is unknown.

Genetically Engineered "Superbugs"

Due to the unique nature of GEMs, they deserve a separate discussion with respect to improving biodegradation rates of hazardous organic compounds. Although the potential benefits of GEMs in bioremediation are anticipated to be numerous, public resistance to the environmental release of GEMs will not subside quickly, if at all. Therefore, it appears that those who advocate the development of GEMs for bioremediation should be prepared to wage extensive, time-consuming legal battles with organizations such as the Foundation of

Economic Trends in order to gain permission to release their specially designed GEMs to the environment. Although President Bush was recently introduced to GEM technology for bioremediation and was ostensibly impressed by its potential, it is difficult to imagine that regulatory permission to release bioremediation GEMs to the environment will be forthcoming in the near future. Therefore, although GEM research for bioremediation should continue to produce advances in microbial capabilities, it appears that we will not be able to directly receive the benefits of the research for some time. Consequently, it will be more fruitful in the near-term to devote a larger portion of our limited research funds to studies that will increase our knowledge of microbial ecology and its influence on bioremediation.

The Microbial Ecology Approach

As discussed previously, this approach for bioremediation endeavors to stimulate indigenous microorganisms to degrade the organic contaminants at elevated rates by making adjustments to certain physical and chemical conditions of the contaminated site. Because it does not generally include the application of microorganisms, it appears that the approach will not be subject to the proposed TSCA rule. In addition, since it involves stimulation of the existing microbial community in the contaminated soil or water, it avoids many of the drawbacks discussed for the microbiological approach (e.g., competition, microbial fitness, and persistence). Furthermore, it appears to be the most useful way to initiate in situ aquifer treatment using bioremediation because it generally only involves the addition of chemicals to stimulate the indigenous organisms.[4,18] These chemicals, such as oxygen, hydrogen peroxide, or inorganic nutrients, more easily disperse throughout the contaminated zone of the aquifer compared to injected microorganisms.

However, there are limitations to the approach. First, in cases where a spill has just occurred, the indigenous microbial community may have been destroyed or inactivated. Consequently, it would take some period of time for recolonization or reactivation to occur. Also, it may take a period of time for the microbial community to develop a metabolic capability to degrade the spilled contamination. Finally, the initial concentration of the contamination may be at such an elevated level that effective degradation by naturally occurring microorganisms is precluded because the contaminant concentration is toxic for natural microorganisms.

At the other end of the spectrum of contaminant concentrations, contaminant levels may be too low for effective microbial reduction and yet still exceed regulatory action levels. For example, it has been observed that natural microbial communities in laboratory microcosms can reduce the concentrations of a number of contaminants in water to the low parts-per-billion range.[14] This range may be the lower practicable limit for biological reduction using natural microorganisms. Yet for some contaminants, current or proposed regulatory standards are much lower than this range. For example, the regulatory standard recommended by the EPA for dioxin in water is 0.013 parts-per-quadrillion. Furthermore, there is also concern that as advances in analytical methods produce lower and lower chemical detection limits, regulatory actions levels will follow the limits down; in effect, creating a situation where the remedial goal is an ever-lower, moving target. If this moving target extends to the sub-parts-per-trillion range, the microbial ecology approach may lose its usefulness.

Another limitation of the microbial ecology approach is that the time period to achieve satisfactory contaminant reduction is uncertain. Although this is also true for the microbiological approach, presumably the time period will be longer for the microbial ecology approach because it does not involve artificially increasing the densities of contaminant-specific degraders. This uncertainty can be important in a situation where rapid site cleanup is desired. Therefore, if the contaminated site poses immediate and unacceptable hazards to human health or the environment, the uncertainty involved in the time requirement for the microbial ecology approach would tend to preclude its use.

CASE HISTORIES IN SUPPORT OF THE MICROBIAL ECOLOGY APPROACH

Despite the above-described limitations, the microbial ecology approach has been successfully applied in a number of contaminant situations. Two such situations are presented below and provide support for expanded use of the microbial ecology approach.

Case History 1: Biological Upgrade of the Alyeska Ballast Water Treatment Facility

The Alyeska Ballast Water Treatment (BWT) Facility, Valdez, Alaska is operated by the Alyeska Pipeline Service Company (Alyeska) and treats the ballast water off-loaded from crude oil tanker ships before the water is discharged to Port Valdez. The ballast water is derived from surface waters and collected in areas adjacent to the numerous ports-of-call where the crude oil shipments had been unloaded. During transit to Valdez the ballast water becomes contaminated with oil, grease, and dissolved aromatic hydrocarbon compounds. Ballast water batches collected from both freshwater and saltwater areas are often off-loaded to the BWT facility at the same time. Consequently, the salinity of the ballast water passing through the BWT facility can vary from brackish to seawater conditions.

The BWT facility processes an average flow rate of 10,000 gallons-per-minute (14.4 million gallons-per-day). Flow rates may double during the winter due to elevated demands for heating oil in the "lower 48" and the need to maintain ship stability with large amounts of ballast water during the voyages through the storm-stricken Gulf of Alaska at those times. Ballast water temperatures will be lowest during these times as well, typically less than 40°F. Alternatively, flow rates decline below the average during the summer months and ballast water temperatures may exceed 50°F.

Up until 1986, the facility treatment process for the water phase consisted of an oil/water separation stage, a dissolved air floatation stage, a pH adjustment stage, and passage through two large impound basins before discharge to the Port

Valdez ecosystem. While this system was effective in reducing the oil and grease content of the ballast water from perhaps 4000 to 5000 mg/L to ~2 mg/L, the concentrations of dissolved petroleum hydrocarbons (especially benzene, ethylbenzene, toluene, xylenes, or BETX) were not appreciably reduced (average effluent concentrations were roughly 7 mg/L). Due to concerns over the dissolved contaminants entering the Port Valdez ecosystem, the EPA mandated in the early 1980s that Alyeska perform an evaluation of various means to upgrade treatment performance of the BWT facility. Once the most appropriate treatment upgrade option was identified, it had to be installed.

Six upgrade options including resin adsorption, granular and powdered activated carbon, and three forms of biological treatment were evaluated both for feasibility and economic impact. Among the biologically based options considered was simple aeration of the impound basins, and this option was projected to incur an economic impact many millions of dollars lower than any of the other options. However, there was uncertainty over the level of effectiveness of biological treatment of the ballast water since the salinity and temperature of the water would vary considerably over time and because microbial activity was anticipated to be limited by the near-arctic climatic conditions characteristic of Valdez, Alaska.

Therefore, the first aspect of the existing treatment system that was examined in this regard was dissolved oxygen concentrations in the impound basins. The rationale for this action was that if there were contaminant-specific microbial degraders already present within the impound basins and the microbes were metabolizing a portion of the dissolved contaminants, dissolved oxygen concentrations might be reduced in portions of the basins. This assessment, essentially a feasibility study based on the microbial ecology approach, was carried out in early 1986.

A survey of dissolved oxygen concentration profiles in each basin found that, although the average residence time of ballast water in the basins was roughly 12 hours and the basins appeared to be well mixed, the lower two-thirds of the water column in each basin was anaerobic (devoid of oxygen). This indicated that microbial consumption of oxygen in the basins exceeded the rate of replenishment from the atmosphere in the lower portions of the water column in each basin. From a microbial ecology aspect, it was theorized that supplementation of oxygen throughout the water columns of the impound basins would produce significant reductions in the concentrations of the dissolved contaminants because microbial biodegradative activity would be enhanced.

The results of the dissolved oxygen profiling study encouraged Alyeska to conduct a bench-scale biological treatment study onsite that was designed and overseen by Woodward-Clyde Consultants in 1986. The results of the study indicated that oxygen addition to the ballast water in the impound basins may produce a 90% average reduction in BETX concentrations under normal flow conditions.[19] In response to the findings of the bench-scale study, Alyeska installed aeration equipment that was specifically designed to maximize oxygen injection without producing an excessive amount of air stripping of the contaminants.

The reductions in the concentrations of dissolved petroleum hydrocarbon compounds in the effluent of the BWT facility after installation of the aeration units are summarized in Figures 14.3 and 14.4. As can be seen, impound basin aeration produced a significant decrease in dissolved concentrations of the petroleum hydrocarbon compounds. Additional studies were conducted and concluded that biodegradation was the major mechanism of contaminant reduction.[20,21] The average yearly decrease has been estimated to be ~85% compared to typical pre-aeration performance.[22] Seasonal fluctuations do occur, with increases during the winter when flowthrough rates are highest, water temperatures are lowest, and microbial metabolic activity is at a minimum. Nevertheless, effluent concentrations of the dissolved contaminants at those times of the year remain significantly lower than typical effluent concentrations measured before impound basin aeration was implemented. During the summer months, total dissolved concentrations of the contaminants typically decrease below 0.5 mg/L.

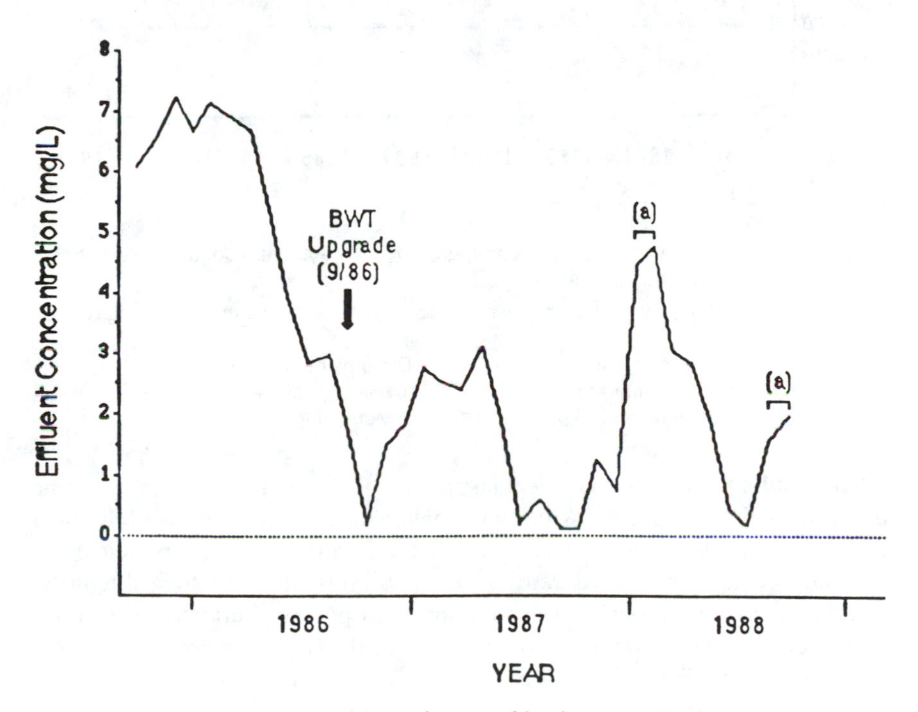

Figure 14.3. Total aromatic hydrocarbons* (mg/L) in BWT effluent. (Concentrations below detection limits set to zero.)

*Total aromatic hydrocarbons include:

Benzene	ortho-Xylene
Ethylbenzene	meta-Xylene
Trimethylbenzene	para-Xylene
Toluene	

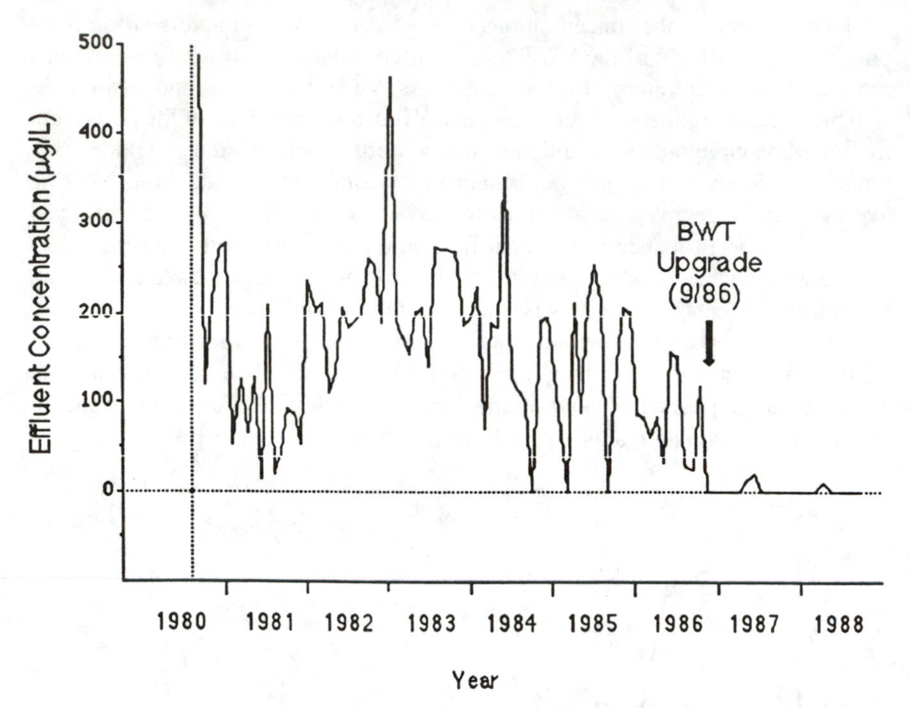

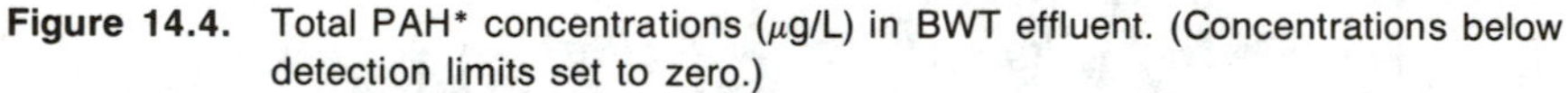

Figure 14.4. Total PAH* concentrations (μg/L) in BWT effluent. (Concentrations below detection limits set to zero.)

*PAH (polynuclear aromatic hydrocarbons) monitored under existing NPDES permit:

Naphthalene	Dimethylnaphthalene
Phenanthrene	1-Methylnaphthalene
Anthracene	2-Methylnaphthalene

The results of these studies demonstrated that the biological upgrade could produce significant reductions in toxic contaminant concentrations even under relatively adverse environmental conditions. Furthermore, the principles of microbial ecology were used to assess the feasibility of the biological upgrade and the microbial ecology remediation approach produced effective treatment. Implementation of this treatment option reduced the economic impact of the upgrade by many millions of dollars compared to the costs projected to be associated with the other upgrade options.

Case History 2: Biological Treatment of Groundwater and Soil Contaminated by Wood Preservatives

Activities at a former wood-treating facility in northwest Montana led to severe contamination of soil and groundwater by the wood preservative chemicals creosote and pentachlorophenol. The contaminant plume in the groundwater extended over three-quarters of a mile offsite and beneath a town located just downgradient

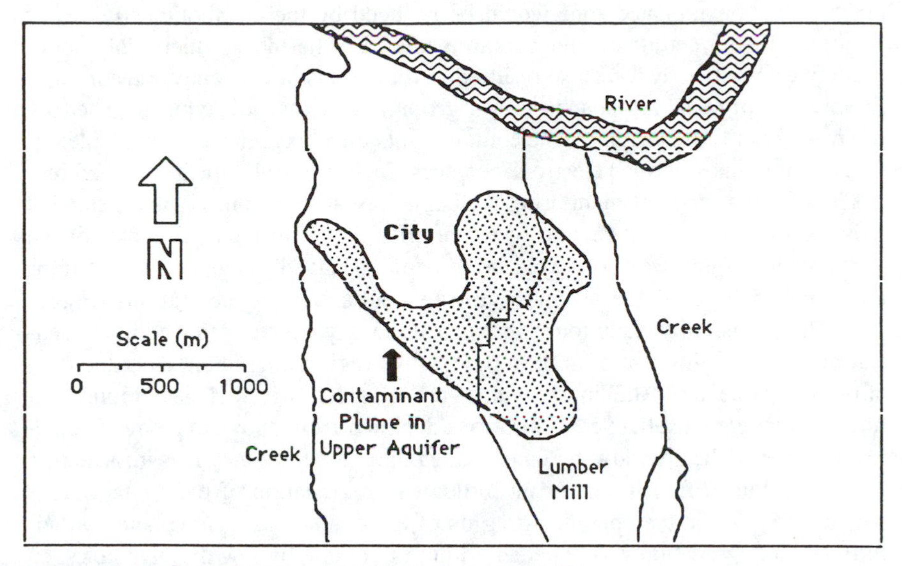

Figure 14.5. Groundwater contaminant plume in aquifer.
Source: Reference No. 28.

from the facility (Figure 14.5). The groundwater contamination was detected in 1979 and the site was placed on the EPA's National Priority List in 1983.

In Situ Bioremediation Pilot Study for Aquifer Treatment

The severity of contamination and the unconfined nature of the contaminated aquifer immediately below the site indicated that pump-and-treat approaches would have little effect on the contaminant plume and would probably be necessary for many years. As a result, the pump-and-treat approach was deemed unsatisfactory by the EPA for this site. Therefore, innovative approaches were considered.

One of the few innovative approaches potentially feasible for aquifer treatment involved in situ bioremediation of the contaminant plume. It was theorized that since this form of treatment could act on both the dissolved and sorbed contamination in the affected region of the aquifer, aquifer cleanup may be achieved in a more timely manner compared to pump-and-treat approaches. The fundamental question was whether or not the treatment approach could be applied effectively at this site.

Therefore, a preliminary feasibility study based on the microbial ecology approach was implemented in the fall of 1986 to evaluate the potential for in situ bioremediation.[5,23] It was theorized that since the contamination had been present in the aquifer for many years, there were already contaminant-specific microbial degraders present in the affected regions of the aquifer. It was also theorized that since the components of creosote and pentachlorophenol degrade most quickly under aerobic conditions,[24,26] dissolved oxygen concentrations in the groundwater as it entered the contaminated zone would be reduced by the metabolic activities of microorganisms acting on the contaminants. Furthermore, microbial biodegradative activity may have also resulted in reductions of concentrations of other dissolved inorganic compounds in the groundwater that are typically used by microorganisms to metabolize the contaminants once oxygen had been depleted. These compounds, termed electron acceptors, include nitrate, nitrite, and sulfate.

Therefore, a series of monitoring wells that lay along a transect that extended from pristine upgradient regions of the aquifer down through the center of the contaminant plume were surveyed in situ for dissolved oxygen concentrations and sampled for analyses of nitrate, nitrite, and sulfate concentrations (Figure 14.6). The feasibility study found that dissolved oxygen was reduced in concentration from roughly 4 to 6 mg/L in the aquifer region upgradient from the contaminated zone to <0.2 mg/L within the plume.[5,23] It was also found that nitrate, nitrite, and sulfate concentrations declined from above to below detectable levels once the groundwater entered the contaminated zone. This information convinced the EPA that significant biological degradation of the contaminants was taking place at the upgradient region of the contaminant plume and that additional biodegradation was impeded within the plume by low dissolved oxygen concentrations. Therefore, in the spring of 1987, the EPA instructed Woodward-Clyde Consultants to design and conduct a 12-month, in situ bioremediation pilot-scale study at the site. Details of the pilot study follow.

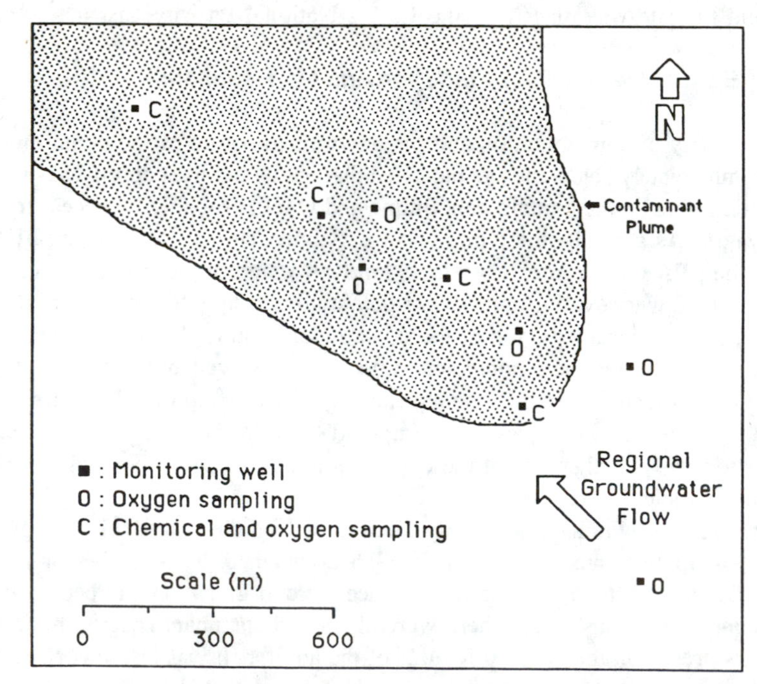

Figure 14.6. Locations of wells monitored for dissolved oxygen (O) and sampled for chemical analyses as part of bioremediation feasibility study.

Four existing monitoring wells located within the heart of the contaminant plume were converted to injection wells and connected to a hydrogen peroxide injection system (Figure 14.7). Two of the wells were located at each of two injection sites, and were screened at two depths (~15 and 30 feet below ground surface at each injection site) within the contaminated aquifer. Therefore, injection would take place at two depths in two locations within the plume.

The peroxide injection system supplied hydrogen peroxide at a concentration of 100 mg/L and at a combined flow rate of 40 gallons-per-minute. Hydrogen peroxide decomposes in water to dissolved oxygen at a ratio of 2 to 1.[27] Therefore, the initial oxygen concentration in the injected water was anticipated to be 50 mg/L.

Before peroxide injection was started, two 4-inch diameter monitoring wells with extended screened intervals (15- to 20-foot long screens) were installed ~200 feet downgradient from the pair of injection sites (Figure 14.7). Extended screen intervals were installed on these wells so that sampling from the wells would integrate groundwater conditions over extended vertical distances in the aquifer. These wells (Wells 25 and 26) along with a monitoring well (Well 12) located just upgradient from the injection sites were used as the primary monitoring wells for the pilot study.

A bromide tracer study was then performed and demonstrated that the injection wells were hydrologically connected to the primary monitoring wells and also to a number of monitoring wells located further downgradient. The tracer study also indicated that injected water was reaching the primary downgradient monitoring wells within two days. Therefore, groundwater flow rate in the treatment zone under the influence of the injection system was >100 ft/day.

Continuous peroxide injection began in July 1987. The primary monitoring wells were sampled generally biweekly for dissolved oxygen, creosote components (16 priority pollutant polynuclear aromatic hydrocarbons or PAHs), and pentachlorophenol (PCP). Since the study was conducted under EPA scrutiny, strict sampling, analytical, and quality control/quality assurance (QA/QC) procedures were followed. With respect to the latter, no exceedences of the QA/QC goals occurred during the study.

For the first 125 days, dissolved oxygen concentrations in the two downgradient monitoring wells remained below 0.5 mg/L (Figure 14.8). This occurred despite the fact that high oxygen concentrations were being injected just 200 feet upgradient from the wells. However, by Day 148, dissolved oxygen concentrations in Well 26 had increased above 12 mg/L and remained between 12 and 20 mg/L for the remainder of the 12-month study. Dissolved oxygen concentrations remained low in Wells 12 and 25.

Dissolved contaminant concentrations in Well 12 were quite variable throughout the study (Figure 14.9). Short-term increases in contaminant concentrations took place roughly 100 and 250 days after the pilot study began.

Dissolved contaminant concentrations were also variable in Well 25 (Figure 14.9). A short-term increase in concentrations in samples from Well 25 also took place approximately 100 days after the study began, and the increase was roughly the same order of magnitude as was observed in Well 12. These observations

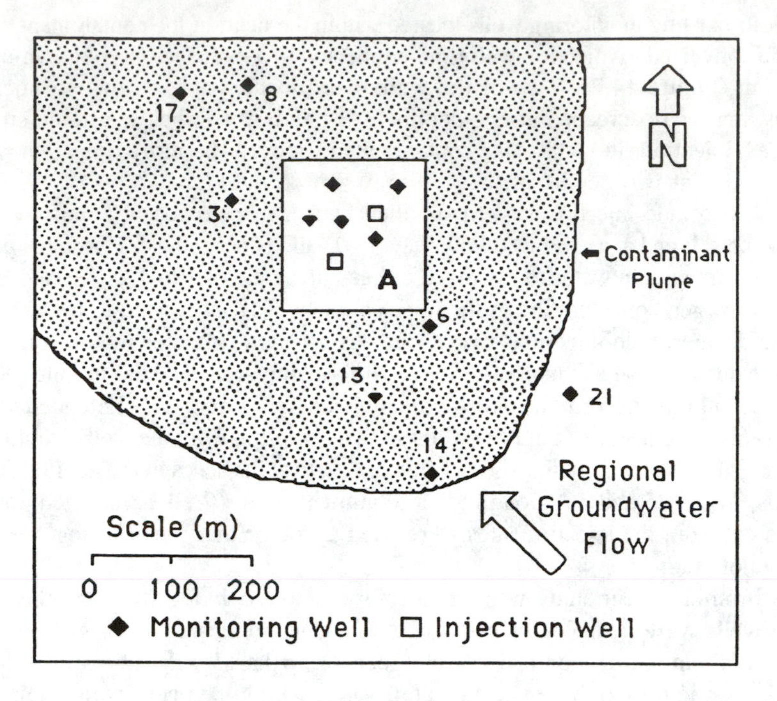

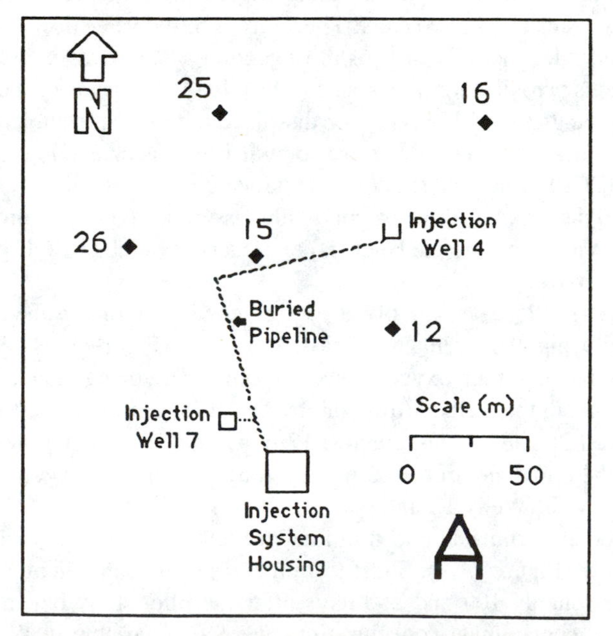

Figure 14.7. Location of injection sites in contaminant plume. Inset A in top figure is expanded in lower diagram to show positions of the injection wells, the near-field monitoring wells, and the injection system housing.

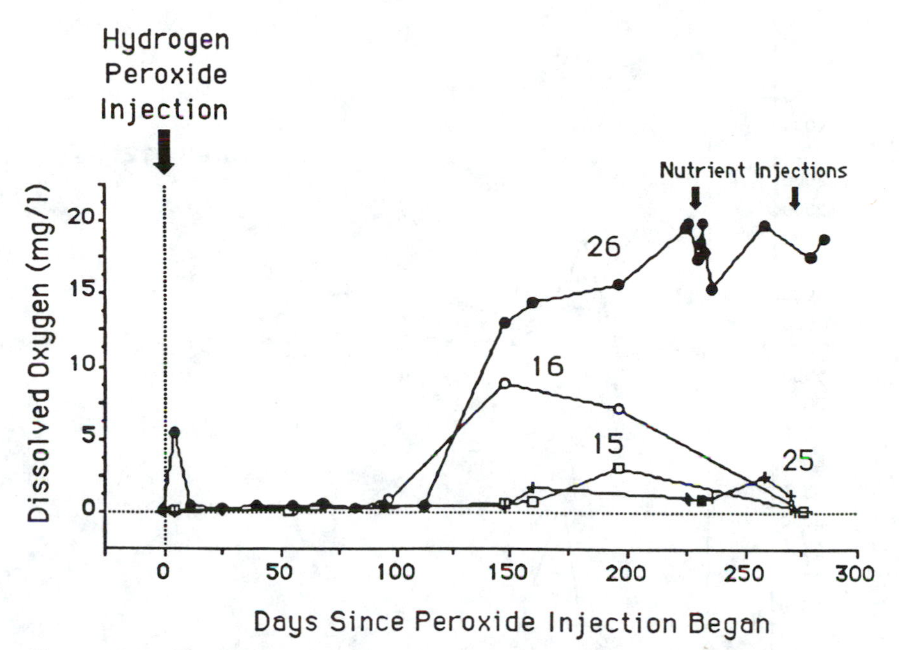

Figure 14.8. Dissolved oxygen concentrations in groundwater over time in wells located downgradient from the injection system.

suggested that a regional increase in concentrations of the dissolved contaminants had occurred at this time. Such fluctuations in dissolved contaminant concentrations are typically observed in aquifers contaminated with wood preservatives.[28] Interestingly enough, however, the second, short-term increase in dissolved contaminant concentrations observed in Well 12 approximately 250 days after the study began was not observed in Well 25.

Dissolved contaminant concentrations in Well 26 were extremely variable for the first 100 or so days (Figure 14.9). As was seen for Wells 12 and 25, dissolved concentrations in Well 26 increased for a short time approximately 100 days after the study began to roughly the same order of magnitude as was observed in the other two wells, providing additional evidence that the increases in dissolved contaminant concentrations were regional in nature. However, after Day 148, total dissolved contaminant concentrations in samples collected from the well were below 20 μg/L and continued to decline for the remainder of the study. By Day 454, all of the PAH concentrations were below detectable levels and the PCP concentration was 1 μg/L. The marked reduction in contaminant concentrations in the well coincided with the marked increase in dissolved oxygen concentrations (Figure 14.10). The results of the pilot study indicated that dissolved oxygen breakthrough had occurred at Well 26 and breakthrough coincided with dramatic reductions in the concentrations of dissolved contaminants. Interpretation of these results follows.

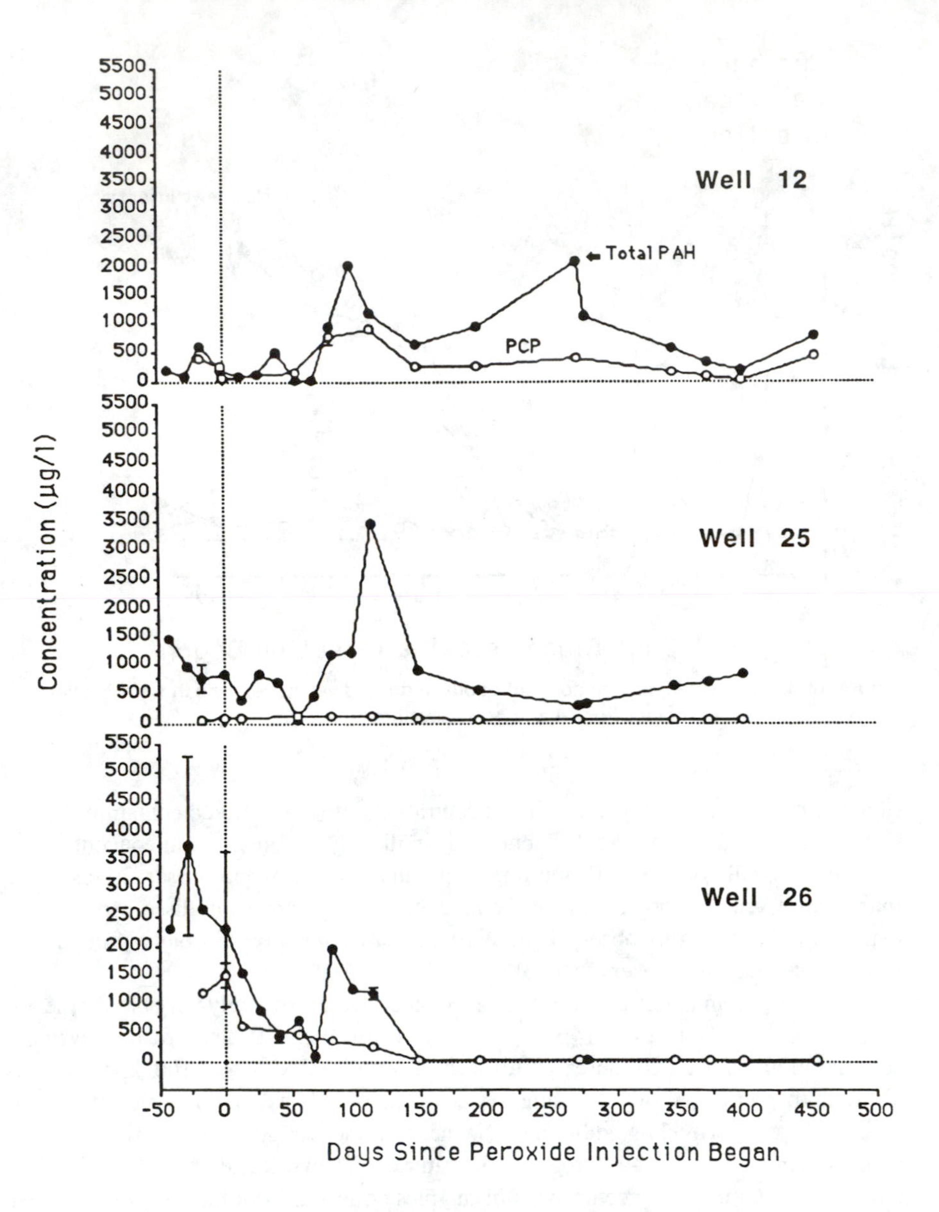

Figure 14.9. Organic contaminant concentrations in groundwater samples from monitoring Wells 12, 25, 26 over time. Only detected PAH compounds were summed for total PAH calculation; concentrations below detection limits were set to zero. Error bars represent ±1 standard error for occasional collections of replicate samples from a well.

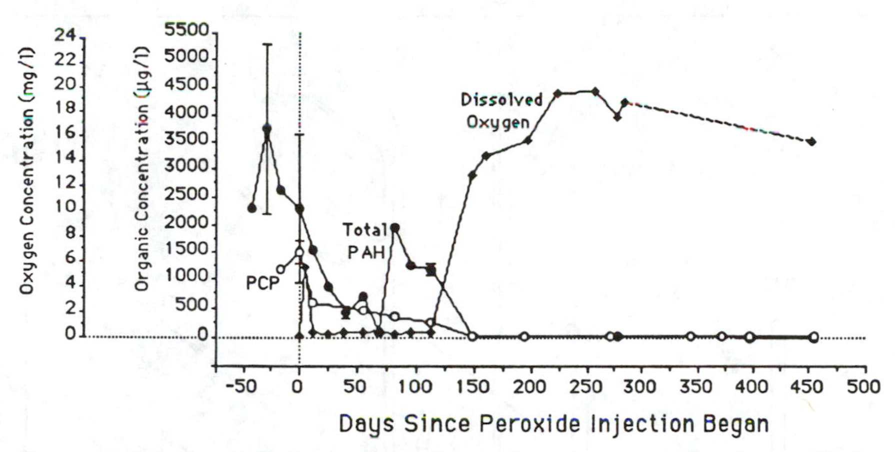

Figure 14.10. Organic contaminant and dissolved oxygen concentrations in groundwater from Well 26 over time.

Peroxide injection had created zones of high dissolved oxygen conditions (oxic zones) in the vicinity of each injection well (Figure 14.11). As peroxide injection continued, the microbial communities residing in the vicinity of each well screen received high oxygen concentrations and began to metabolize the dissolved and sorbed contaminants present at those locations at elevated rates. As microbial degradation proceeded, the concentrations of the contaminants in the vicinities of the injection screens declined. Once the contaminant concentrations decreased, microbial oxygen consumption declined and allowed the injected oxygen to persist greater distances downgradient. In effect, the oxic zones began to expand. Oxic zone expansion from Injection Site 7 took place at a higher rate and expanded more extensively than the oxic zone created by Injection Site 4. We (Woodward-Clyde Consultants) believe that the reasons expansion was more rapid and extensive from Injection Site 7 were that the total injection rate through Site 7 was greater than the rate for Site 4, and that Site 4 was located in a much more contaminated region of the aquifer. Due to these conditions, oxic zone expansion from Site 4 was reduced relative to Site 7.

By Day 148, oxic zone expansion from Injection Site 7 had passed monitoring Well 26 (Figure 14.12). Since the concentrations of dissolved contaminants within the oxic zone had been reduced by enhanced microbial activity, once the front of the oxic zone from Site 7 passed the location of Well 26, the water within the zone contained significantly lower concentrations of the contaminants. We believe that this is the reason why marked reductions of contaminant concentrations were observed in the well once oxygen concentration increased. We also believe

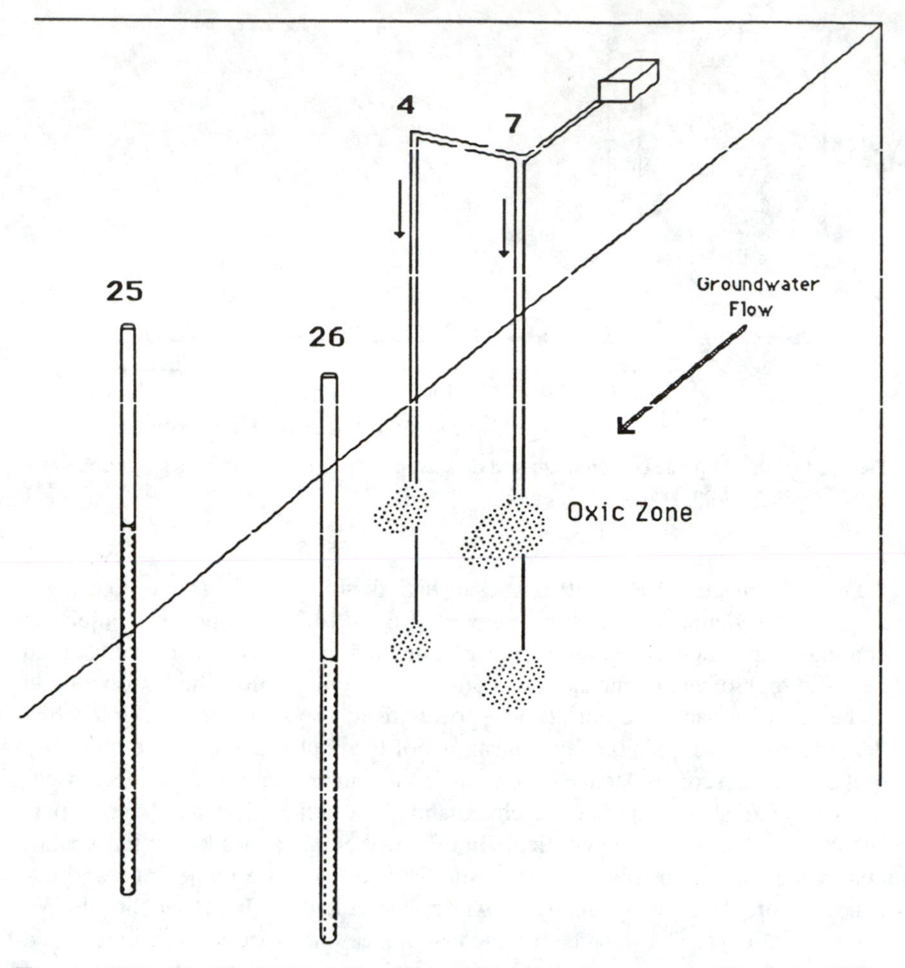

Figure 14.11. Hypothetical three-dimensional view of injection system and Monitoring Wells 25 and 26: Initial stage of peroxide injection.

that the rate of oxic zone expansion (i.e., the rate of downgradient movement of the oxic front) was controlled by the rate of microbial reduction of the sorbed contaminants in the groundwater pathway leading to the vicinity of Well 26. No other mechanism can be invoked to explain why the injected oxygen did not reach Well 26 more quickly than it did nor why dissolved oxygen failed to reach Well 25 in elevated concentrations.

This pilot study was the first Superfund study to demonstrate that injection of hydrogen peroxide could create an extended oxic zone within a contaminant plume

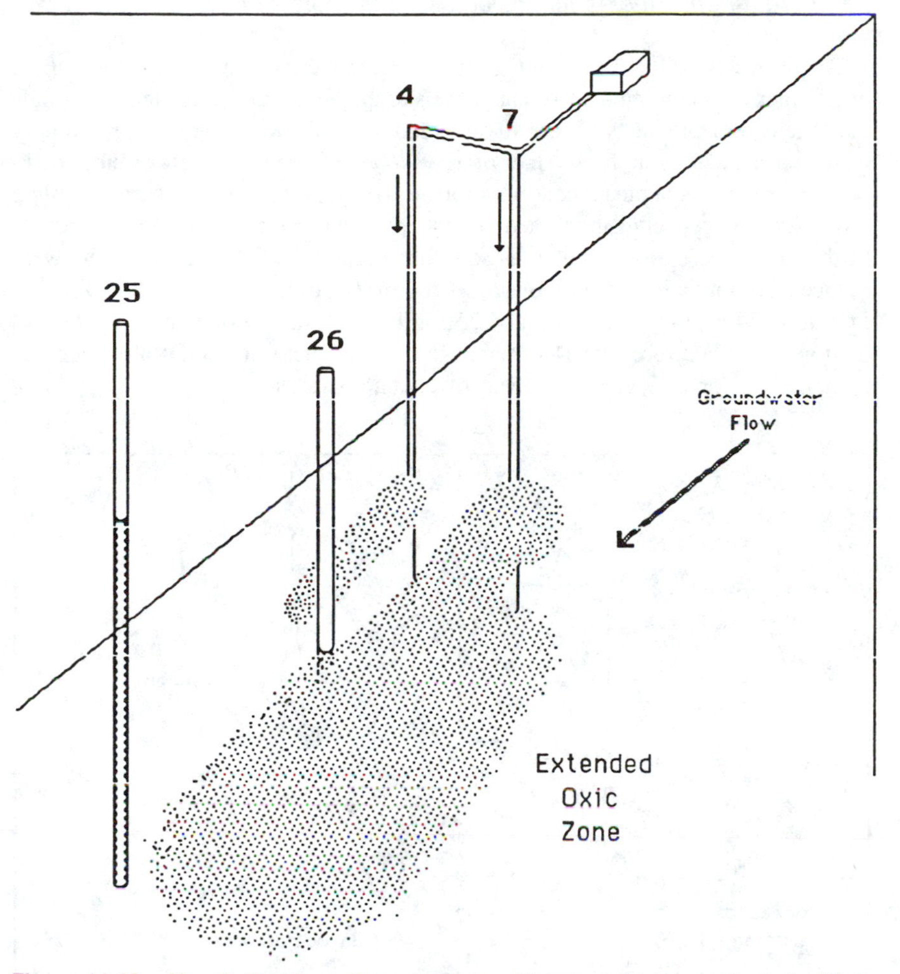

Figure 14.12. Hypothetical three-dimensional view of injection system and Monitoring Wells 25 and 26: view after 148 days of injection.

and that creation of the oxic zone would coincide with marked contaminant reductions within the zone. We presented the results of the study to the EPA in the summer of 1988,[29] and after a review of the results by bioremediation experts at the EPA's Robert S. Kerr Environmental Research Laboratory in Ada, Oklahoma, the EPA concurred with our interpretation and handed down a record of decision in December 1988 that stipulated implementation of in situ bioremediation for aquifer treatment. This was the first time in the Superfund program that such a decision had been handed down.[30]

Land Treatment Studies for Biological Treatment of Contaminated Soils

The positive results of the pilot study for aquifer treatment indicated that there was a microbial community present at the site that could be stimulated to degrade the organic contaminants. These findings led to implementation of a preliminary land treatment study in the summer of 1988 to assess the usefulness of bioremediation to reduce contaminant concentrations in affected soils. Again, strict sampling procedures, analytical methods, and QA/QC procedures were followed. The results of this study were also positive, as soil concentrations of the contaminants were reduced from initial concentrations of ~786 mg/kg of total PAHs and 750 mg/kg PCP to ~74 mg/kg total PAHs and 25 mg/kg PCP in approximately 80 days of treatment.[31] Therefore, the December Record of Decision (ROD) also included a stipulation for biological treatment of contaminated soils.

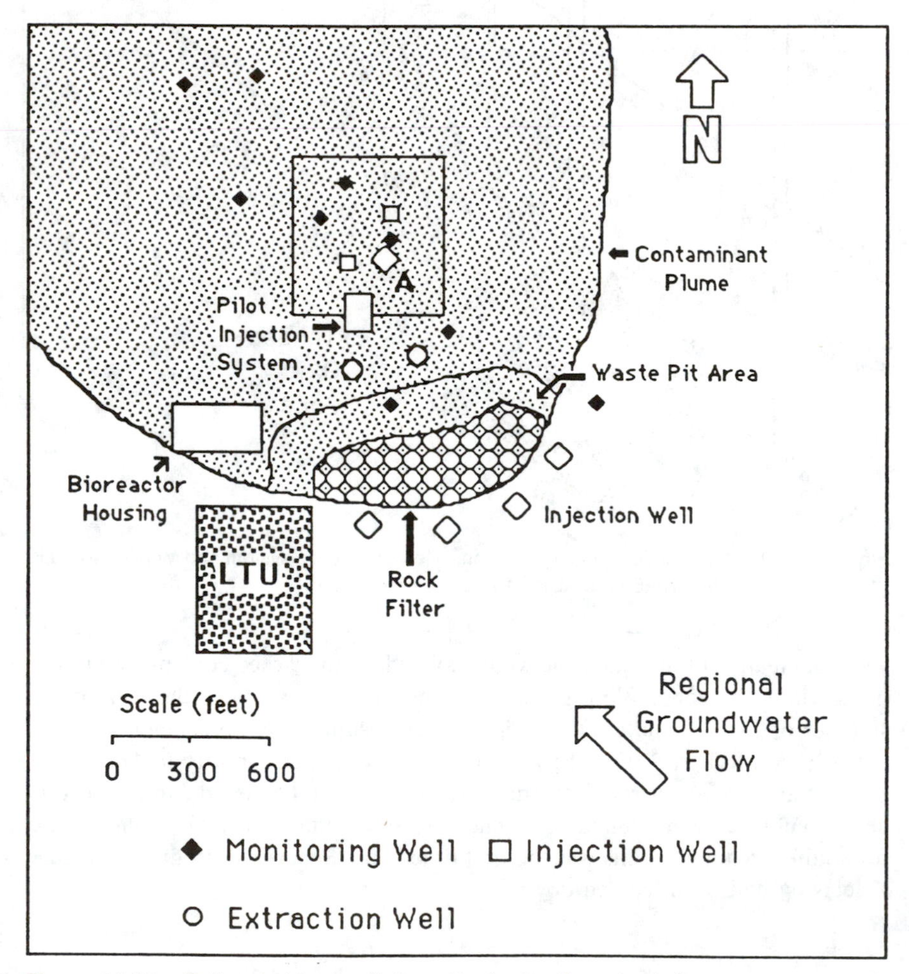

Figure 14.13. Full-scale bioremediation plan for the Superfund site.

The full-scale bioremediation design for the site was prepared in 1988[31] and approved in 1989. Figure 14.13 depicts a summary of the approved remedial activities and they are summarized below:

- continued operation of the pilot-scale peroxide injection system, since it apparently serves as a biological "sponge" to adsorb and detoxify dissolved contamination that enters the oxic zones from aquifer regions upgradient from the pilot treatment area
- full-scale land treatment of contaminated soils in a one-acre, RCRA-type, land treatment unit (LTU)
- extraction of heavily (two-phase) contaminated groundwater from the vicinity of the contaminant source area. The organic phase will be separated from the aqueous phase and stored for recycling. The aqueous phase will be treated in an aboveground bioreactor treatment system and trickled over a rock filter composed of rocks separated from the contaminated soils. The filter will be located over the head of the contaminant plume and the water that trickles down to the water table in this area is anticipated to enhance contaminant degradation rates in this region
- installation and operation of a full-scale injection well in the contaminant plume to assess the hydrological influence of a full-scale injection well on the aquifer. The results of this assessment will be used to design the distribution of injection wells to be deployed along the upgradient rim of the contaminant source area in 1991.

During the summer of 1989, a number of these full-scale bioremediation activities were implemented at the site including full-scale biological treatment of contaminated soil in the LTU. Figures 14.14 and 14.15 present a summary of the data collected in 1989 for the LTU. Two separate lifts of contaminated soil were applied during the 1989 treatment season. Contaminant concentrations in the first lift were reduced below EPA-stipulated levels within 40 days. The second lift was applied in August, and the higher initial contaminant concentrations coupled with the generally cooler ambient temperatures during treatment of the second lift extended the time period necessary for satisfactory cleanup to approximately 90 days.

Summary of the Case Histories

The case histories demonstrated the usefulness of the microbial ecology approach both in the conductance of the feasibility studies and in the actual treatment efforts. Furthermore, in Case History 2, because the feasibility study was conducted without time-consuming laboratory investigations and assessments of the microbiological content of the affected groundwater, we were able to rapidly develop and present evidence to the EPA of the potential usefulness of in situ bioremediation. This in turn allowed us to proceed directly to the pilot-scale

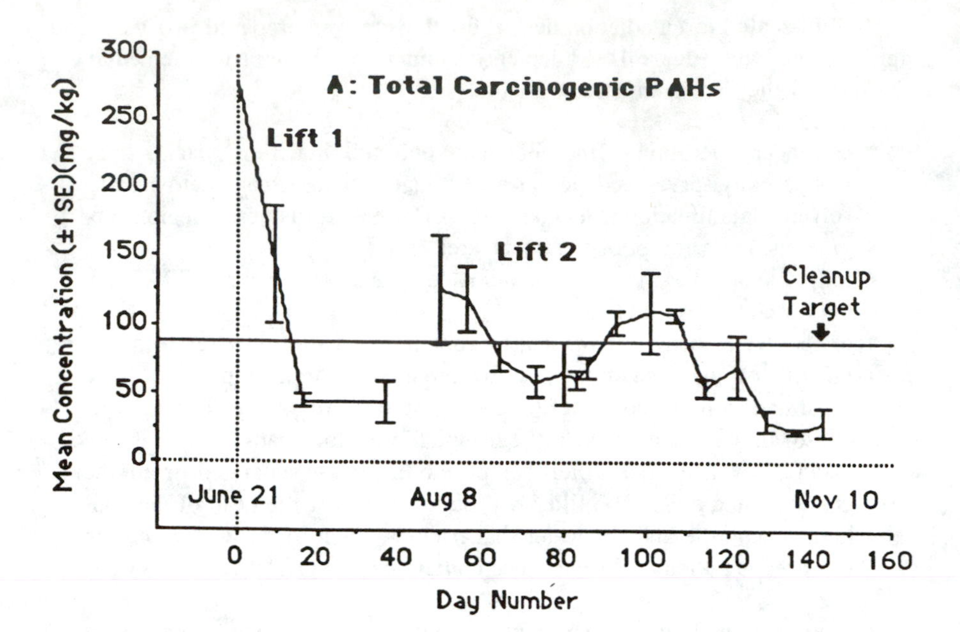

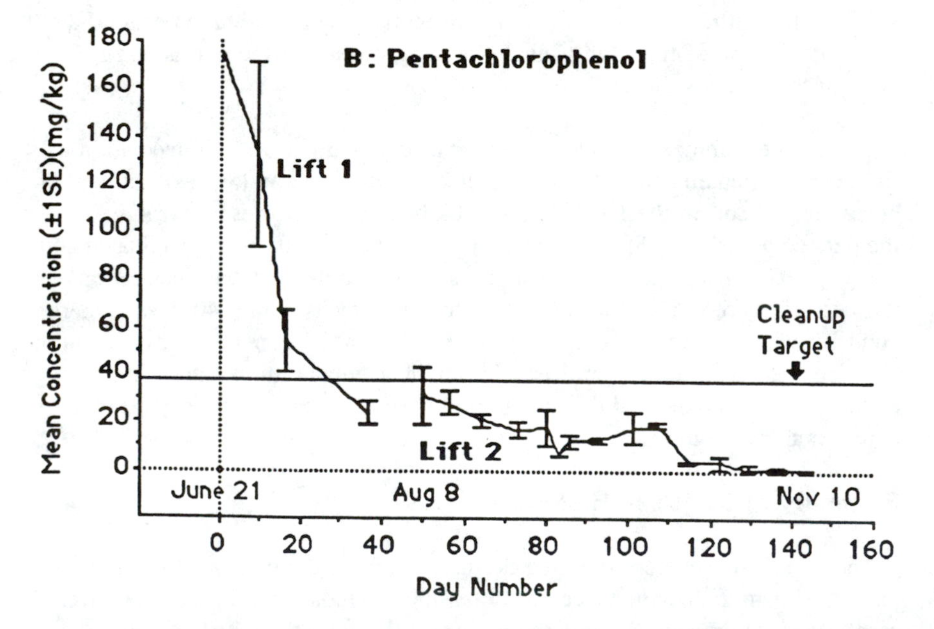

Figure 14.14. Mean concentrations of organic contaminants in soils treated in the Land Treatment Unit, 1989: (A) Total carcinogenic PAHs (polynuclear aromatic hydrocarbons; (B) Pentachlorophenol. Error bars = ±1 Standard Error (SE). Units: mg/kg. Cleanup Targets: 88 mg/kg for total carcinogenic PAHs, 37 mg/kg for pentachlorophenol.

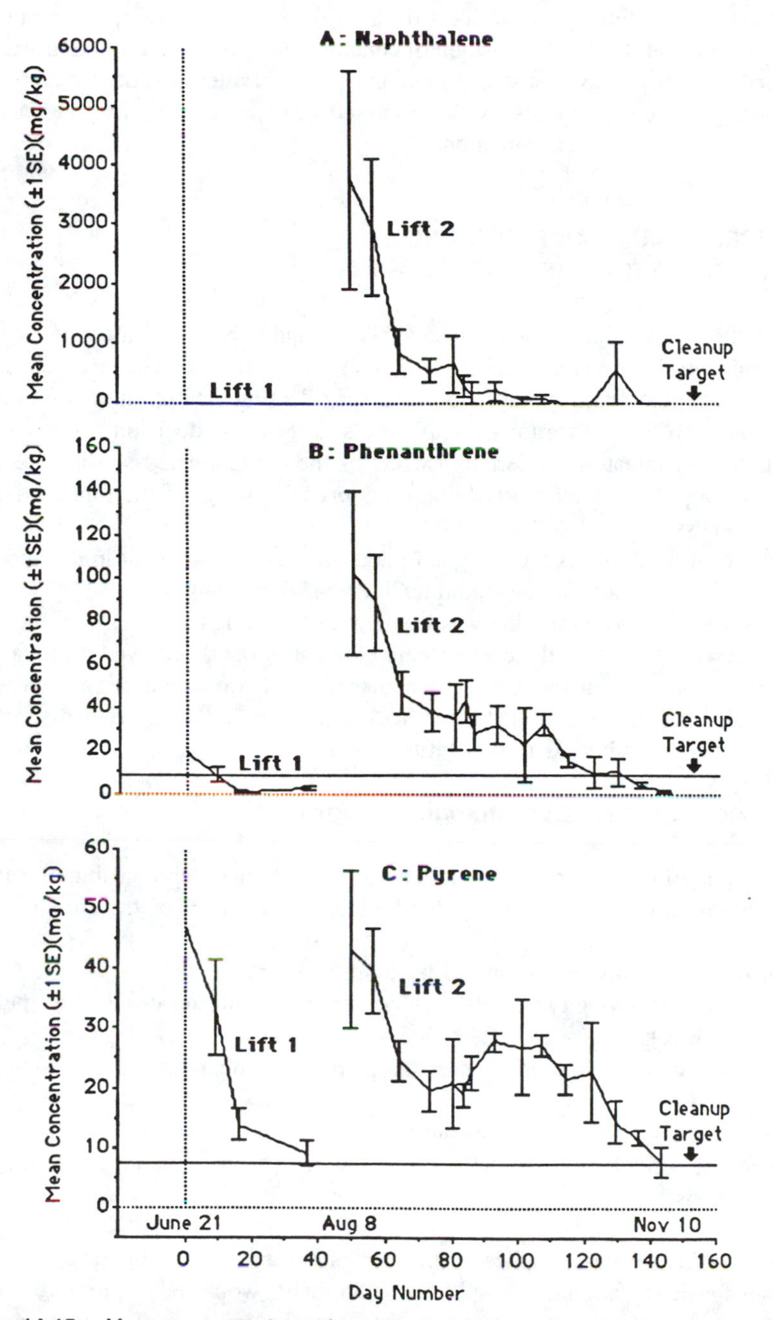

Figure 14.15. Mean concentrations of organic contaminants in soils treated in the Land Treatment Unit, 1989: (A) Naphthalene (Cleanup Target: 8 mg/kg); (B) Phenanthrene (Cleanup Target: 8 mg/kg for total carcinogenic PAHs, 37 mg/kg for pentachlorophenol.

treatment study. Finally, the success of the microbial ecology approach in these studies demonstrated that application of contaminant-specific microorganisms was not necessary to achieve satisfactory enhancements of microbial degradation rates of the organic contaminants. As is discussed below, the same may be true for a number of contaminant situations.

CONCEPTUAL FRAMEWORK FOR BIOREMEDIATION DECISIONMAKING

Both bioremediation approaches have been found to be useful for some organic contaminant situations while having limitations in others. The preceding discussions suggest a conceptual framework for selecting the optimum bioremediation approach for restoration of a specific site. For this decision framework, a preliminary evaluation of certain aspects of the contaminated site is necessary to decide a priori which bioremediation approach may have the highest degree of usefulness.

Figure 14.16 presents a conceptual diagram designed to assist in the decision process for selection of the optimum bioremediation approach. The decision process is based on a preliminary assessment of how long the contaminant(s) have been present at the site, the contaminant concentration(s) involved, how rapidly site cleanup must be achieved, and whether or not toxic metals are also present. Depending on the results of this evaluation process, either bioremediation approach may be indicated to be useful.

Indicators for the Microbiological Approach

The following site aspects would tend to indicate that the microbiological approach would be the most useful for biological treatment of the site:

- A site recently contaminated by a spill event
- A site containing extremely high contaminant concentrations (e.g., percent levels)
- A site containing high concentrations of toxic metals and organic compounds
- A site requiring rapid cleanup
- A site containing extremely low, but unacceptable contaminant concentrations.

For the first situation, application of microbial strains possessing metabolic capabilities for degradation of the spilled contaminants would be required to initiate appreciable contaminant degradation because the indigenous microorganisms would be anticipated to be absent or nonfunctional.

Time Aspects:

"Fresh" Spill ------------> Microbiological Approach Using Strains Adapted to High Organic Concentrations

"Old" Spill --------------> Microbial Ecology Approach

Contaminant Aspects:

Concentration:

High (toxic) ---------> Microbiological Approach Using Strains Adapted to High Contaminant Concentrations

"Moderate" ----------> Microbial Ecology Approach

Low with Risk -----> Microbiological Approach Using Strains Adapted to Low Contaminant Concentrations

High Concentrations of Toxic Metal(s):

Present --------------> Microbiological Approach Using Strains Adapted to Metal(s)

Absent --------------> Microbial Ecology Approach

Site Aspects:

Immediate Cleanup Required ---------> Microbiological Approach

Immediate Cleanup Not Required -----> Microbial Ecology Approach

Aquifer Involved ----------------------> Microbial Ecology Approach

Figure 14.16. Conceptual diagram for deciding on bioremediation approach.

For the second situation, application of strains capable of degrading the contaminants at elevated concentrations would probably be required because the existing concentrations would be anticipated to be toxic to the majority of natural microorganisms.

For the third situation, microbial strain(s) with the ability to withstand and function effectively in the presence of the metals would be required since natural microorganisms would not be anticipated to possess this capability.

The fourth situation requires application of microbial strains that can achieve satisfactory contaminant reductions more rapidly than could be anticipated to be possible by the indigenous microorganisms.

For each of the preceding situations, microbial strains possessing such capabilities are already available or could be developed for site-specific application. The last situation, however, poses a special circumstance, because it appears that it will require the application of microorganisms capable of metabolizing the contaminants of concern at very low concentrations (< 1 part-per-billion) and reducing them to concentrations perhaps several orders-of-magnitude lower in order to meet regulatory goals.

The literature indicates that many natural microorganisms tend to switch "off" or to more abundant carbon substrates once the concentrations of the substrates they have been using decline below certain minimal levels. As was discussed earlier, these minimal levels may be typically in the low parts-per-billion range[32] and for some contaminant compounds, this level of reduction will not be sufficient to meet regulatory action levels. Although microorganisms with capabilities to survive and grow at very low concentrations of organic substrates (termed oligotrophs or oligocarbophiles) have been isolated from the environment[33,34] I am unaware of specific applications of oligotrophs to reduce such low concentrations of hazardous contaminants to acceptable levels in affected soils and waters. Oligotrophic research using microbiological approaches may produce important advances for bioremediation and it remains a research area ripe for discovery and application.

Indicators for the Microbial Ecology Approach

The following site aspects would tend to indicate that the microbial ecology approach would be the most useful bioremediation approach:

- A site that has been contaminated for some time
- A site contaminated for some time with multiple organic contaminants
- A site contaminated with "moderate" concentrations of contamination
- A site that does not have to be cleaned up quickly
- A site containing a contaminated aquifer.

For the first situation, the longer the contamination has been present at the site, the more likely it will be that a natural microbial community has developed in the contaminated soil or water with the capability to degrade the contamination. Although the time period necessary for the development of an indigenous microbial community with the capability to degrade one or more contaminants is not precisely known, it appears to be a period no longer than several months. For example, within months of the EXXON Valdez oil spill, bioremediation teams from the EPA and EXXON observed indications that numerous indigenous microbial strains possessed or quickly developed a capability to degrade the spilled crude oil.[35]

Direct stimulation of an indigenous, contaminant- and site-adapted microbial community would be anticipated to be more effective than application of a non-site-specific, prepackaged microbial strain that could be outcompeted by the existing indigenous microorganisms. It would also tend to be less time-consuming and potentially as effective as the second form of the microbiological approach (development of high densities of site- and contaminant-specific strains in the laboratory for eventual reapplication to the site).

For a site contaminated in the past by a hazardous substance containing multiple organic contaminants (such as crude oil, many petroleum products, and creosote), the indigenous microbial community can be anticipated to have developed the capacity to degrade many of the organic constituents. This capability would be difficult to simulate using added microorganisms. Furthermore, it can be anticipated that the indigenous community would possess the interactive nature necessary to achieve essentially complete biotransformation of a number of the contaminants. Biotransformation in this context refers to mineralization (conversion to carbon dioxide), incorporation into microbial biomass, or incorporation into humic substances.

With respect to situations where contaminant concentrations are moderate, the definition of what is "moderate" is important. Based upon Woodward-Clyde Consultants studies, contaminant concentrations up to several thousands of parts-per-million did not preclude the indigenous community from effectively degrading the contaminants. It may be that even percent levels can be dealt with using the microbial ecology approach; however, this remains to be seen. The lower bound is also poorly known. In our studies, dissolved contaminant concentrations were reduced below the detection limits for a number of pollutants, and these detection limits were <1 part-per-billion. Therefore, the term "moderate" concentration may be applicable to a range in concentration extending over 8 orders-of-magnitude.

From a practical viewpoint, the majority of reported sites are contaminated with "moderate" organic concentrations, typically less than 100,000 parts-per-million. This suggests that the microbial ecology approach may be the most useful bioremediation approach for immediate application to most sites. However, due to uncertainty over the final contaminant concentrations achievable using this approach, it should be recognized that for some organic contaminants, the regulatory action level may be unreachable. In those cases, it may be necessary to apply microbial strains adapted to very low concentrations of the contaminant(s) once the indigenous community has reduced concentrations to minimum levels achievable by the community.

For sites which pose little direct risk to human or environmental health, the microbial ecology approach may be useful since immediate cleanup may not be required. One of the uncertainties involved with the microbial ecology approach is the time period required for satisfactory cleanup. If cleanup time is not an issue, the microbial ecology approach may be the most cost-effective means to restore the site, since the approach is generally less expensive to assess and apply than the microbiological approach.

If a site contains a contaminated aquifer, the microbial ecology approach appears to be the most capable means of remediating the aquifer in situ. This is because the approach characteristically takes the form of adjusting the physical and chemical characteristics (such as pH, redox) by injection of the proper chemicals. The chemicals will usually disperse throughout the contaminated aquifer more readily than injected microorganisms. However, the physical characteristics of the aquifer will be the fundamental determinants of how successful any in situ remedial approach will be for the aquifer. If the nature of the aquifer does not allow easy manipulation of the chemical or physical characteristics of the aquifer, remedial action will take longer to succeed (if at all) no matter which approach is applied.

LONG-TERM IMPLICATIONS OF BIOREMEDIATION FOR SITE CLEANUP

Regardless which bioremediation approach is espoused, advocates of bioremediation face an important uncertainty with respect to its future usefulness. While few would deny that bioremediation will be useful for reducing the volumes and concentrations of organic contaminants in soils and waters, uncertainty over what contaminant concentrations will be deemed protective of human health and the environment in the future will have crucial bearing on the long-term usefulness of the remedial technology.

When a site is contaminated by hazardous organic compounds, the contamination exists in two basic forms: the unbound (free) state and the bound state. The bound state in this context refers to organic contaminants sorbed to mineral surfaces or complex organic matrices (humic substances). When bioremediation techniques are applied, the overall concentration of organic contaminant declines (Figure 14.17a). Portions of the contamination may be mineralized, converted to microbial biomass, or incorporated (bound) in humic substances. Concurrent with the reduction in contaminant concentration is a reduction in the risk to human health and the environment (Figure 14.17b).

However, after some period of time, the rate of contaminant reduction will decrease, and eventually a semisteady state will be reached in which low concentrations of the contaminant exist in both the bound and unbound states. Risk reduction will also become asymptotic. The fundamental question at this point is whether or not the level of risk posed by the residual contamination is acceptable or not. This issue is currently a matter of some conjecture and considerable debate and largely depends on risk factors, extrapolation models, and our current understanding of potential hazards posed by low levels of contaminants. If the resultant levels of contamination are deemed protective, the site will have been ostensibly restored. If the resultant levels exceed regulatory goals, the site will require additional remediation by technologies which are anticipated to be extremely costly and perhaps not presently available.

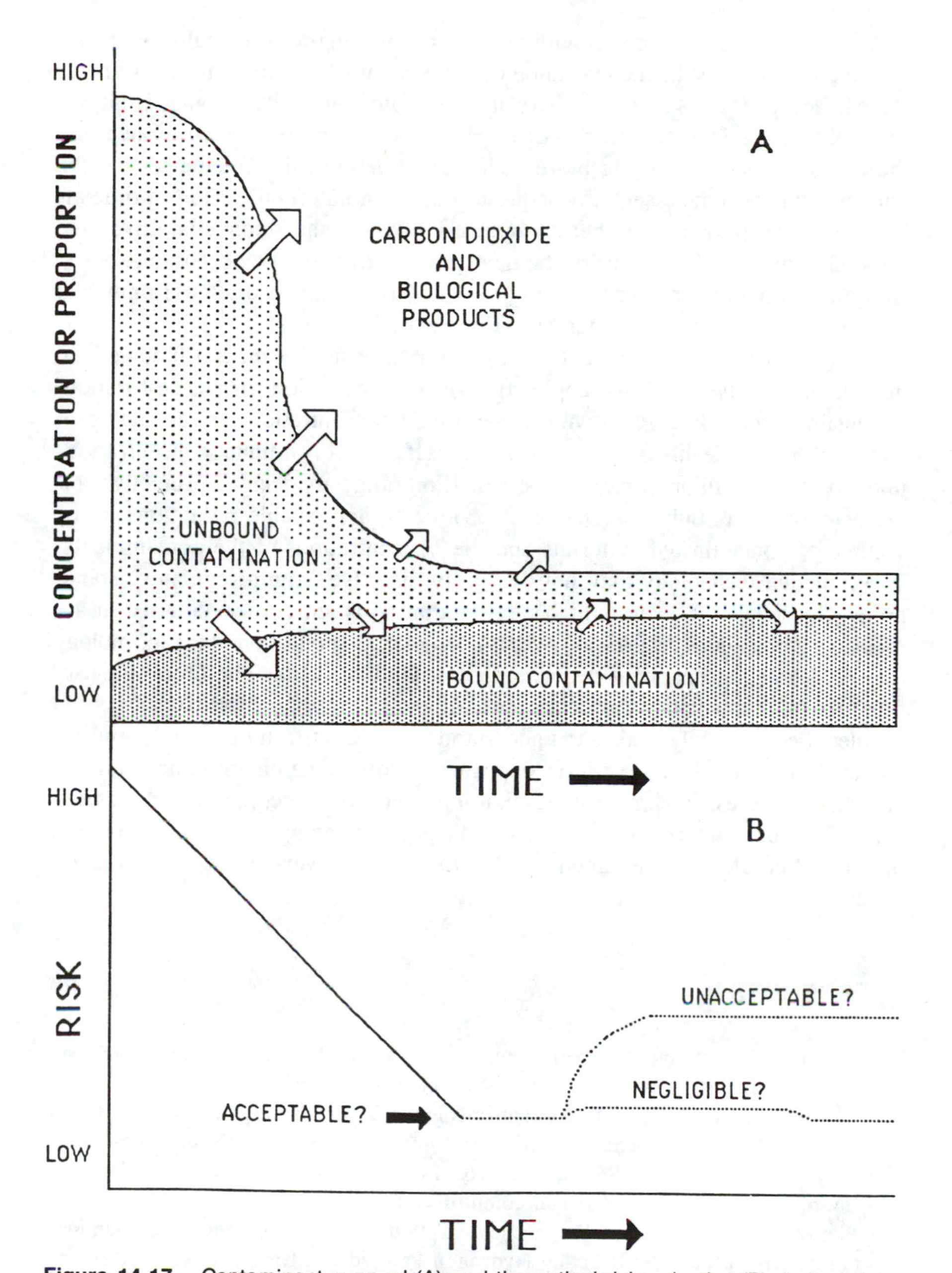

Figure 14.17. Contaminant removal (A) and theoretical risk reduction (B) during a bioremediation program. Open arrows extending out of shaded areas depict conversions of contamination to carbon dioxide or biological products. Open arrows extending between shaded areas depict exchanges of contamination from bound to unbound conditions. Sizes of arrows indicate relative rate of process. *Source:* Reference No. 3.

A second aspect of concern with respect to the long-term usefulness of bioremediation involves the incorporation of the contaminants into humic substances. Recent isotope studies by researchers at Utah State University, Logan, Utah, indicate that a considerable fraction of PAH compounds become incorporated into the humic component during bioremediation.[36] While it is anticipated that the contaminants become essentially sequestered for an indefinite period of time after humic incorporation and thereby reduced in toxicity and bioavailability,[37] the possibility exists that portions of the incorporated fraction may become released (become unbound) at a later time (Figure 14.17a). If that occurs, environmental risks would be expected to increase (Figure 14.17b).

Whether or not the subsequent increase in risk is significant is a fundamental uncertainty involved with the application of bioremediation for site restoration, and again depends largely on what contaminant concentrations are deemed protective of human health and the environment. If only contaminant concentrations in the parts-per-trillion to parts-per-quadrillion range are deemed protective, bioremediation may only be capable of producing appreciable reductions in the volume of contaminated materials and the concentrations of the contaminants. The remedial technology may not be totally effective from a regulatory standpoint. Consequently, even though bioremediation is anticipated to assist in appreciable contaminant reductions, over the long-term, application of the technology may be deemed incapable of restoring the contaminated site. Therefore, the usefulness of bioremediation for helping us solve our environmental contamination problems rests largely with our understanding of risks to human and environmental health posed by long-term exposure to low contaminant concentrations and the proper establishment of regulatory goals for the contaminants of concern. Hopefully, future assessments of risks posed by contaminants will be based on more accurate risk assessments and not simply on ever-descending detection limits.

REFERENCES

1. Zitrides, T. G. "Mutant Bacteria Overcome Growth Inhibitions in Industrial Waste Facility," Industrial Wastes 24: 42–44 (September/October 1978).
2. Chakrabarty, A. M. "Genetic Mechanisms in the Dissimulation of Chlorinated Compounds," In: A. M. Chakrabarty, ed., *Biodegradation and Detoxification of Environmental Pollutants* (Boca Raton, FL: CRC Press, 1982).
3. Bourquin, A., ECOVA, Personal communication.
4. Piotrowski, M. R. "In Situ Biogeochemical Reduction of Hydrocarbon Contamination of Groundwater by Injecting Hydrogen Peroxide: A Case Study in a Montana Aquifer Contaminated by Wood Preservatives," Ph.D. Dissertation, Boston University, Boston, MA, 1989a. (University Microfilms, Inc. Order No. 8913768).
5. Piotrowski, M. R. "Improving Feasibility Studies," *HAZMAT WORLD* 2(6):42–45 (1989b).

6. Barles, R. W., C. G. Daughton, and D. P. H. Hsieh. ''Accelerated Parathion Degradation in Soil Inoculated with Acclimated Bacteria Under Field Conditions,'' *Arch. Environ. Contam. Toxicol.* 8: 647–660 (1979).
7. Kilbane, J. J., D. K. Chatterjee, and A. M. Chakrabarty. ''Detoxification of 2,4,5-Trichlorophenoxyacetic Acid from Contaminated Soil by Pseudomonas cepacia,'' *Appl. Environ. Microbiol.* 45:1697–1700 (1983).
8. Goldstein, R. M., L. M. Mallory, and M. Alexander. ''Reasons for Possible Failure of Inoculation To Enhance Biodegradation,'' *Appl. Environ. Microbiol.* 50:977–983 (1985).
9. Lewandowski, G., S. Salerno, N. McMullen, L. Gneiding, and D. Abramowitz. ''Biodegradation of Toxic Chemicals Using Commercial Preparations,'' *Environmental Progress* 5(3):212–217 (1986).
10. Atlas, R. M., and R. Bartha. ''Effects of Some Commercial Oil Herders, Dispersants, and Bacterial Inocula on Biodegradation of Oil in Seawater,'' Louisiana State University Center for Wetland Resources, LSU-SG-73-01, 1973.
11. Lehtomaki, M., and S. Niemela. ''Improving Microbial Degradation of Oil in Soil,'' *Ambio* 4:126–129 (1975).
12. Hill, D. L., T. J. Phelps, A. V. Palumbo, D. C. White, G. W. Strandberg, and T. L. Donaldson. ''Bioremediation of Polychlorinated Biphenyls: Degradation Capabilities in Field Lysimeters,'' *Appl. Biochem. Biotech.* 20/21:233–243 (1989).
13. Forrester, I., Gemini Biochemical Research Ltd., Calgary, Canada, Personal communication.
14. Alexander, M. *Introduction to Soil Microbiology.* (New York, NY: John Wiley & Sons, 1977).
15. Hanson, W., U.S. EPA, Superfund '89 presentation, Washington, DC, November, 1989.
16. Fleirmans, C., Department of Energy, Deep Subsurface Microbiology Program, Personal communication.
17. Moreland-Day, S., former Executive Director of the ABTA, Personal communication.
18. Jamison, V. W., R. L. Raymond, and J. O. Hudson, Jr. ''Biodegradation of High Octane Gasoline in Groundwater,'' *Dev. Ind. Microbiol.* 16: 305–311 (1975).
19. Woodward-Clyde Consultants. ''Alyeska Bench Test for Removal of Aromatic Hydrocarbons,'' Report prepared for Alyeska Pipeline Service Company, 1986.
20. Woodward-Clyde Consultants. ''Biodegradation of Volatile Aromatic Hydrocarbons in Aqueous Environments,'' Report prepared for Alyeska Pipeline Service Co. and the U.S. EPA, 1987a.
21. Woodward-Clyde Consultants. ''Evidence for Biodegradation of BETX in Bench-Scale and Prototype Aerated Lagoon Wastewater Treatment Systems,'' Report prepared for Alyeska Pipeline Service Co., 1987b.
22. Marrs, D., SOHIO, Personal communication.
23. Woodward-Clyde Consultants. ''Preliminary Feasibility Assessment of In Situ Biological Treatment,'' Appendix A in: *Draft Feasibility Study for Site Remediation, Groundwater Contamination Site, Libby, Montana.* Prepared for Champion International, Stamford, CA, 1987c.
24. DeLaune, R. D., W. H. Patrick, Jr., and M. E. Casselman. ''Effects of Sediment pH and Redox Conditions on Degradation of Benzo(a)pyrene,'' *Mar. Pollut. Bull.* 12(7):251–253 (1981).

25. McGinnis, G. D. ''Biological and Photochemical Degradation of Pentachlorophenol and Creosote,'' Mississippi State University, Forest Products Laboratory. Prepared for the U.S. Forest Products Laboratory, U.S. Department of Agriculture, Madison, WI, 1982.
26. Rochkind, M. L., J. W. Blackburn, and G. S. Sayler. ''Microbial Decomposition of Chlorinated Aromatic Compounds,'' U.S. EPA, Hazardous Waste Engineering Research Laboratory, Office of Research and Development, Cincinnati, OH. EPA/600/2-86/090, 1986.
27. Texas Research Institute, Inc. ''Final Report: Enhancing the Microbial Degradation of Underground Gasoline by Increasing Available Oxygen,'' Prepared for the American Petroleum Institute, Washington, DC, 1982.
28. Godsy, E. M., U.S. Geological Survey, Personal communication.
29. Woodward-Clyde Consultants. ''Pilot-Scale In-Situ Biodegradation Stimulation Study,'' Presented to the U.S. EPA July 1988, 1988a.
30. Wallace, K., U.S. EPA, Region VIII, Personal communication.
31. Woodward-Clyde Consultants. ''Feasibility Study for Site Remediation, Groundwater Contamination Site, Libby, Montana,'' Prepared for Champion International, Stamford, CA, 1988b.
32. Alexander, M. ''Biodegradation of Organic Chemicals,'' *Environ. Sci. Technol.* 18(2):106–111 (1985).
33. Poindexter, J. S. ''Oligotrophy: Fast and Famine Existence,'' *Adv. Microbial Eco.* 5:63–89 (1981).
34. Valiela, I. *Marine Ecological Processes* (New York, NY: Springer-Verlag, 1984).
35. ''Oil Spill Bioremediation Project Status Report and Addendum,'' U.S. Environmental Protection Agency Office of Research and Development, Bioremediation Project Office, Valdez, AK. July and September Reports, 1989.
36. Radtke, C., Utah State University, Personal communication.
37. McCarthy, J. F. ''Bioavailability and Toxicity of Metals and Hydrophobic Organic Contaminants,'' In: I. H. Suffet and P. MacCarthy, Eds. *Aquatic Humic Substances: Influence on Fate and Treatment of Pollutants.* American Chemical Society Series No. 219. American Chemical Society, 1989.

CHAPTER 15

Recycling of Petroleum Contaminated Soils in Cold Mix Asphalt Paving Materials

Terry C. Sciarrotta, Southern California Edison Company, Environmental Research, Rosemead, California

PURPOSE

The purpose of the study was to evaluate the hazardous characteristics of cold mix asphalt pavements containing both petroleum contaminated sands and clean sands. Research sponsored by the United States Environmental Protection Agency (USEPA), the Electric Power Research Institute (EPRI) and several state agencies has suggested that asphalt incorporation may be useful for treating petroleum contaminated soils and other wastes. However, the most often cited reservation about this technology is the potential for polynuclear aromatic hydrocarbon (PNAH) releases and hydrocarbon air emissions. In most cases these studies were based upon the use of hot batching technologies.

While cold mix technology produces a lower grade pavement, these pavements may be suitable for light duty use. The asphalt grade employed in this technology is characterized by a low PNAH content and, in the case of slow curing blends, are relatively nonvolatile. Hydrocarbon emissions are also reduced because the mixing occurs at ambient temperatures avoiding the potential emissions from fuel combustion and volatilization of the lighter asphalt fractions.[1,2] For this reason, cold mix asphalt incorporation is not subject to permitting by the South Coast Air Quality Management District in California.

ASPHALT CHEMISTRY

Asphalts are bituminous materials which occur naturally or are derived from nondestructive separation of petroleum fractions. Typically, this is achieved through fractional distillation or solvent de-asphalting. Asphalt should not be confused with tar, which is obtained through destructive processing of coal, wood, or petroleum. Asphalt contains aliphatic, mononuclear aromatics and PNAHs, and differs from tars which contain primarily mononuclear and PNAHs.

Structurally, asphalt is considered to be either a colloid consisting of asphalt micelles suspended in oils or an intermicellar resin phase; or a solution of asphaltenes dissolved in the oil-resin phase. Asphalt materials are categorized as asphalt cements or liquid asphalts. These materials are described as follows:[1]

- *Asphalt cement* is the heaviest fraction of asphalt. The consistency of these materials range from between solid to semiliquid at room temperature. At least five grades of asphalt cement are produced and are categorized based upon their degree of hardness or resistance to penetration. In general the heavier fractions of asphalt are characterized by high concentrations of aromatic (both mononuclear and polynuclear rings), nitrogen, sulfur, oxygen, and trace amounts of metals or organo-metallic compounds. Asphalt cements are typically used in hot mix technologies.
- *Liquid asphalts* are produced from the lighter fractions of the residual asphalt, or by dissolving asphalt cements in solvent or emulsifying asphalt cements in water. These materials are used in cold mix technologies. The liquid asphalts are graded by their viscosity. The four grades of liquid asphalt are 3000, 800, 250, and 70. The numbers indicate the minimum viscosity for the established grade. The acceptable viscosities within each grade expand across a specified range. For instance, 3000 grade liquid asphalts may exhibit a viscosity of anywhere between 3000 and 6000 centistokes. The different types of liquid asphalts are described as follows:
  - Rapid curing (RC) liquid asphalts are produced by dissolving relatively hard asphalt cements in a kerosene or naphtha solvent. These materials may be expected to contain PNAH compounds originating from the heavy asphalt solute and are characterized by a high VOC content due to the solvent carrier.
  - Medium curing (MC) liquid asphalts are produced by dissolving softer asphalt cements in a kerosene type solvent. These materials may also contain PNAH compounds, but in a lower concentration than the RC liquid asphalts, because they are derived from the lighter fractions of asphalt cement. These materials may also contain a high VOC content due to the solvent carrier.
  - Slow curing (SC) liquid asphalts (or road oils) may be derived from distillation, being the lightest fraction of the asphalts, or by fluxing the lightest fraction of the asphalt cements with lighter oils. These materials are most similar to residual fuel oils, which also are derived

from the lighter ends of the petroleum distillate (see Figure 15.1). SC liquid asphalts are expected to be composed primarily of chain-type aliphatics rather than aromatic compounds. However, the potential for PNAH contaminants cannot be discounted and, because some PNAH compounds have been designated as potential carcinogens, even trace concentrations may be of concern. The SC liquid asphalts are semivolatile or nonvolatile, and therefore the potential for VOC emissions is not a major concern.

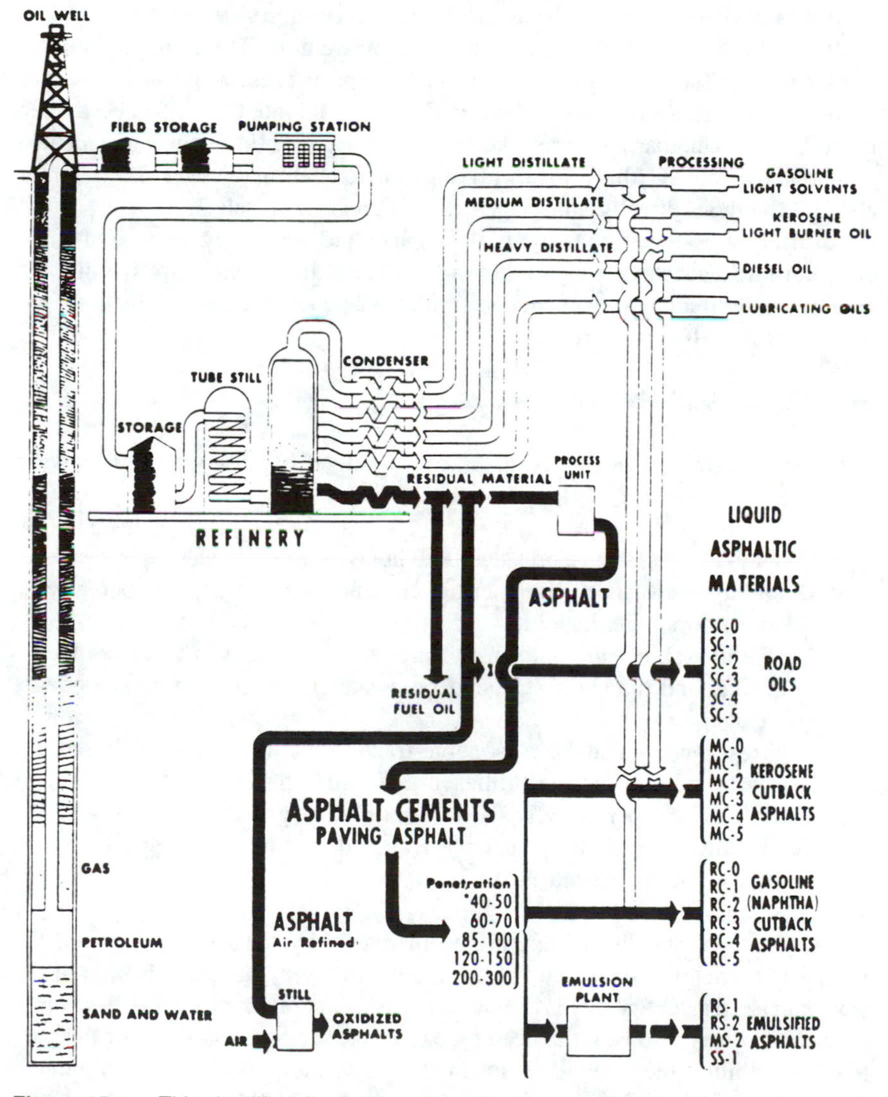

Figure 15.1. This simplified flow chart shows the inter-relationships of processed crude oil with asphalts, refined oils, and gasoline. (After Asphalt Institute)

— Asphalt emulsions are produced by using surfactants to emulsify asphalt cements in water. Anionic, cationic and nonpolar asphalt emulsions are available. These materials may contain PNAH compounds, depending upon the grade of asphalt cement from which they are derived. Emulsions are relatively nonvolatile.

ASPHALT PAVING MIXTURES

Asphalt paving mixtures typically are composed of aggregate and/or sand (90% to 95% by weight) and asphalt (5% to 10% by weight). The aggregate and/or sand is responsible for the primary load-bearing properties, while asphalt serves as the binder and as a protective coating. The asphalt binder functions best when the aggregate/sand particles are "wetted" with asphalt. If the particle is coated with water or a clay film prior to mixing, the asphalt may cover the water or clay film without directly adhering to the aggregate particle.[1]

To ensure good adhesion between the asphalt and the aggregate or sand particle, moisture and clays are to be avoided. Thus, this technology has limited application for treating petroleum contaminated soils with high clay fractions and a high capacity for water retention.

Integrity of Asphalt Pavement

The suitability of asphalt paving mixes is judged based upon the following properties:

- Stability—resistance to deformation; stability is measured in units of Hveems.
- Durability—resistance to weathering, crushing, and degradation; partially dependent upon permeability.
- Flexibility—ability to conform to long-term variations in the base due to settling and the ability to bend repeatedly without fracture (fatigue resistance).
- Skid resistance—to allow reasonable traction in all weather conditions.
- Permeability—low permeabilities are desired to resist the impact of weathering and permeability is an important factor in durability.
- Workability—the ability to achieve a smooth finish when the pavement is placed and compacted in the road.

The pavement stability is determined primarily by the friction between the aggregate particles, the viscosity of the asphalt mixture, and the mix ratio of the asphalt to aggregate. Stability is enhanced by using aggregates with rough-textured surfaces and may also be influenced by particle size and gradation. For the purposes of stability, the amount of asphalt should be minimized because too much asphalt will act as a lubricant and cause the mixture to flow. This is especially true for "critical" sand mixtures or aggregate mixtures containing a high percentage of fines.[1] As seen in this study, a slight excess of asphalt in a critical mixture can reduce the pavement's stability.

Durability is primarily dependent upon the aggregate/sand resistance to crushing, abrasion, and weathering, and the asphalt's resistance to weathering and aging. It is assumed that the impact of weathering on the particle will be minimal as long as it is coated with asphalt. The durability of the asphalt may be impacted by several chemical reactions. Reactions with it that may change the asphalt binder composition and thereby affect its durability include:

- Oxidation—asphalt forms water-soluble oxidation products which can leach from the asphalt, changing its composition. As the more soluble oils are leached from the asphalt mixture, the exposed surface may become hard and brittled. If the pavement cracks, deeper layers may be exposed to air and water, thus compromising the durability of the road.
- Volatilization—resulting in the loss of the lighter oil fractions.
- Separation—the lighter oils are sometimes selectively absorbed by the aggregate.
- Syneresis or staining—the tendency of some asphalt mixtures to exude the lighter oils at the pavement surface. Syneresis may impact the skid resistance of the pavement as well as the durability of the asphalt.

The durability of the asphalt pavement mixture is usually enhanced by high asphalt content and compacted impervious mixtures. Application of a protective top coating after curing may also enhance the pavement's durability.

The long-term flexibility and fatigue resistance of the asphalt pavement and the workability is enhanced by the same factors impacting durability. As can be seen, high asphalt content will increase the durability, flexibility, and workability of the asphalt pavement, but may decrease its stability especially in critical sand mixtures.[1]

FEASIBILITY STUDY

A feasibility study was conducted in the spring of 1989 to determine whether sandy soils contaminated with residual fuel oil could be stabilized with an asphalt binder and used as a paving mixture. Proposed uses for the paving mixture included all weather surfacing of light-duty access roads, storage yards, automobile parking, and possible use as a temporary liner in an emergency run-off reservoir. The soils had been excavated during the repair of residual fuel storage tanks and transport lines, and stockpiled upon plastic tarpaulins spread throughout the facility.

In order to evaluate the soil characteristics, the samples were obtained of the sand and fuel oil for bench-scale testing. Twenty representative samples were obtained by trenching through the stockpiled sands with a backhoe. Each sample was sealed in a plastic bag and labeled for transport to the laboratory. The approximate volumes of each stockpile were estimated by measuring its dimensions with a tape measure to allow proper weighting of the sample analyses.

Each sample was analyzed for oil content and moisture content. The oil content values were weighted based upon the calculated volume of the respective stockpile. The values ranged between 1.3% and 4.7%, with the weighted average being 2.7% (Table 15.1). The moisture values ranged from between 3.1% and 10.7%.

Table 15.1. Moisture and Oil Content of Contaminated Soil

Sample No.	% Moisture	% Fuel Oil
1	5.9	2.0
2	4.7	1.9
3	6.4	1.3
4	4.4	3.3
5	2.2	2.5
6	4.4	2.3
7	10.0	1.3
8	3.2	3.1
9	3.1	3.2
10	7.4	4.2
11	4.3	3.3
12	4.8	2.9
13	4.4	3.7
14	8.0	1.6
15	10.7	4.7
16	6.6	3.2
17	9.0	4.3
18	6.0	1.9
19	7.6	2.5
20	8.1	2.6

Two representative samples were selected for grain size evaluation. The results of these analyses (Table 15.2) were then compared to the desired gradation requirements for road mix asphalt/sand mixtures and aggregate/sand asphalt concrete mixtures. Based upon these comparisons, the sands would have to have been supplemented with imported aggregate in a ratio of 70:30 (aggregate to sand) in order to meet the desired gradation.

Achieving the desired gradation by supplementing the sands with aggregate would more than triple the project costs. In an effort to compensate for the gradation requirements, it was proposed to improve the stability by blending a more viscous liquid asphalt with the sand/fuel oil mixture, thus eliminating the need for aggregate. Also the designated pavement use was downgraded to light duty

Table 15.2. Material Gradation

Sieve Size (inches)	Sample 1 (%)	Sample 2 (%)	Average
1.5	100	100	100
1.0	100	86.4	93.2
0.75	98.7	84.5	91.6
0.50	96.6	84.5	90.6
0.38	93.9	83.9	88.9
No. 4	88.2	79.0	83.6
No. 8	81.5	73.3	77.4
No. 16	72.5	67.0	69.8
No. 30	60.5	60.4	60.5
No. 50	44.3	51.6	48.0
No. 100	28.2	40.4	34.3
No. 200	15.1	29.2	22.2

Theoretical Total Asphalt Demand: 8.26
Residual Oil: 2.73
Theoretical Additional Asphalt Demand: 5.53

pavement. The recommended guidelines for light duty pavement[3] suggest a minimum Hveem stability of 30, but this value was not considered an absolute for the intended application.

With these criteria in mind, bench-scale testing was performed using weighted composite samples of the fuel oil contaminated sands. The samples were mixed with SC-3000 liquid asphalt in amounts equal to 3.5% (by weight) of the mixture, and increasing the amount of liquid asphalt by 1% increments up to 6.5%. The resulting mixture was tested in both a cured and uncured condition in accordance with Caltran's Test Methods 304 and 366. As shown in Table 15.3, and as expected from the literature, the highest Hveem stability was achieved at the lower concentrations of liquid asphalt, and stability decreased rapidly as excess asphalt was added. However, the durability, flexibility, and workability increased with higher asphalt content.

In addition to testing the sand/asphalt mixture, samples of fuel oil were blended with the liquid asphalt and compared with the specifications for commercial liquid asphalt grades. The SC-3000 liquid asphalt exhibited a viscosity of 4910 centistokes, which is in the mid range for this grade (see Table 15.3). When the SC-3000 was blended with samples of residual fuel oil in a ratio of approximately 2:1, the viscosity dropped dramatically, but was still higher than the specification for SC-70. The open cup flash point of the blend was higher than the minimum specification for all of the common liquid asphalt grades, and the total distillate content (i.e., the amount of volatile lost at temperatures 680°F) were lower than the specified maximum. The SC-3000/fuel oil blend exceeded the specifications of SC-70 liquid asphalt grades.

Table 15.3. Feasibility Study Data

Contaminated Sand/Liquid Asphalt Paving Mixture

					Recommended
SC-3000 percentage	3.5	4.5	5.5	6.5	4.0 to 5.0
Hveem stability	32	26	17	<10	30 minimum

Fuel Oil and Liquid Asphalt Viscosities[a]

Sample SC-3000	SC-3000 Specs.	SC-800 Specs.	SC-70 Specs.
4910	3000–6000	800–1600	70–140

SC-3000/Fuel Oil Blend (66.9%/33.1%)

Text	Blend	SC-3000 Specs.	SC-800 Specs.	SC-70 Specs.
Viscosity[a] ASTM D2170	185	3000–6000	800–1600	70–140
Open cup min. flash point (°F), ASTM D92	320	225 min.	200 min.	150
Total distillate (volume %), ASTM D402	0.5	5 max.	2–12	10–30
Viscosity[a] on residue	51.2	40–350	20–160	4–70

[a]Kinematic viscosity at 140°F; measured in centistokes.

Based upon the above described analyses, it was advised that the moisture content of the onsite sands be reduced to 4% or less, and SC-3000 liquid asphalt be added to the mixture in amounts equaling 4% to 5%. The reduced moisture will provide a more stable mixture during the initial handling and should minimize the period of curing required prior to final placement and compaction. Stability was compromised somewhat in the interest of durability, flexibility, and workability. However, a chip seal topping could be added to improve durability after curing.

ENVIRONMENTAL TESTING

The experimental pavement samples were prepared from the stockpiled-sands contaminated with residual fuel oil, and control samples were prepared from "clean" sands. The samples were analyzed to determine whether they contained toxic constituents and specifically whether they would be regulated as hazardous substances in California. The California toxicity criteria are more encompassing than the Resource Conservation and Recovery Act (RCRA) criteria. They include:

- limits on specific carcinogens (including some PNAH compounds)
- limits on heavy metals, similar to the RCRA EP Toxicity test
- bioassay testing

The experimental and control samples were tested for PNAH content and heavy metals content, and were submitted to bioassay testing in accordance with California Code of Regulations (CCR), Sections 66696–666700. The test results indicated:

- The asphalt pavement produced from fuel oil contaminated sands represented no greater environmental risk than pavement produced from clean sands.
- Neither asphalt pavement would be classified as a hazardous waste pursuant to existing California regulations.

PNAH Analyses

The samples were extracted with methylene chloride in a Soxhlet thimble in accordance with EPA procedures for extracting semivolatile and volatile organic compounds from solid materials (SW-846, Method 8.86).

The control sample yielded 78.2 mg of extractable hydrocarbon per gram of sample. The samples were analyzed by high resolution capillary gas chromatography/mass spectroscopy (GC/MS). PNAH compounds were mostly less than the detection limits for each substance. Four organics were just over their limit for detection (Table 15.4).

The experimental sample yielded 80.6 mg of extractable hydrocarbon per gram of sample. The same four PNAH compounds were detected from the GC/MS analyses of this sample. In each case the concentration was less in the experimental than the control. This suggests that the source of PNAH compounds is from the asphalt and less from the fuel oil. All other volatile and semivolatile organic compounds were below their respective detection limits (Table 15.4).

Heavy Metals Analyses

The pavement samples were analyzed for both total and soluble heavy metals content pursuant to California hazardous waste regulations. Unlike the RCRA EP Toxicity criteria, the California Code of Regulations establishes limitations based upon both total content (i.e., Total Threshold Limit Concentration; TTLC) and the extractable or soluble content (Soluble Threshold Limit Concentration; STLC). As one would expect, the established TTLC is always higher than the STLC. A waste is considered hazardous if the total concentration is equal to or greater than the TTLC, and nonhazardous if the total concentration for a given parameter is less than the TTLC but higher than or equal to the STLC, the sample is analyzed further using the California Waste Extraction Test (WET). If the concentration in the WET extract exceeds the STLC, the substance is deemed hazardous.[4]

The subject samples were digested and analyzed for total metals content by EPA Methods 3050 and 3010 as specified in CCR, Title 22, Section 66700 (b). In both cases the heavy metals content was less than the TTLC and the STLC. Therefore, the WET was not performed. The sample results and the regulated

Table 15.4. Volatile and Semivolatile Organic Compounds

Compound	D.L.	μg/Kg Exper.	Cntrl
o-Xylene	0.5	nd	nd
Isopropylbenzene (cumene)	0.5	nd	nd
Indan	0.5	nd	nd
Neopentylbenzene	0.5	nd	nd
1,2,3,4-Tetramethylbenzene	0.5	nd	nd
Naphthalene	0.5	nd	nd
2-Methylnaphthalene	0.5	nd	nd
1-Methylnaphthalene	0.5	nd	nd
1,3,5-Trisopropylbenzene	.05	nd	nd
Biphenyl	0.5	nd	nd
2,6-Dimethylnaphthalene	0.5	0.58	1.41
Hexamethylbenzene	0.5	nd	nd
Acenapthene (1,2-Dihydroacenaphthylene)	0.5	nd	nd
2,3,5-Trimethylnaphthalene	0.5	1.60	2.08
Fluorene	0.5	nd	nd
2-Methylfluorene	0.5	nd	nd
Dibenzothiophene	0.5	nd	nd
Phenanthrene	0.5	0.74	1.66
Anthracene	0.5	nd	nd
Fluoranthene	0.5	nd	nd
Pyrene	0.5	nd	nd
2,3-Benzofluorene	0.5	nd	nd
1,1′-Binaphthalene	0.5	nd	nd
Benzo(a)anthracene	0.5	0.92	1.29
Chrysene/triphenylene	1.0	nd	nd
Benzo(b + k)fluoranthene	1.0	nd	nd
Benzo(e)pyrene	1.0	nd	nd
Benzo(a)pyrene	1.0	nd	nd
Perylene	2.0	nd	nd
9,10-Diphenylanthracene	2.0	nd	nd
1,2,5,6-Dibenzanthracene	2.0	nd	nd
Benzo(g,h,i)perylene	4.0	nd	nd
Anthanthrene	4.0	nd	nd
Coronene	5.0	nd	nd
1,2,4,5-Dibenzopyrene	5.0	nd	nd

concentration limits are summarized in Table 15.5. These results indicate that the pavement samples would not be considered hazardous due to heavy metals content.

Aquatic Bioassay

Static aquatic bioassay tests were performed on both the experimental and control samples. The samples were milled to pass through a number 10 sieve and suspended in the test solution in accordance with procedures prescribed in *Standard Methods for the Examination of Water and Wastewater.* Test mixtures containing

Table 15.5. Total Threshold Limit Concentration

Element	Experimental (mg/kg)	Control (mg/kg)	TTLC (mg/kg)	STLC (mg/kg)
Antimony	<1.00	<1.00	500	15
Barium	1.20	2.50	10,000	100
Beryllium	0.01	0.01	75	0.75
Cadmium	<0.02	<0.02	100	1.0
Chromium VI	[a]	[a]	500	5
Chromium (total)	0.13	0.12	2,500	560
Cobalt	<0.30	<0.30	8,000	80
Copper	0.08	0.08	2,500	25
Lead	<0.50	<0.50	1,000	5.0
Molybdenum	<0.70	<0.70	3,500	350
Nickel	<0.20	0.25	2,000	20
Silver	<0.02	<0.02	500	5
Vanadium	0.57	0.56	2,400	24
Zinc	0.97	0.37	5,000	250
Iron	150	120	—	—
Arsenic	0.18	0.18	500	5
Selenium	<0.01	<0.01	100	1.0
Mercury	.202	.143	20	0.2
Thallium	<0.70	<0.70	700	7.0

[a]The total chromium was less than the hexavalent limit.

250 milligrams per liter (mg/L) and 750 mg/L of each sample were prepared from tap water and introduced into 10 L tanks. Ten fathead minnows, *Pimephales promelas,* were placed into each tank approximately 20 minutes after the samples were introduced. Temperature, pH, and dissolved oxygen were measured every 24 hours. A zero mortality rate was reported after 96 hours for each tank.

REGULATORY REVIEW

A review of the California Health and Safety Code, Sections 25250.1(a) (2) and (5), and (b) revealed that the fuel oil contaminated soil qualified as "used oil" and as a "recyclable oil." Finally, Section 25250.1(e) states "Used oil which meets the standards set in subdivision (c), is not hazardous pursuant to the criteria adopted pursuant to Section 25141 for constituents other than those listed in subdivision (c), and is not mixed with any waste listed as hazardous in Part 261 of Chapter 1 of Title 40 of the Code of Federal Regulations (RCRA) *is not regulated by the Department.*"

The fuel oil contaminated soil met the above criteria as an unregulated recyclable product which could be used for other beneficial purposes. The tests performed in this study demonstrated that fuel oil contaminated soil could be safely mixed with cold mix asphalt for road construction.

ROAD BUILDING

Over a period of several days, the stockpiled sands were mixed and aerated to reduce the moisture content and to obtain a homogeneous blend. The road bed was graded and the blended sands were mixed by blading and dragging over the road bed surface, graded to a uniform thickness, sprayed with the asphalt, and then bladed into a windrow. The soil was picked up from the windrow with a travelling pugmill which mixed the slow curing (SC) liquid asphalt, spread across the surface at a uniform 3 inches thickness and compacted (Figure 15.2). Curing occurred over 6 to 12 months. The durability of the pavement was evaluated in the spring of 1990, and it was determined that there was no need for a top coat.

Figure 15.2. Blading of the newly mixed slow cure liquid asphalt with fuel oil contaminated soil.

CONCLUSIONS AND FUTURE STUDIES

The study results indicate that sandy soils contaminated with residual fuel oil can be successfully stabilized with liquid asphalts, and the resultant pavement does not pose a significant environmental risk. Study results indicate that both experimental and control asphalt pavements would not be classified as hazardous under current California hazardous waste testing protocol. The PNAH content observed in this study may reflect the characteristics of the SC liquid asphalt and may not be characteristic of other asphalt grades.

After one year of use, the demonstration road performed well within specifications with no sign of deterioration. Although classified as a light duty road, heavy equipment has been using the road on a routine basis. However, the purpose of the road does not require much turning, which can be quite damaging to SC pavements. A need has not been demonstrated yet for pavement upgrading with a chip seal or other improved surfaces. However, the thermal loading of the summer months may significantly plasticize the pavement. We feel it is prudent to continue to monitor the performance of the demonstration road and to conduct additional research on other petroleum contaminates mixed with asphalts.

The technology cannot be universally applied to all petroleum contaminated soils, and is not yet applicable to contaminated clay soils. The applicability of this technology to other petroleum contaminates such as gasoline was not determined by this study. The importance of bench-scale testing for particular site conditions cannot be overemphasized. Tests need to be performed on other hydrocarbons such as diesel oil, gasoline, and mineral oils for asphalt mixing suitability. Less suitable petroleum products could be blended with more suitable materials to create pavements that meet design criteria. Blending experiments will also be conducted with less suitable soils such as silts and clays that are contaminated with a host of petroleum products.

Air Emissions

As was the case in this study, the contaminated soils may have to be aerated to reduce the moisture content to acceptable levels. Although the cold-mix asphalt paving process may not be subject to Air Pollution Control District requirements, the excavation and aeration of petroleum contaminated soils may be regulated. Since the subject study was conducted, the South Coast Air Quality Management District in Los Angeles has implemented Rule 1166 which requires a 90% reduction in VOC emissions from the excavation of regulated contaminated soils. Regulated soils include those which emit VOCs in concentrations exceeding 50 ppm when measured at a distance of up to 3 inches from the surface with an organic vapor analyzer (calibrated to hexane). Exemptions are provided for soils contaminated with organic compounds that have an initial boiling point of 302°F or greater, provided the soil is not heated. The #6 fuel oil contaminated soils met the provisions for exemption.

REFERENCES

1. Monismith, C. L. ''Asphalt Paving Mixtures, Properties, Design, and Performance,'' Institute of Transportation and Traffic Engineering, Course Notes, Berkeley, 1961–1962.
2. ''Bituminous Emulsions for Highway Pavements,'' Transportation Research Board, National Cooperative Highway Research Program, Synthesis 30, 1975.
3. Asphalt Institute Manual Series No. 14 (MS-14), Specification RM-3, 1969.
4. California Code of Regulations, Title 22, Article 11.

CHAPTER 16

Soil Venting at a California Site: Field Data Reconciled with Theory

Paul C. Johnson, Curtis C. Stanley, and **Dallas L. Byers,** Shell Development/ Shell Oil Company, Westhollow Research Center, Houston, Texas

David A. Benson and **Michael A. Acton,** Applied Geosciences, Inc., Tustin, California

INTRODUCTION

The development and evaluation of soil treatment technologies is being driven by regulatory demands, which often require or suggest that residual total petroleum hydrocarbon (TPH) concentrations in soil be reduced below 1000 mg/kg. In some areas this limit is as low as 100 mg/kg TPH; lower concentrations are mandated for other specific compounds (i.e., PCBs, metals, etc.). Thermal desorption, incineration, in situ soil venting, solidification, stabilization, biodegradation, and soil washing are examples of the many processes currently being studied.

In situ soil venting, or vacuum extraction, is the focus of this chapter. As illustrated in Figure 16.1, typical components include: vapor extraction (recovery) wells installed within the vadose (unsaturated) zone, blowers or vacuum pumps, and vapor treatment systems. By applying a vacuum at the vapor extraction well, contaminant vapors are removed, the natural rate of volatilization in the soil is enhanced, and residual hydrocarbon concentrations are reduced. More complex systems may utilize surface seals (polymer liners, asphalt, etc.), horizontal wells, and forced or passive air injection wells.

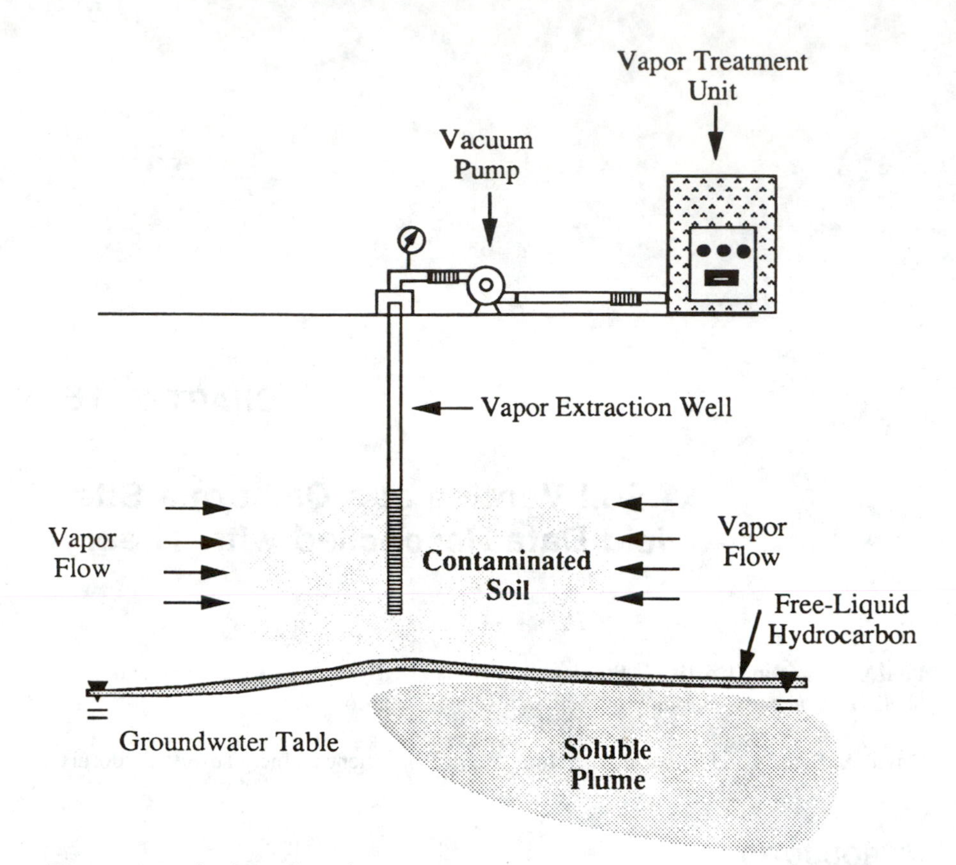

Figure 16.1. In situ soil venting system.

While soil venting-related articles began to appear in peer-reviewed journals and conference proceedings in the early 1980s, soil venting is no different in principle than the vapor extraction processes practiced much earlier for the control of contaminant vapor migration into buildings. The majority of articles are anecdotal reports of field applications; many of these have been summarized by Hutzler, et al.[1] In general, many of the sites can be characterized as gasoline or solvent spills in relatively permeable (sandy) soils, the process is reported to have been "successful," although few report the final cleanup levels achieved, and not enough information is presented to allow one to generalize to other spill sites or contaminant types. Most reports are of the type described above; however, several authors have attempted to elucidate the governing phenomena, through modeling and controlled laboratory- and field-scale experiments. Some of the more noteworthy studies include: the controlled gasoline spill field-scale experiments reported by Thornton and Wootan,[2] Marley and Hoag's soil column venting experiments,[3] Terra Vac's Superfund Innovative Technology Evaluation (SITE) Program demonstration,[4] laboratory experiments conducted by Parker, et al. to assess mass transfer related limitations,[5] and the modeling studies by Johnson, et al.,[6] Wilson, et al.,[7] Massmann,[8] and Marley, et al.[9]

In 1987, Shell Oil Company initiated a study of this remediation process. The goal of this study was to answer the following questions:

- *What processes govern the behavior observed during soil venting applications?*
- *What are the limitations/advantages of this process?*
- *Can remediation by soil venting achieve regulated target cleanup levels?*
- *What impact, if any, will remaining residual levels have on the environment?*
- *What information/data is needed to answer these questions?*

In addition, it was envisioned that answers to these questions could be used in the development of an approach to (a) identify sites at which soil venting could be effectively practiced, and (b) design efficient soil venting systems. This study, which is still in progress, consists of three phases: (1) mathematical modeling to identify the relevant phenomena and formulate simple predictive equations to be used as screening tools for predicting process variables (flowrates, contaminant removal rates, etc.), (2) laboratory-scale experiments, and (3) field applications at two service station field sites. Modeling results have been previously reported,[6,10] as well as a suggested approach for designing, monitoring, and operating soil venting systems.[10] In this chapter we will focus on a comparison of field observations with screening model predictions. Results of the laboratory studies will be the subject of another report.

The organization of this chapter follows the decision process flow diagram suggested by Johnson, et al.,[10] which is reproduced in Figure 16.2. While studies were conducted at two service stations, for clarity, only one is discussed in this chapter.

BACKGROUND REVIEW AND SITE CHARACTERIZATION

In February 1986, underground steel gasoline storage tanks were excavated and replaced at a service station located in Costa Mesa, California. Soil samples from beneath the tanks contained elevated levels of gasoline-range hydrocarbons, although the excavated tanks showed no signs of failure. Additional sampling of the soil and groundwater in May 1987 indicated that the soil contamination extended down into the saturated zone, and high levels (>50 mg/L) of total gasoline-range hydrocarbons were detected in the groundwater.

This site was identified as being appropriate for the in situ soil venting field study. It will be shown that site characteristics differ significantly from those

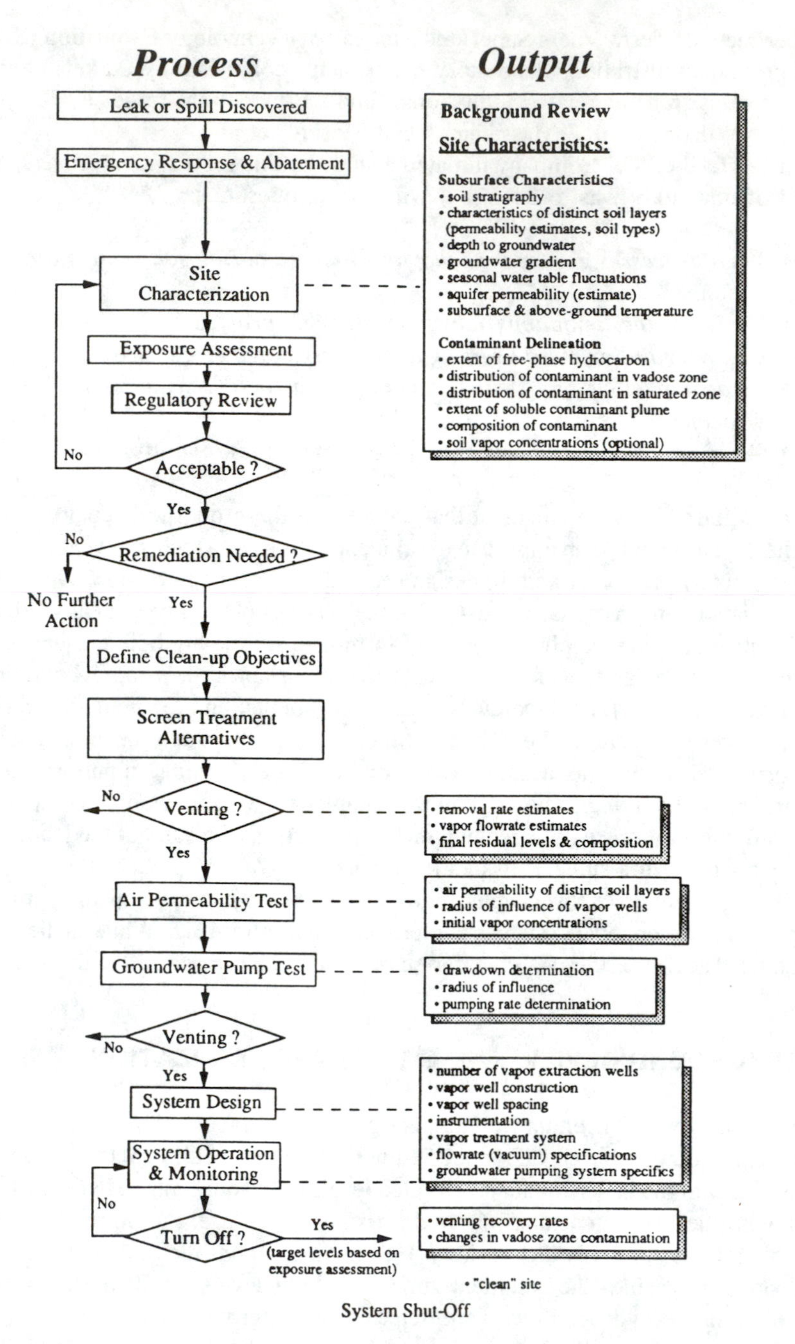

Figure 16.2. In situ soil venting design process (from Johnson et al., 1990).

typically reported for "successful" soil venting applications. Subsequently, in January to March 1988, an extensive site characterization program was conducted. During this phase several field analytical methods were tested, and the following site-specific parameters were defined:

Subsurface Characteristics

- soil stratigraphy
- characteristics of distinct soil layers (soil types, permeability estimates)
- depth to groundwater
- groundwater gradient
- aquifer permeability estimate
- temperature
- seasonal water table fluctuations

Contaminant Delineation

- extent of free-phase contamination
- contaminant distribution in vadose zone
- contaminant distribution in saturated zone
- extent of soluble contaminant plume
- composition of contaminant
- soil vapor concentrations

A site plan illustrating the former and present tank locations, soil boring locations, and other relevant features of this site is given in Figure 16.3. Figure 16.4 presents the soil types encountered in each distinct soil layer, and measured total petroleum hydrocarbon (TPH) concentrations along the HB–17–HB–3 cross section. Residual TPH concentrations were measured in the field by analyzing the vapors in a heated-headspace cell with a portable gas chromatography instrument equipped with a flame ionization detector (GC-FID). Residual TPH levels determined in this way compared well with duplicate samples analyzed by a certified laboratory. An unusually large number of soil borings were drilled with the hope that the effects of soil venting could be quantified at the completion of the study.

The information collected during the site investigation can be summarized as follows:

- as illustrated in Figure 16.4, there are four distinct geologic units at this site: sandy clay is found from 0 to 10 ft below ground surface (BGS); fine to coarse sand occupies 10 to 30 ft BGS; a silty clay and clayey silt layer lies between 30 and 42 ft BGS; and fine to medium sand is detected from 42 ft to the deepest boring depth (60 ft). Groundwater was observed at 48 ft BGS prior to venting.

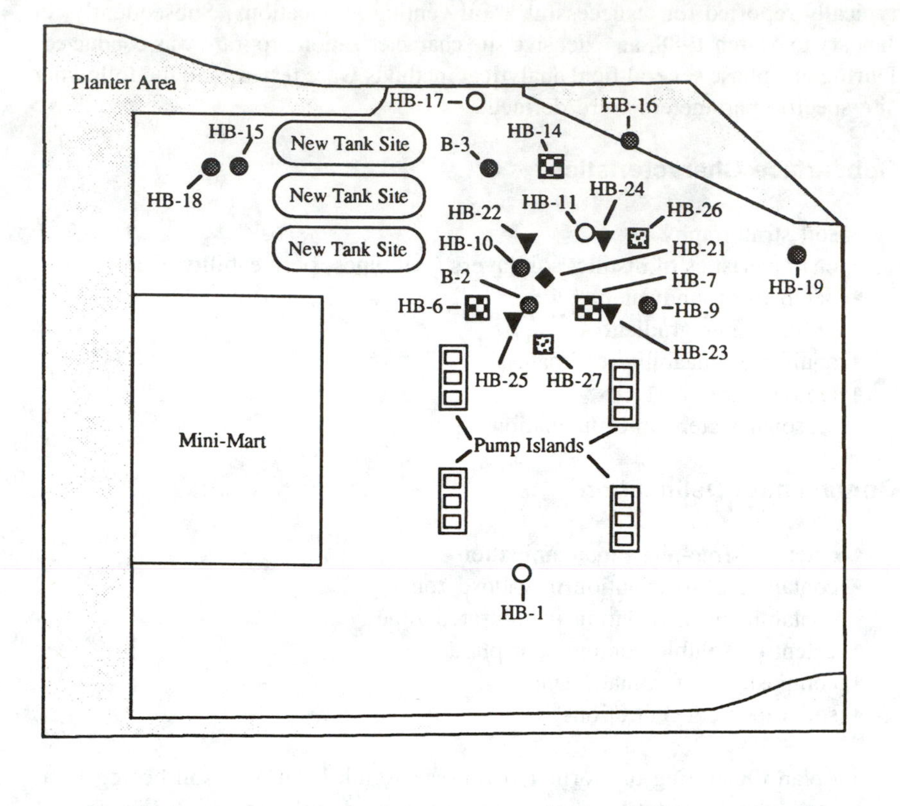

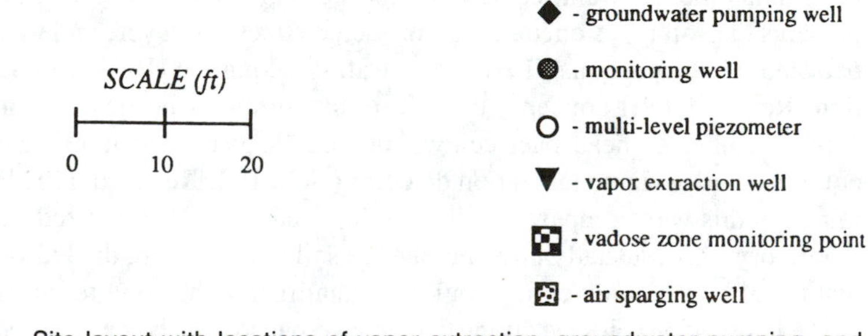

Figure 16.3. Site layout with locations of vapor extraction, groundwater pumping, and monitoring wells.

- hydrocarbons were detected from ground surface down to 60 ft BGS, with the highest levels of hydrocarbons located in the region 35 ft to 60 ft BGS (encompassing the clayey and sandy soil zones). Based on the data given in Figure 16.4, it is estimated that ~4000 kg (1500 gal) of gasoline was present near the water table, and ~150 kg (60 gal) was present in the 0 to 30 ft region below ground surface.

- free-liquid gasoline was observed floating on the groundwater, and was subsequently removed by pumping. Table 16.1 presents results of a GC analysis of a sample of the gasoline. The composition given in Table 16.1 should be regarded as an estimate of the true composition because approximately 20% of the compounds could not be positively identified. The contributions of the "unknown" compounds were subsequently added to known, positively identified compounds with retention times close to those of the unknown compounds. For comparison, the composition of a "fresh" gasoline (i.e., that taken from a gas pump) is also listed in Table 16.1. As expected, the fresh gasoline contains a larger fraction of lower boiling point compounds (propane-isopentane) than the free product sample from this site.

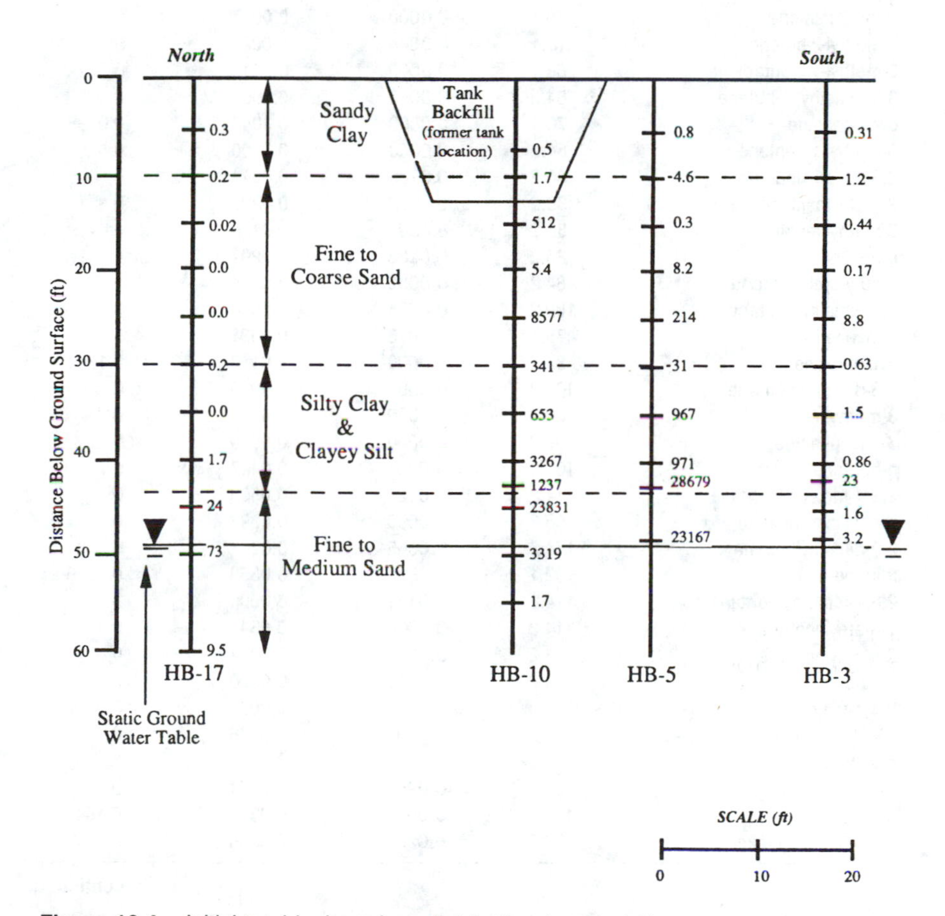

Figure 16.4. Initial total hydrocarbon distribution (mg/kg-soil).

Table 16.1. Composition of the Costa Mesa Gasoline Sample and a Fresh Gasoline

Compound Name	Mw (g)	Fresh Gasoline	Costa Mesa Sample Gasoline	Approximate Composition
propane	44.1	0.0001	0.0000	0
isobutane	58.1	0.0122	0.0000	0
n-butane	58.1	0.0629	0.0000	0
trans-2-butene	56.1	0.0007	0.0000	0
cis-2-butene	56.1	0.0000	0.0000	0
3-methyl-1-butene	70.1	0.0006	0.0000	0
isopentane	72.2	0.1049	0.0069	0.0177
1-pentene	70.1	0.0000	0.0005	0
2-methyl-1-butene	70.1	0.0000	0.0008	0
2-methyl-1,3-butadiene	68.1	0.0000	0.0000	0
n-pentane	72.2	0.0586	0.0095	0
trans-2-pentene	70.1	0.0000	0.0017	0
2-methyl-2-butene	70.1	0.0044	0.0021	0
3-methyl-1,2-butadiene	68.1	0.0000	0.0010	0
3,3-dimethyl-1-butene	84.2	0.0049	0.0000	0
cyclopentane	70.1	0.0000	0.0046	0.0738
3-methyl-1-pentene	84.2	0.0000	0.0000	0
2,3-dimethylbutane	86.2	0.0730	0.0044	0
2-methylpentane	86.2	0.0273	0.0207	0
3-methylpentane	86.2	0.0000	0.0186	0
n-hexane	86.2	0.0283	0.0207	0
methylcyclopentane	84.2	0.0083	0.0234	0
2,2-dimethylpentane	100.2	0.0076	0.0064	0
benzene	78.1	0.0076	0.0021	0
cyclohexane	84.2	0.0000	0.0137	0.1761
2,3-dimethylpentane	100.2	0.0390	0.0000	0
3-methylhexane	100.2	0.0000	0.0355	0
3-ethylpentane	100.2	0.0000	0.0000	0
n-heptane	100.2	0.0063	0.0447	0
2,2,4-trimethylpentane	114.2	0.0121	0.0503	0
methylcyclohexane	98.2	0.0000	0.0393	0
2,2-dimethylhexane	114.2	0.0055	0.0207	0
toluene	92.1	0.0550	0.0359	0.1926
2,3,4-trimethylpentane	114.2	0.0121	0.0000	0
3-methylheptane	114.2	0.0000	0.0343	0
2-methylheptane	114.2	0.0155	0.0324	0
n-octane	114.2	0.0013	0.0300	0
2,4,4-trimethylhexane	128.3	0.0087	0.0034	0
2,2-dimethylheptane	128.3	0.0000	0.0226	0
ethylbenzene	106.2	0.0000	0.0130	0
p-xylene	106.2	0.0957	0.0151	0
m-xylene	106.2	0.0000	0.0376	0.1641
3,3,4-trimethylhexane	128.3	0.0281	0.0056	0

continued

Table 16.1. Continued

Compound Name	Mw (g)	Fresh Gasoline	Costa Mesa Sample Gasoline	Approximate Composition
o-xylene	106.2	0.0000	0.0274	0
2,2,4-trimethylheptane	142.3	0.0105	0.0012	0
n-nonane	128.3	0.0000	0.0382	0
3,3,5-trimethylheptane	142.3	0.0000	0.0000	0
n-propylbenzene	120.2	0.0841	0.0117	0.1455
2,3,4-trimethylheptane	142.3	0.0000	0.0000	0
1,3,5-trimethylbenzene	120.2	0.0411	0.0493	0
1,2,4-trimethylbenzene	120.2	0.0213	0.0705	0
n-decane	142.3	0.0000	0.0140	0
methylpropylbenzene	134.2	0.0351	0.0170	0
dimethylethylbenzene	134.2	0.0307	0.0289	0.0534
n-undecane	156.3	0.0000	0.0075	0
1,2,4,5-tetramethylbenzene	134.2	0.0133	0.0056	0
1,2,3,4-tetramethylbenzene	134.2	0.0129	0.0704	0.1411
1,2,4-trimethyl-5-ethylbenzene	148.2	0.0405	0.0651	0
n-dodecane	170.3	0.0230	0.0000	0
naphthalene	128.2	0.0045	0.0076	0
n-hexylbenzene	162.3	0.0000	0.0147	0.0357
methylnapthalene	142.2	0.0023	0.0134	0
Total		1.0000	1.0000	1.0000

VADOSE/SATURATED ZONE INSTRUMENTATION

Vadose zone monitoring wells were installed in three of the soil borings during the site characterization phase. The locations are shown in Figure 16.3, and details of construction are shown in Figure 16.5a. Each "well" is a 50 ft long 1-in PVC pipe to which are attached three Teflon® tubing lines, and three thermocouple lines. The lines extend from the ground surface to depths of roughly 10 ft, 25 ft, and 40 ft. At ground surface these lines are attached to quick-disconnect couplings, and are used to monitor the subsurface pressure, temperature, and vapor concentrations.

Several groundwater monitoring wells were also fitted with special caps so that the groundwater table level and subsurface pressure could be measured during operation of the venting system. Special caps were fabricated because the monitoring wells must remain sealed during the water level measurement. Uncapping a well releases the vacuum and any effect that it has on the groundwater level at that point. Figure 16.5b is a schematic of the "AGI Custom Cap," which is constructed from a commercially available monitoring well cap, and allows one to use a standard electronic water level sensor.

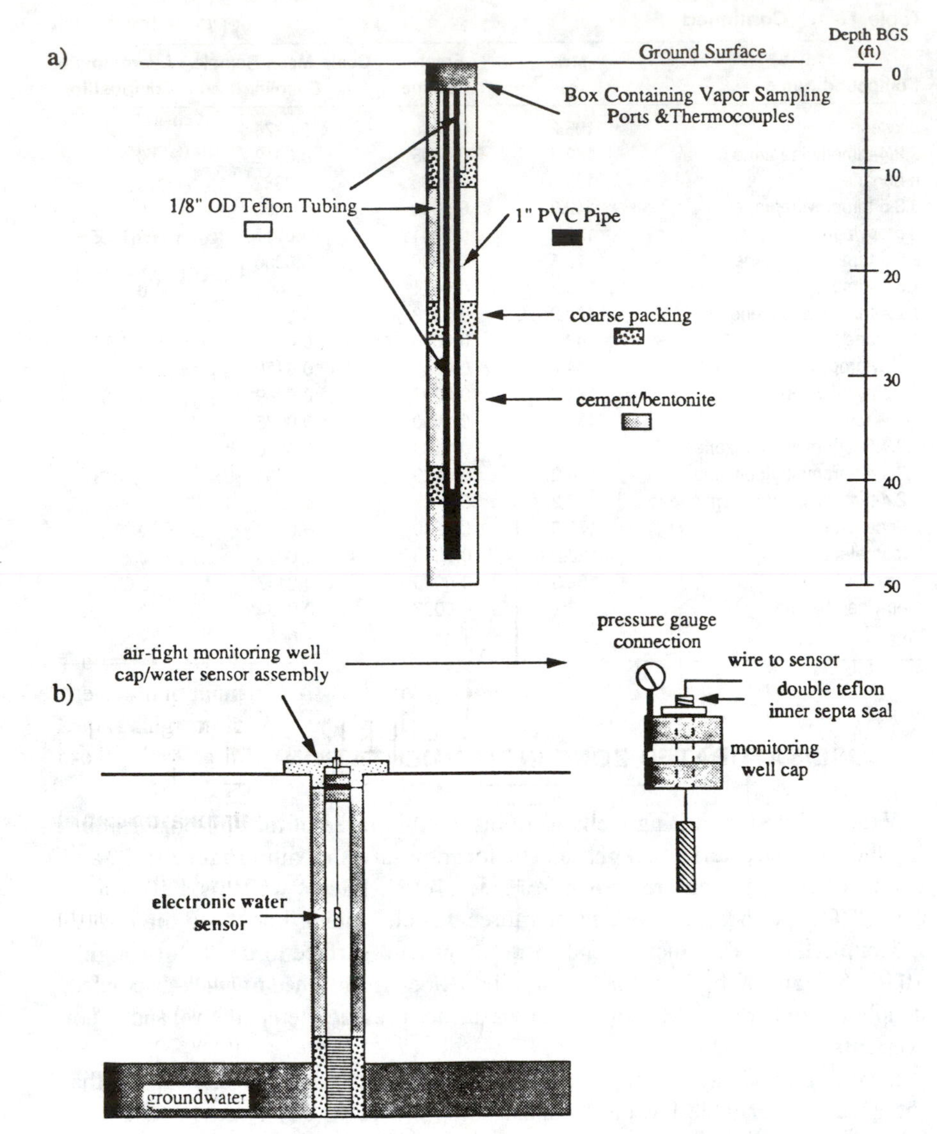

Figure 16.5. (a) Vadose zone monitoring installation, and (b) air-tight groundwater level measuring system.

REVIEW OF REMEDIATION OPTIONS

Based on the site characterization results presented in Figure 16.4 it should be clear that remediation was required at this site. Of the available options, an in situ method was preferable, given the depth of contamination. Site characteristics, however, differ significantly from those of the soil venting "success" stories found in the literature. At this site we encounter three significantly different situations. The upper fine sandy zone (10 to 30 ft BGS) is likely to be ideal for venting due to its high air permeability, the low permeability silty clay zone (30 to 40 ft BGS) is not a good candidate, and cleanup of the sandy zone beneath the clay at ~45 ft BGS is a challenge because the unsaturated soil (which is expected to be permeable to vapor flow) is confined within a 5 ft region between the clay zone and groundwater. Given the variety of distinct soil zones and their associated challenges, this site was a good candidate for the soil venting study. From practical considerations, it is also likely that soil venting was the only feasible alternative among the set of available in situ technologies (biodegradation, soil washing, steam stripping, etc.).

In general, it is suggested that the thought procedure outlined in Johnson et al.[10] be used to help decide if soil venting is applicable.

COMPARISON OF PREDICTED AND MEASURED SOIL GAS CONCENTRATIONS

Soil gas samples were obtained from the monitoring wells prior to operation of the venting system. The vapor concentrations and compositions measured on 1/24/89, as determined by GC-FID analysis, are given in Table 16.2. These values were compared with the expression:[10]

$$C_{est} = \sum_i \frac{x_i P_i^v M_{w,i}}{RT} \tag{1}$$

where:

C_{est} = estimate of contaminant vapor concentration [mg/L]
x_i = mole fraction of component in liquid-phase residual (x_i = 1 for single compound)
P_i^v = pure component vapor pressure at temperature T [atm]
$M_{w,i}$ = molecular weight of component i [mg/mole]
R = gas constant = 0.0821 1-atm/mole-°K
T = absolute temperature of residual [°K]

Utilizing the measured composition given in Table 16.1:

$$C_{est} = 220 \text{ mg/L } (20°C)$$

As can be seen, the predicted value agrees well with the measured values in Table 16.2, for sampling points located close to the zone of contamination. While the "exact" composition was used in this prediction, it should be noted that the value predicted for the nine-component "approximate" distribution given in

Table 16.2. Vadose Zone Monitoring Well Vapor Concentrations and Compositions

Well ID	Date	Distance from Venting Well HB-25 (ft)	Depth (ft)	Total Hydrocarbons (mg/L)	% Vapor Composition Boiling Point Fraction 1	2	3	4	5
HB-6S	1/24/89	8	10	13	15.4	34.6	30.8	19.2	0.0
HB-6S	12/9/89	8	10	0.15	13.3	18.7	25.3	26.0	16.7
HB-6D	1/24/89	8	40	308	32.0	49.1	17.6	1.2	0.0
HB-6D	1/25/89	8	40	168	37.8	50.0	11.8	0.4	0.0
HB-6D	2/28/89	8	40	91.8	27.7	52.9	18.0	1.4	0.0
HB-6D	12/9/89	8	40	37.2	13.1	39.1	38.1	9.2	0.6
HB-7S	1/24/89	15	10	32.7	45.6	33.3	15.9	5.2	0.0
HB-7S	1/25/89	15	10	28.9	51.9	34.6	10.7	2.8	0.0
HB-7S	2/28/89	15	10	11.5	27.0	25.2	32.2	15.7	0.0
HB-7S	12/9/89	15	10	3.2	14.2	38.5	10.7	51.3	20.0
HB-7M	1/24/89	15	25	298	36.2	50.8	12.3	0.7	0.0
HB-7M	1/25/89	15	25	47.4	34.8	26.8	27.0	11.4	0.0
HB-7M	2/28/89	15	25	20.2	17.3	32.7	38.6	11.4	0.0
HB-7M	12/9/89	15	25	16.6	0.5	5.6	46.7	44.7	2.4
HB-7D	1/24/89	15	40	239	43.2	48.3	8.2	0.3	0.0
HB-7D	2/28/89	15	40	241	28.6	42.4	25.5	3.5	0.0
HB-7D	12/9/89	15	40	106	26.9	41.5	27.9	3.4	0.2
HB-14S	1/24/89	30	10	6.3	58.7	27.0	12.7	1.6	0.0
HB-14S	12/9/89	30	10	0.2	27.1	33.3	12.4	14.3	12.9
HB-14M	1/24/89	30	25	18.2	47.8	29.7	21.4	1.1	0.0
HB-14M	12/9/89	30	25	0.42	17.9	34.2	27.6	13.4	6.8
HB-14D	1/24/89	30	40	251	40.0	45.9	13.2	0.8	0.0
HB-14D	1/25/89	30	40	199	43.4	45.0	10.8	0.8	0.0
HB-14D	2/28/89	30	40	170	6.4	44.3	40.4	8.1	0.8
HB-14D	12/9/89	30	40	44	14.3	40.7	36.2	7.7	1.1

Boiling Point Fractions:
1: methane-isopentane (<28°C)
2: isopentane-benzene (28–80°C)
3: benzene-toluene (80–111°C)
4: toluene-xylenes (111–144°C)
5: >xylenes (>144°C)

Table 16.1 is 270 mg/L. This illustrates that one needs only an estimate of the composition (a "boiling point distribution") to reasonably predict the maximum soil gas concentrations.

AIR PERMEABILITY TEST

To confirm the assumption that soil venting could be practiced at this site, and to gain information needed for design considerations, an air permeability test was conducted. Figure 16.6 illustrates the basic equipment setup. A vacuum pump

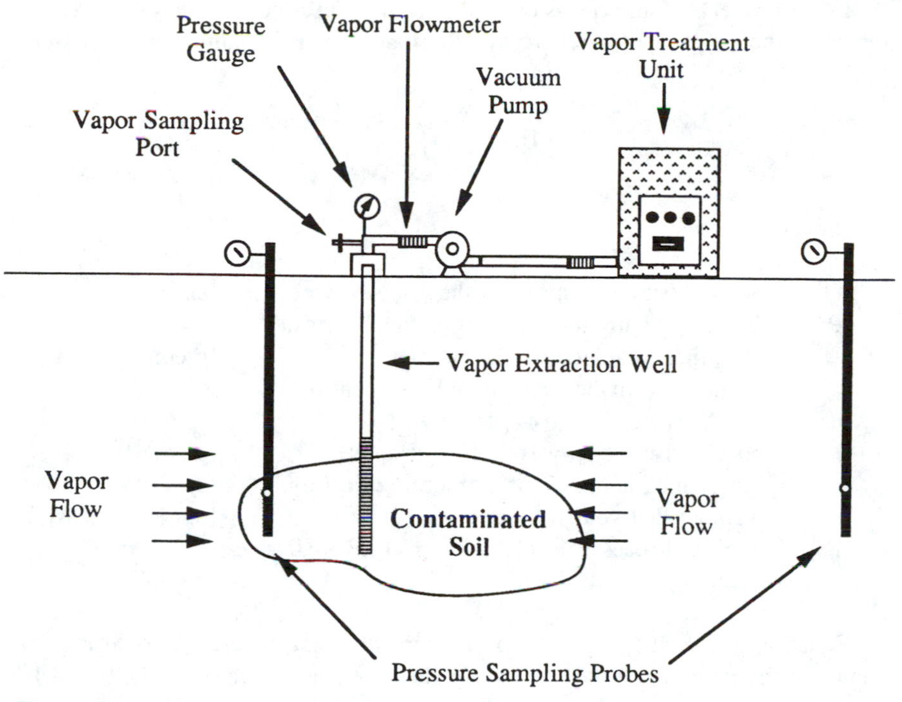

Figure 16.6. Air permeability test setup.

is connected to a well screened through the vadose zone. When the vacuum pump is turned on, vapor flowrates and subsurface vacuums at a few distances away from the extraction well are monitored. The test can also be conducted by injecting air into the subsurface and monitoring the corresponding pressure increase. Data is analyzed, as described below, to obtain estimates of soil permeability and the effective "radius of influence" of the vapor extraction wells.

The air permeability test began on the evening of 8/24/88 and continued through the early morning of 8/25/88. The vacuum pump was connected to an existing groundwater monitoring well (HB-9), which had a screened interval that extended into the unsaturated fine sand zone. For the first 210 min air was withdrawn from the well. The vacuum at the wellhead was equivalent to 45 in H_2O (= 0.89 atm absolute pressure). After 210 min, the vacuum pump was turned off and the vadose zone pressure was monitored for 65 min. After this relaxation period, the pump was reversed and air was injected into the ground at a gauge pressure of 83 in H_2O (= 1.20 atm absolute pressure). During each phase of the experiment the pore vapor pressure was recorded with time at vadose monitoring wells HB-14, HB-7, and HB-6. During the injection test, the vapor pressure in monitoring well HB-10 was also monitored. Unfortunately, the total vapor flowrate during the vacuum extraction test was not measured accurately, but was estimated to be 15 scfm.

In Johnson et al.,[6] an expression is derived that describes the transient development of the pore vapor pressure distribution about a vacuum extraction well:

$$P' = \frac{Q}{4\Pi m(k/\mu)} \int_{\frac{r^2\epsilon\mu}{4kP_{Atm^t}}}^{\infty} \frac{e^{-x}}{x}\, dx \qquad (2)$$

In Equation 1:

Q	=	volumetric flowrate from the vacuum well	$[cm^3/s]$
P′	=	gauge pressure measured at distance r from the injection/extraction well	$[g/cm\text{-}s^2]$
r	=	distance from the well where P′ is measured	[cm]
k	=	soil permeability to vapor flow	$[cm^2]$
μ	=	vapor viscosity ($= 1.8\times10^{-4}$ g/cm-s)	[g/cm-s]
t	=	time since initiation of extraction/injection	[s]
ϵ	=	vapor-filled void fraction	[dimensionless]
P_{Atm}	=	ambient pressure (1 atm $= 1.01\times10^6$ $g/cm\text{-}s^2$)	$[g/cm\text{-}s^2]$

Equation 2 is equally valid for an air injection test, where Q would then be equal to the negative of the volumetric flowrate injected into the well, and P′ would be equal to the gauge pressure measured at the distance r away from the injection point. For $(r^2\epsilon\mu/4ktP_{atm})<0.1$, Equation 2 can be approximated by:

$$P' = \frac{Q}{4\Pi m(k/\mu)}\left[-0.5772 - 1n\left(\frac{r^2\epsilon\mu}{\propto\kappa\Pi_{A\tau\mu}}\right) + 1n(t)\right] \qquad (3)$$

When P′ is plotted as a function of 1n(t), therefore, the slope A, and y-intercept B, are equal to:

$$A = \frac{Q}{4\Pi m(k/\mu)} \qquad B = \frac{Q}{4\Pi m(k/\mu)}\left[-0.5772 - 1n\left(\frac{r^2\epsilon\mu}{4kP_{Atm}}\right)\right] \qquad (4)$$

The permeability, k, can be calculated from one of two methods. In the first, the value of the slope of the P′-vs-1n(t) curve is used along with known values of Q, m, and μ. The second method involves the values of the slope and the y-intercept:

$$\text{Method 1: } k = \frac{Q\mu}{4A\Pi\mu}$$

$$\text{Method 2: } k = \frac{r^2\epsilon\mu}{4P_{Atm}} \exp\left[\frac{B}{A} + 0.5772\right] \qquad (5)$$

At steady-state, the radial pressure distribution is expected to obey the following relationship:[6]

$$P'(r) = P_w \left[1 + (1 - (P_{Atm}/P_w)^2) \frac{\ln(r/R_w)}{\ln(R_w/R_I)} \right]^{1/2} - P_{Atm} \qquad (6)$$

where:

$P'(r)$ = gauge pressure at a distance r from vacuum/injection well
P_w = absolute pressure at vacuum/injection well
P_{Atm} = absolute ambient pressure
R_I = radius of influence
R_w = radius of vacuum/injection well

Figure 16.7 presents the field data for the vacuum and injection tests. Only the pressure readings from the lowest pressure sensor on each vadose monitoring well are presented, because no response was detected at any other zone. Recall that the lowest pressure sensor lies in the permeable fine sand zone just above the water table, and this is the only zone where we expect vapor flow to occur. Because the vapor flowmeter was inadvertently not placed at the wellhead, Method 2 (see Equation 5) was used to calculate permeabilities. This approach does not require values for the vapor flowrate (Q), or the thickness of the zone through which air flows (m). The calculated k values appear in Table 16.3. Most values are between 2 darcys and 10 darcys, with the average being about 8 darcys (excluding the k = 270, darcy HB-7D vacuum test data). These values are typical of medium to fine sands, which is consistent with the well boring logs from this site.

While all field data points were used to obtain the permeability values, it is useful to check if the restriction $(r^2\epsilon\mu/4ktP_{atm}) < 0.1$ is satisfied. Based on a permeability value of k = 8 darcys, the time $t^* = (10r^2\epsilon\mu/4kP_{Atm})$ required to

Table 16.3. Permeability Values Derived from Field Data at the Costa Mesa Site

Test	Well	Distance from Vacuum/Injection Well (ft)	k (darcys)	t* (min)
Vacuum	HB-7D	11	270	3
Vacuum	HB-6D	53	2	75
Vacuum	HB-14D	32	23	27
Injection	HB-7D	11	2	3
Injection	HB-6D	53	9	75
Injection	HB-14D	32	7	27
Injection	HB-10	25	8	16

(Vacuum/injection well was HB-9.)
$t^* = 10(r^2\epsilon\mu/4kP_{atm}.)$

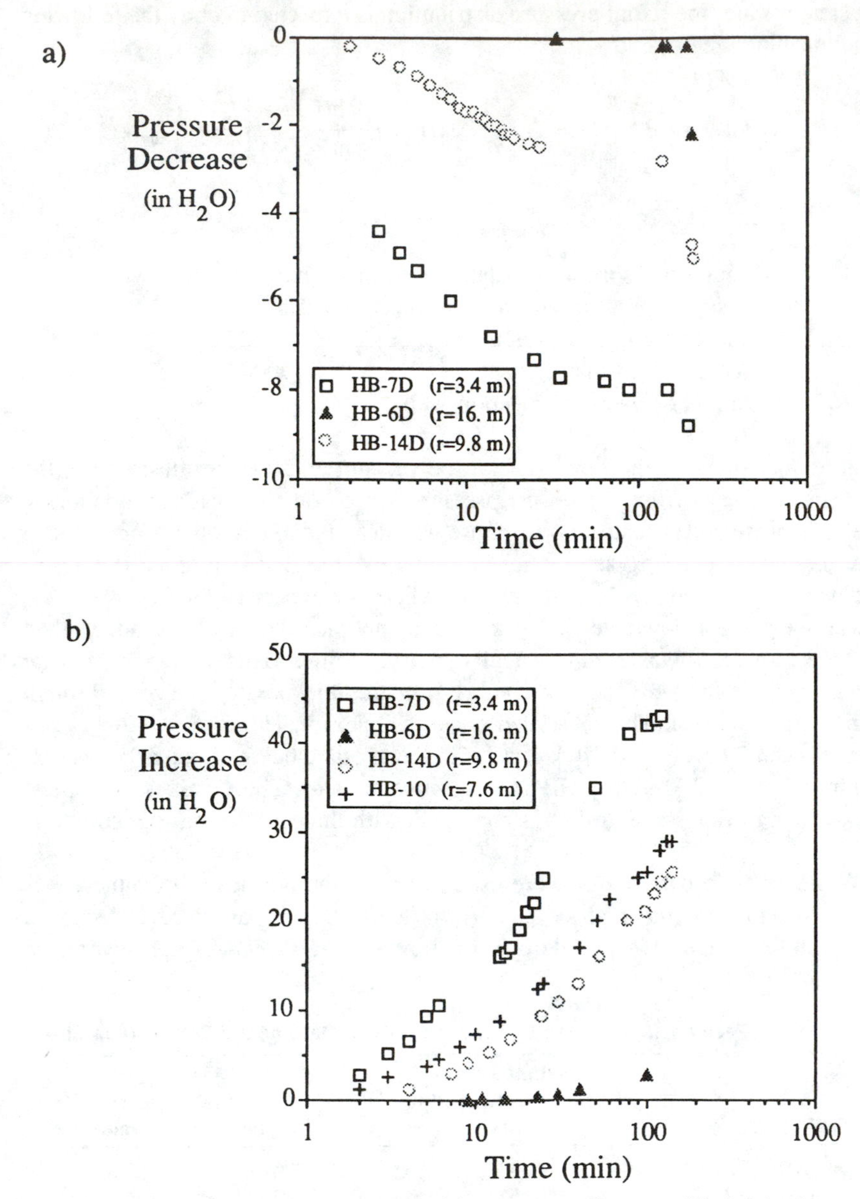

Figure 16.7. Air permeability test results: (a) vapor extraction test; (b) air injection test. *Note:* (in H_2O) denotes vacuums expressed as equivalent water column heights.

satisfy the restriction at each vadose well was calculated. The results are also listed in Table 16.3. With the exception of well HB-6D, which is located 53 ft from the vacuum/injection well, all t* values are less than the duration of the vacuum/injection tests.

Figure 16.8 presents the radial pressure distribution measured at the end of the vacuum and injection tests. These are compared with radial pressure distributions calculated from Equation 6 with R_I = 60 ft. As can be seen, the predicted and measured distributions agree reasonably well.

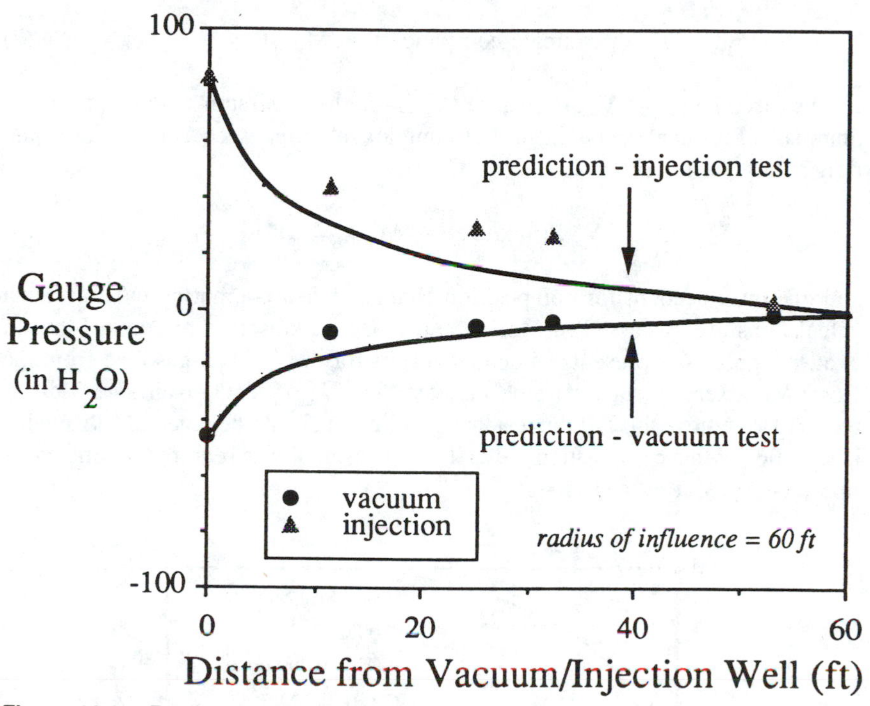

Figure 16.8. Predicted and measured radial pressure distributions.

SYSTEM DESIGN

In Johnson et al.,[10] it is suggested that soil venting system designs be based on site geology, soil characteristics, location of the saturated zone, location of the contaminant, and contaminant type. In brief, the following guidelines are to be followed:

- vapor extraction wells should be placed in the center of the contaminated zone
- vapor extraction well screened intervals should only encompass the zone of contamination
- groundwater extraction wells should be incorporated in the design, if the extraction wells, or contaminated soil, are in the vicinity of the saturated zone

Johnson et al.[10] estimate the minimum number of extraction wells required based on the concept that a minimum volume of air is required to remove a given mass of contaminant. For example, with a pure component contaminant, the maximum vapor concentration that can be extracted is predicted by Equation 1, with x_i = 1. If the estimated spill mass is m_{spill}, then the minimum volume of air, V_{min}, is:

$$V_{min} \text{ (single component)} = M_{spill}/C_{est} \tag{7}$$

In this case, $1/C_{est} = (V_{min})^*$, where $(V_{min})^*$ is the minimum volume of air per unit mass of residual contaminant. For complex mixtures, such as gasoline, Equation 7 must be rewritten as:

$$V_{min} \text{ (mixture)} = M_{spill} (V_{min})^* \tag{8}$$

and one must account for composition changes when computing $(V_{min})^*$. Such calculations are discussed in Marley and Hoag,[3] Johnson et al.,[6] and Johnson et al.[10] Figure 16.9 presents computed results for the residual gasoline from the Costa Mesa service station which indicate that $(V_{min})^* \sim 100$ 1-air/g-residual is required to remove 95% of the residual gasoline. It should be noted that this value is specific for this composition; ~25 1-air/g-residual was required for the gasoline used by Marley and Hoag.[3]

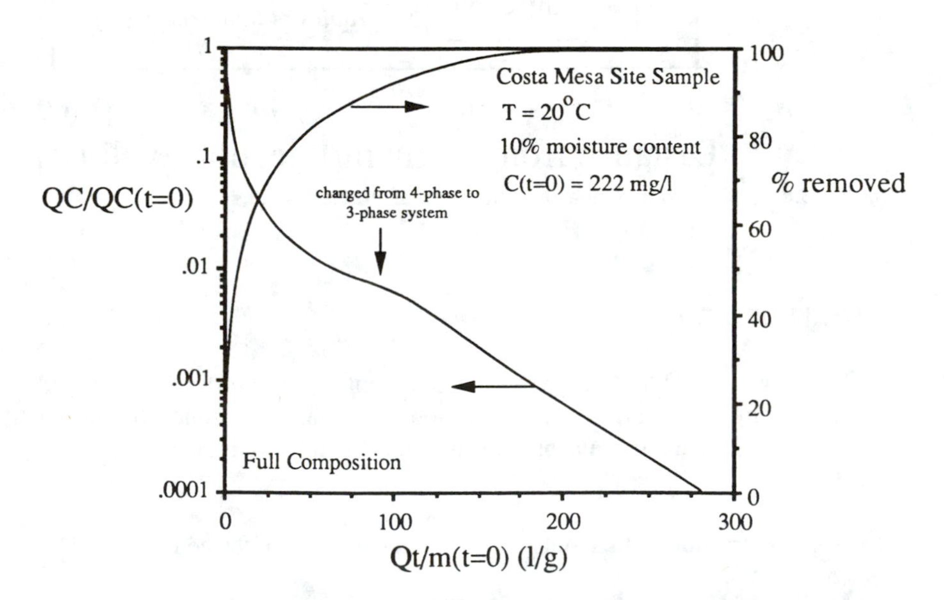

Figure 16.9. Equilibrium model predictions for the free-liquid gasoline sample obtained from the Costa Mesa site.

Given $(V_{min})^*$, M_{spill}, the estimated flowrate from a single vapor extraction well Q_{well}, and the desired duration of remediation, $\tau_{clean\text{-}up}$, the minimum number of required vapor extraction wells N_{well}, is computed from:

$$N_{well}\ (\text{minimum}) = \frac{M_{spill}\ (V_{min})^*}{Q_{well}\ \tau_{clean\text{-}up}} \tag{9}$$

where Q_{well} is estimated, for vertical wells, from (Johnson, et al.)[6]:

$$Q_{well} = H \frac{\pi \kappa}{\mu} P_w \frac{[1 - (P_{Atm}/P_w)^2]}{1n(R_w/R_I)} \tag{10}$$

where:

P_w = absolute pressure at vacuum/injection well
P_{Atm} = absolute ambient pressure
R_I = radius of influence
R_w = radius of vacuum/injection well
H = thickness of screened interval/permeable zone
κ = soil permeability
μ = viscosity of air (= 1.8×10^{-4} g/cm-s)

Consider the upper (10–30 ft BGS), middle (30 to 40 ft BGS), and lower (40 to 45 ft BGS) unsaturated contaminated zones. Based on boring logs and results of the air permeability test, the parameters shown in Table 16.4 characterize these zones.

Table 16.4. Parameters Shown by Results of Air Permeability Test

	Soil Zone		
Parameter	**Upper**	**Middle**	**Lower**
Subsurface interval (ft BGS)	10–30	30–40	40–45
Soil type	medium sand	clayey silt	fine sand
Permeability (darcy)	50	0.1	8
M_{spill} (kg)	150	600	3400
Radius of influence (ft)	40	40	40
Extraction well radius (in.)	2	2	2
Extraction well screen thickness (ft)	10	5	5
Q_{well} (scfm at P_w = 0.90 atm)	220	0.2	17

If a six month remediation period is acceptable, then the minimum number of wells required for each zone can be calculated from Equation 9, using $(V_{min})^* \sim 100$ 1-air/g-residual. The results are shown in Table 16.5.

Table 16.5. Calculation of Minimum Number of Wells Required

Zone	N_{well} (minimum)
Upper	0.010
Middle	40
Lower	2.6

The N_{well} values listed in Table 16.5 illustrate the effect that soil type has on the usefulness of venting. The upper zone can likely be cleaned with a single vapor extraction well, and the lower zone requires more than three. The middle clayey silt zone, however, requires a minimum of 40 vapor extraction wells. Given the areal extent of the contaminated zone, it is clear that the number of wells required to clean the middle zone is not practical.

At the Costa Mesa site the following were installed:

- one vapor extraction well (HB-22) screened from 25 to 30 ft BGS in the upper zone
- two extraction wells (HB-23, HB-24) screened from 33 to 38 ft BGS in the middle zone
- one extraction well (HB-25) screened from 35 to 45 ft BGS
- two air sparging wells (HB-26, HB-27) screened from 55 to 60 ft BGS in the unsaturated zone
- one groundwater pumping well (HB-21)

Locations of each are shown in Figure 16.3, while Figure 16.10 indicates the screened intervals. Well HB-22 was installed for remediating the upper zone; HB-23 and HB-24 were installed to determine if the low permeability soil zone could be remedied by direct venting; HB-25 was installed for remediating the lower zone; HB-26 and HB-27 (air injection wells) were installed to study the effect of air sparging on the residual gasoline trapped beneath the groundwater table; HB-21 was installed to prevent groundwater upwelling induced by a vacuum applied to HB-25.

All vapor extraction wells were manifolded to a single vacuum pump, and the extracted vapors were treated by a thermal oxidizer unit.

SYSTEM MONITORING

In addition to being able to monitor the vapor concentration, temperature, and pore vapor pressure at each vadose zone monitoring location, the venting system installed at the Costa Mesa site allows the following to be measured:

- total system flowrate
- vacuum at each vapor extraction and groundwater monitoring well

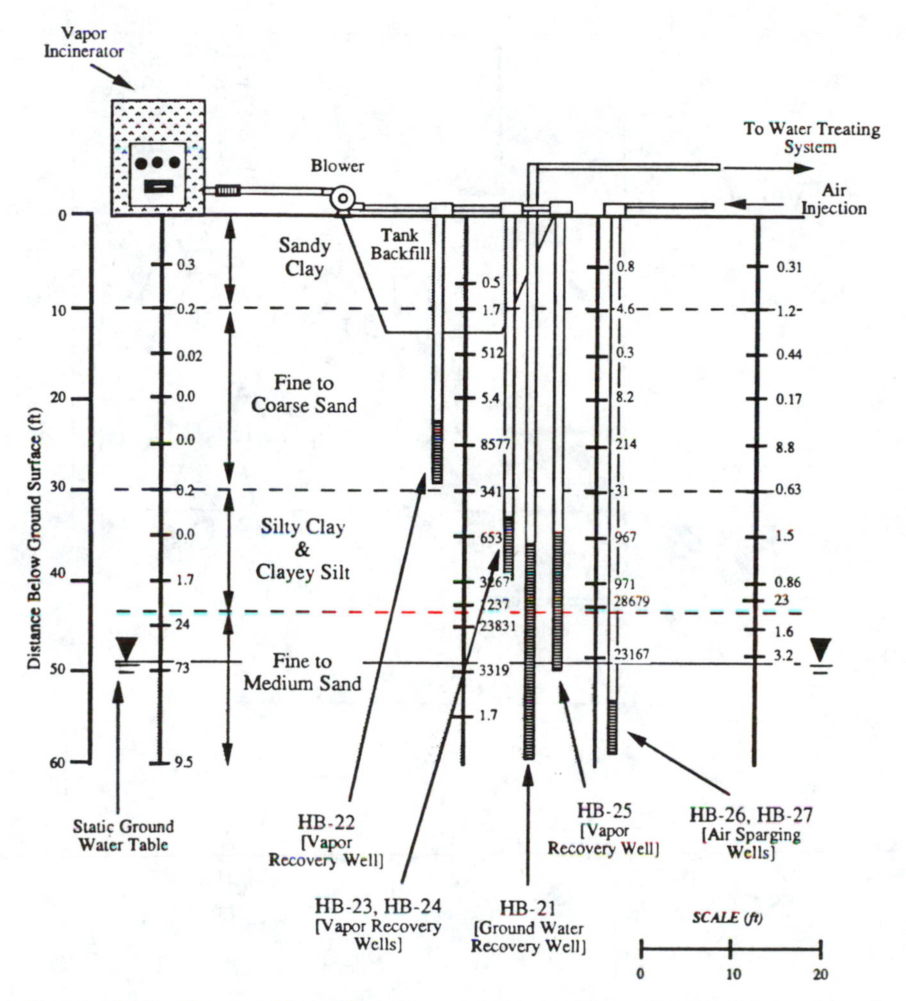

Figure 16.10. Screened intervals and approximate locations of vapor recovery, groundwater recovery, and air sparging wells.

- total system vapor concentration (on-line FID total hydrocarbon detector)
- vapor concentration and composition from each vapor extraction well (as determined by a GC-FID analysis)

SYSTEM OPERATION

Figures 16.11 through 16.14 summarize data obtained during the first 370 days of operation. The operation, monitoring, and improvements to the system are

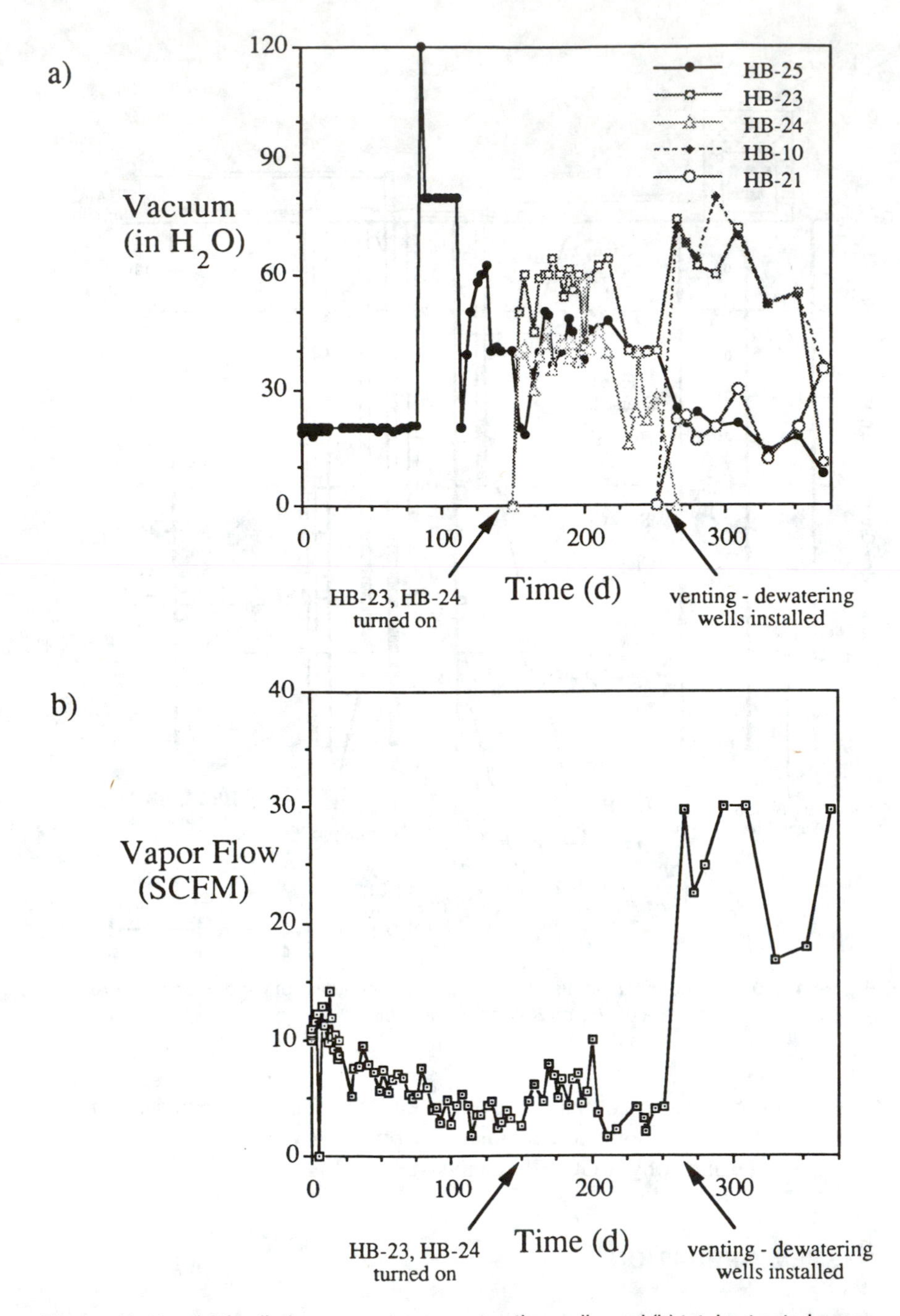

Figure 16.11. (a) Applied vacuum at vapor extraction wells, and (b) total extracted vapor flowrate.

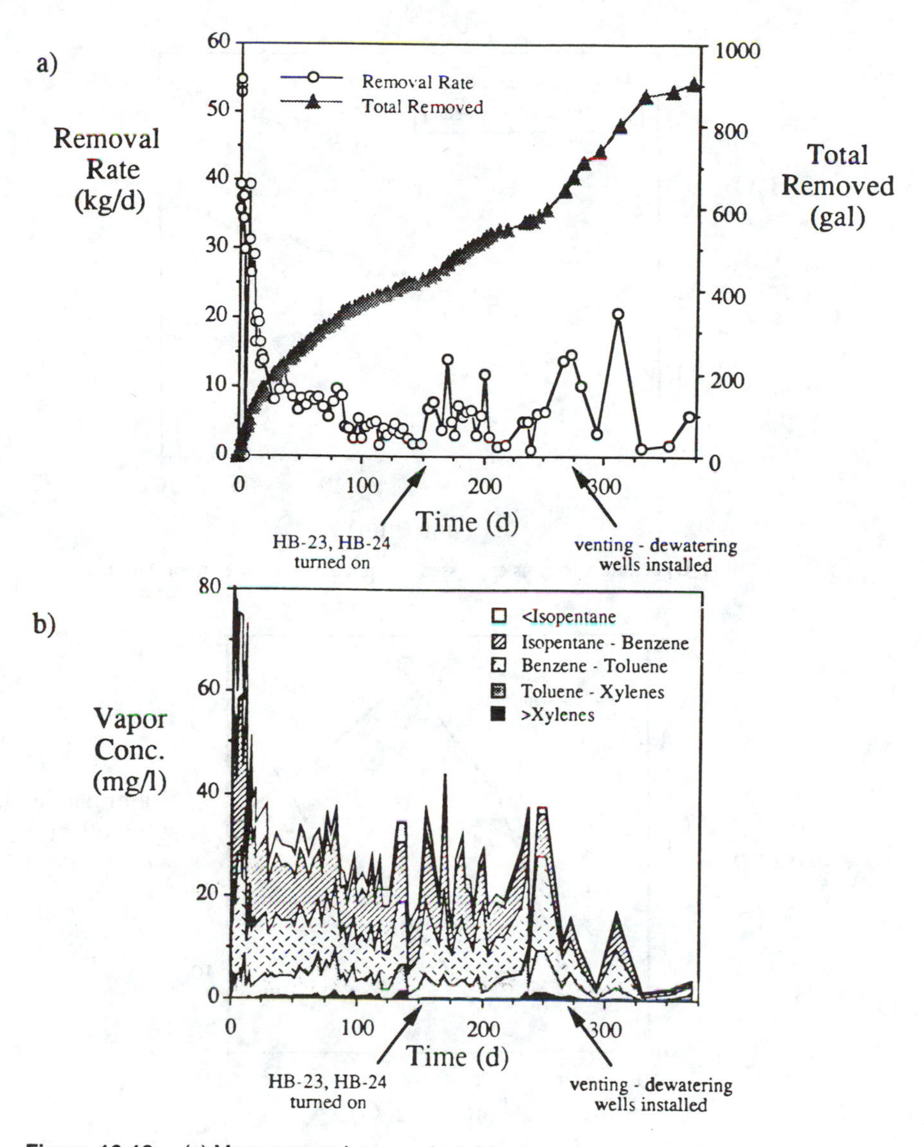

Figure 16.12. (a) Mass removal rate and total hydrocarbon removed, and (b) vapor concentration and composition changes with time.

described below. It should be noted that the goal of the field test was to gain a better understanding of the processes governing the effectiveness of soil venting, and hence, the system was not necessarily operated in the most efficient manner. All quoted "vacuum" values refer to gauge values, and not absolute vacuums. The values are given in in. H_2O, which is equal to the height of a column of

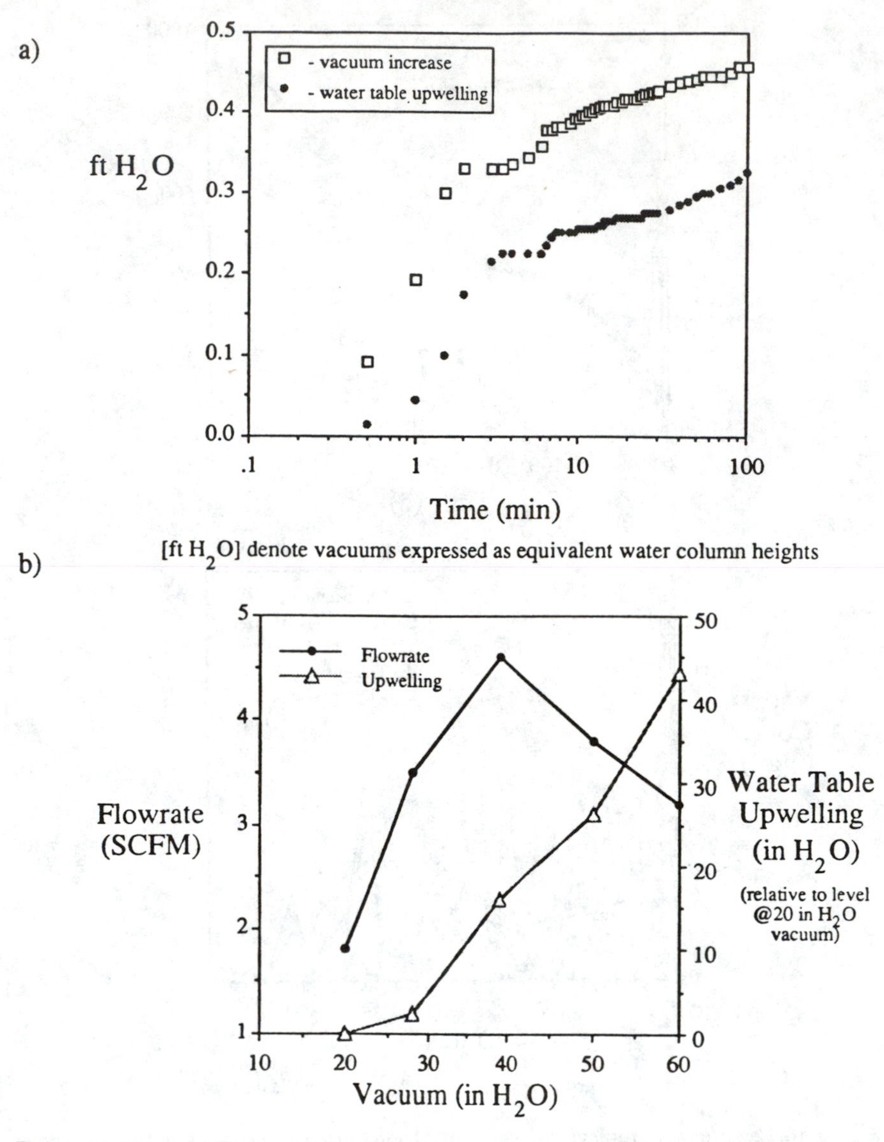

Figure 16.13. (a) HB-25 vapor flowrate and groundwater upwelling dependence on applied vacuum, and (b) transient water table and subsurface vacuum response.

water that would give such a gauge vacuum (472 in H_2O vacuum is equivalent to an absolute vacuum).

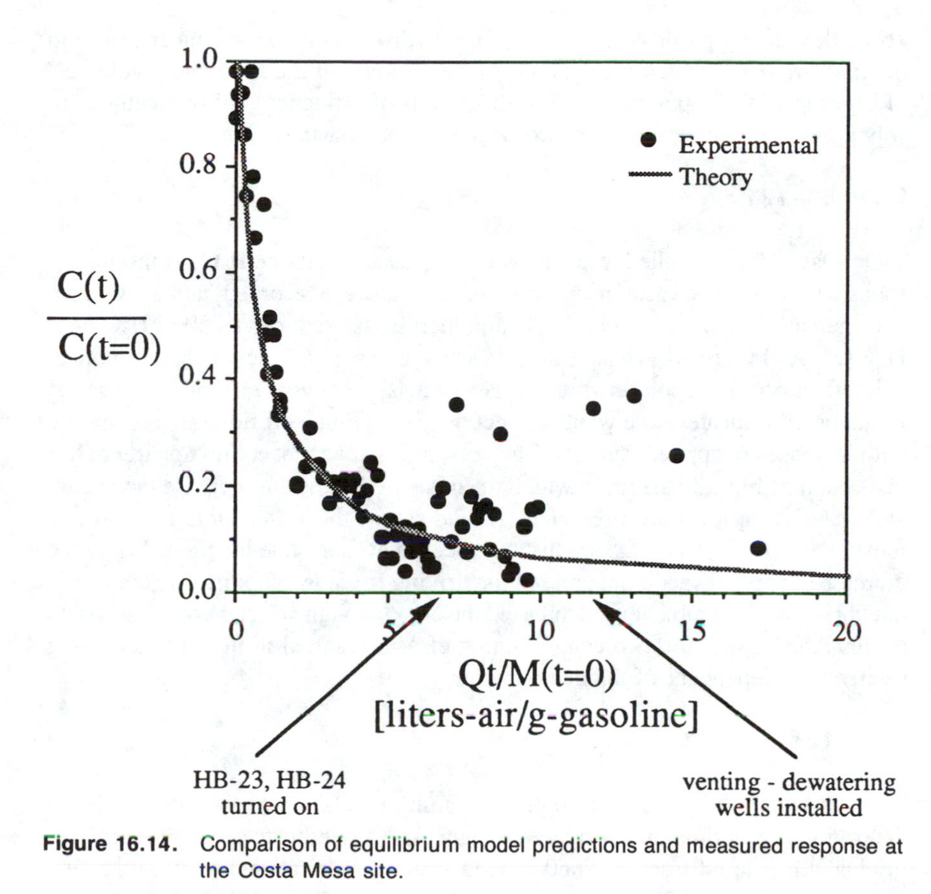

Figure 16.14. Comparison of equilibrium model predictions and measured response at the Costa Mesa site.

Days 1–83

At start-up, the vacuum blower was connected only to vapor extraction well HB-25, which is screened through the lower sandy zone. By operating in this manner, any interactions due to other extraction, or air sparging wells could be eliminated. During Days 1 to 83 a 20 in H_2O (0.96 atm absolute) vacuum was applied, as indicated by Figure 16.11a. Figure 16.11b shows that the extraction well flowrate was initially 12 scfm, but gradually decreased to about 5 scfm by Day 83. For reference, Equation 10 predicts that Q_{well} = 12 scfm, for k = 8 darcys. It is not clear why the flow decreased while the applied vacuum was held constant, although a gradual rise in the groundwater table (either due to natural causes or failure of the groundwater pumping system) could be responsible. Figure

16.12 depicts a significant decrease in the hydrocarbon vapor concentration in the first two days, which corresponds to the removal of the first "pore volume" of vapor from the vapor extraction well's radius of influence. Subsequently there was a gradual decline in the hydrocarbon vapor concentration.

Days 83–121

On Day 83 the applied vacuum was increased to 80 in H_2O, transient responses in subsurface vacuum and water table level were recorded, and the vacuum was increased to 120 in H_2O briefly, and then it was reduced to 80 in H_2O until Day 121. As Figure 16.11b indicates, the change in applied vacuum had no noticeable effect on the vapor flowrate. This result is somewhat curious because, as Equation 10 indicates, one would expect the extraction well flowrate to increase with increases in applied vacuum. The reason, however, becomes apparent after inspection of Figure 16.13a, in which the measured transient increase in vacuum and water table level are presented. As indicated, the water table level in any location is influenced by the subsurface vacuum at that same location. While the increase in applied vacuum increased the driving force for vapor flow, the subsequent rise in water table level decreased the area through which vapor flow could occur. Apparently, the two compensating effects resulted in the flowrate being relatively independent of applied vacuum.

Days 121–135

During this time period the applied vacuum at HB-25 was varied in order to determine the optimum flowrate. As Figure 16.11 indicates, the vacuum was gradually increased from 20 to 60 in H_2O, and at each vacuum the flowrate and water table level in HB-25 were measured. Figure 16.13b presents the measured responses, and it can be seen that the maximum vapor flowrate corresponded to 40 in H_2O vacuum. Subsequently, the vacuum was maintained at 40 in H_2O.

Days 149–266

On Day 149, the two extraction wells screened through the silty clay layer (HB-23, HB-24) were connected to the vacuum pump, and vacuums of 40–60 in H_2O were applied to each. As indicated in Figures 16.11 and 16.12, this had little effect on the total vapor flowrate, while a slight increase in total vapor concentration was detected. This confirms the expectation that remediation of the silty clay layer is not likely to occur due to convective transport, and will be limited by the rate at which hydrocarbon vapors can diffuse into the zones where significant vapor flow occurs.

Days 266–370

It should be clear from the results prior to Day 266 that the rate of hydrocarbon removal was limited by the low vapor extraction well flowrate, which itself was affected by the interaction between the applied vacuum and water table level. As discussed previously, equilibrium-based models predict that the degree of remediation is a function of the volume of vapor drawn through the contaminated soil and the mass of contaminant present in this zone. Recall that model predictions for a sample of gasoline from this site indicate that about 100 1-air/g-gasoline is required to achieve a desired degree of remediation. Figure 16.14 compares model predictions with data obtained at this site. It should be noted that in order to account for some bypassing of air through clean soil zones, the total volume of withdrawn vapor is multiplied by the ratio of (vapor concentration in the extracted vapors/vapor concentration in the vadose monitoring wells) to obtain the abscissa of this plot. Prior to Day 266 this ratio was approximately 1/3, and only 11 liters-air/g-gasoline had been withdrawn.

In order to increase the vapor flow, and decrease the remediation time, modifications were made to the system by Day 266. It was clear that having only a single well, combined with water table-vacuum interactions was limiting the effectiveness of the system. To remedy the problem, monitoring well HB-10 and groundwater recovery well HB-21 were converted to dual groundwater pumping/vapor recovery wells. Each well was connected to the vacuum blower manifold and also equipped with a level-control groundwater pump. Because the water table upwelling will be strongest close to a vapor extraction well, it is logical to construct these dual groundwater pumping/vapor recovery wells whenever soils close to the water table are to be remediated.

Figures 16.11 and 16.12 indicate that when the system modifications were completed, the total vapor flowrate increased to 30–40 scfm, and the total hydrocarbon recovery increased to 900 gal (3 kg hydrocarbon = 1 gal).

Figure 16.12b presents the change in vapor composition with time. Rather than report the large number of compounds detected in the GC-FID vapor analyses, the compounds have been grouped by boiling point:

methane-isopentane (<28°C)
isopentane-benzene (28–80°C)
benzene-toluene (80–111°C)
toluene-xylenes (111–144°C)
>xylenes (>144°C)

In the first 20 days there was a shift in composition to less volatile compounds, then until Day 266 the vapor composition and concentration remained relatively unchanged. Following the installation of dual vapor extraction/groundwater recovery wells, the vapor concentration decreased, but the relative composition did not change significantly through Day 370.

CONCLUSIONS AND FUTURE WORK

This chapter describes an attempt to gain a better understanding of the processes governing the effectiveness of soil venting as a soil remediation process, through the results from a field study. The data obtained thus far confirms previously reported modeling efforts of this study,[6] and supports the use of the structured design and monitoring approach suggested by Johnson et al.[10]

Even though a sufficient degree of remediation has not been achieved by Day 370, this site has been an ideal location for illustrating the influence of site characteristics on the efficiency of soil venting. As predicted, one must be careful not to apply soil venting in situations where other remediation options are likely to be more efficient and cost-effective. It should be clear that not all sites can be remediated in less than 3 to 6 months, as implied by the many of the cases reported in the literature.

The overall goal of this project is still to determine the effectiveness of venting at this site. As of Day 370, questions still remain unanswered concerning level of residual hydrocarbons that can be reasonably obtained. In addition, the potentially positive effect of venting on other processes (i.e., biodegradation), and the effect of possible system enhancements (i.e., air sparging in the saturated zone) on the removal of residual hydrocarbons still remain to be investigated.

ACKNOWLEDGMENTS

The authors would like to acknowledge the following people who contributed to this investigation: Marian W. Kemblowski, James D. Colthart, and Ira J. Dortch of Shell Development Co., Frank Fossati and Carl Grimmer of Shell Oil Co., and Sharon Boltinghouse of Applied Geosciences, Inc.

REFERENCES

1. Hutzler, N. J., B. E. Murphy, and J. S. Gierke. State of Technology Review: Soil Vapor Extraction Systems, U.S. EPA, CR-814319-01-1, 1988.
2. Thornton, J. S., and W. L. Wootan. "Venting for the Removal of Hydrocarbon Vapors from Gasoline Contaminated Soil," *J. Environ. Sci. Health,* A17(1):31–44 (1982).
3. Marley, M. C., and G. E. Hoag. "Induced Soil Venting for the Recovery/Restoration of Gasoline Hydrocarbons in the Vadose Zone," NWWA/API Conference on Petroleum Hydrocarbons and Organic Chemicals in Groundwater, Houston, TX, 1984.

4. U.S. Environmental Protection Agency, Terra Vac In Situ Vacuum Extraction System: Applications Analysis Program, EPA/540/A5-89/003, July 1989.
5. Parker, H. W., K. A. Rainwater, B. J. Clayborn, and M. R. Zaman. "Diffusion Control Model for Vaporization of Hydrocarbon Contamination from the Vadose Zone," 1989 AICHE Spring National Meeting, Houston, TX, April 2–6, 1989.
6. Johnson, P. C., M. W. Kemblowski, and J. D. Colthart. "Practical Screening Models for Soil Venting Applications," NWWA/API Conference on Petroleum Hydrocarbons and Organic Chemicals in Groundwater, Houston, TX, 1988.
7. Wilson, D. J., A. N. Clarke, and J. H. Clark. "Soil Clean-up by in situ Aeration. I. Mathematical Modeling," *Sep. Science Tech.,* 23:991–1037 (1988).
8. Massmann, J. W., "Applying Groundwater Flow Models in Vapor Extraction System Design," *J. Environ. Eng.* 115(1):129–149 (1989).
9. Marley, M. C., A. L. Baehr, and M. F. Hult. "Evaluation of Air-Permeability in the Unsaturated Zone Using Pneumatic Pump Tests: 1. Theoretical Considerations," In review, 1990.
10. Johnson, P. C., C. C. Stanley, M. W. Kemblowski, D. L. Byers, and J. D. Colthart. "A Practical Approach to the Design, Operation, and Monitoring of In Situ Soil Venting Systems," *Ground Water Monitoring Review,* Spring 1990.

CHAPTER 17

Human Health Risks Associated with Contaminated Sites: Critical Factors in the Exposure Assessment

Jayne M. Michaud, Alan H. Parsons, ChemRisk™—
A Division of McLaren/Hart, Portland, Maine

Stephen R. Ripple, Dennis J. Paustenbach, ChemRisk™—
A Division of McLaren/Hart, Alameda, California

INTRODUCTION

The widespread use of petroleum in the United States has created the potential for contamination of soil and groundwater. The environmental, health, and economic implications of soil contamination have become a topic of interest to many in the past several years. The application of risk assessment to evaluations of petroleum-contaminated sites will help prioritize sites, focus resources, and develop cost-effective remediation strategies.[1,2]

Risk assessment is an important tool for evaluating the potential hazards of human exposure to industrial chemicals, such as petroleum hydrocarbons, in air, water, and soil. Health risk assessments have become so widely adopted in the United States that their conclusions are now major factors in many environmental decisions.[3-5] The risk assessment process has helped the American public to better understand the magnitude of risks posed by naturally-occurring and man-made products, and consequently has helped to reduce unwarranted concern over trivial hazards.[6]

The risk assessment process has evolved over the past two decades from a qualitative and semiquantitative process to one that includes refined methodologies for the quantitative evaluation of hazards posed by exposure to toxicants. The knowledge gained through experience in conducting risk assessments, coupled with the scientific advances of the past few years, will greatly improve our ability to estimate accurately the human health risks of low-level exposure to chemicals in this decade.[7]

The goal of a risk assessment is to estimate the probability of an adverse effect on humans, domestic animals, wildlife, or ecological systems from exposure to a chemical or physical agent.[8] The four components of risk assessments—hazard identification, dose-response assessment, exposure assessment, and risk characterization—have been defined in detail in many publications.[8-10] A comprehensive risk assessment document presents, discusses, and evaluates the information regarding toxicity, environmental transport and fate, exposure, and human experience associated with a potential hazard. The importance of risk assessment is that it provides the bridge between scientific research and risk management.

The steps of hazard identification and dose-response assessment have been the subject of considerable debate. Avenues clearly exist for their improvement.[7] The risk assessment practitioner, however, has only a limited opportunity to enter into this debate in the preparation of assessments for regulatory compliance purposes. The hazard identification step which addresses the question of whether a substance is capable of producing an adverse health effect in humans, and the dose-response step which characterizes the relationship between dose and adverse response, involve debate on such issues as the use and interpretation of animal data, consideration of the "weight of evidence," mechanisms of action for carcinogens, tumor type, pharmacokinetics, and low dose extrapolation models. Debate in these areas, although impacting the science of risk assessment, is largely confined to the regulatory policy-making arena. The primary area where the individual risk assessor can and does have a major impact is in the exposure assessment step. It is here where the risk assessment professional integrates site-specific information into the assessment to accurately characterize potential risks associated with a given contaminant. This is particularly true for the assessment of contaminated soils.

OBJECTIVE OF EXPOSURE ASSESSMENT

The objective of exposure assessment is to measure or estimate the intensity, frequency, and duration of human or animal exposure to an agent present in the environment or estimate hypothetical exposures that might arise from the release of new chemicals into the environment. In its most complete form, an exposure assessment should describe: the magnitude, duration, schedule, and route of exposure; the populations exposed; and the uncertainties in all estimates. It predicts the amount of the chemical that may be absorbed by an individual from various media (water, air, soil, food) following exposure.[8] The exposure assessment can often be used to identify feasible prospective control options and to predict the effects of available technologies for controlling or limiting exposure.

Exposure is dependent upon the concentration of a chemical in a particular medium. Although direct measurement is preferred, environmental concentrations are often below the analytical limit of detection. Consequently, contaminant levels often must be estimated through modeling.[11-13] In order to estimate the distribution of a substance in the environment, it is important to understand its physical and chemical properties.[7] With these data and some knowledge of the source, modeling equations may be used to estimate the emission rate, transport, and degree of exposure to a particular contaminant.[11]

Humans may be exposed to contaminants in soil as a result of their occupation, recreational activities, or through exposure in and around the home.[2,14] Exposure can occur through dermal contact with contaminated soil or dust, ingestion of dust or soil from the hands, food, or objects, and inhalation of contaminated fugitive dusts or vapors. Exposure to indoor air contaminants via inhalation is one exposure pathway that is increasingly of public concern. Contaminants in soil and groundwater may volatilize and seep through cracks in building foundations, resulting in elevated indoor air contaminant levels.[15] Also, chemicals that are not strongly bound to soil should be evaluated for their potential to leach into surface or groundwater.

Recent research has shown that indirect routes of exposure such as the ingestion of meat and milk contaminated by particulate emissions falling onto grass and soil with subsequent ingestion by dairy cows, will often be an important route of human exposure.[7] Such indirect routes, i.e., the food chain pathways, become a critical consideration for hydrocarbon contaminants that fall in the semivolatile category.

CRITICAL COMPONENTS OF AN EXPOSURE ANALYSIS

Predicting exposure and subsequent human uptake of chemicals from soil is a complex process, particularly when one considers factors such as bioavailability, environmental degradation and accumulation, ingestion of contaminated soil and foliage by cattle, deposition rate of particulates, dermal uptake, biomagnification, pharmacokinetics in humans, exposure schedule, and the dietary changes that occur over a lifetime.[10] Careful selection of exposure assumptions is critical to the generation of reliable estimates of chemical uptake and the prediction of associated health risks. The critical factors that should be considered in an exposure assessment of petroleum contaminated soils include:

- appropriate analysis of data
- estimation of transport and distribution of petroleum hydrocarbons in soil
- identification of all exposure routes and potentially exposed populations
- accurate estimation of exposure parameters such as:
  - soil ingestion rates
  - dermal contact rates
  - inhalation rates
  - bioavailability of chemicals following exposure

Appropriate Analysis of Data

The statistical methodology used to analyze data from air, water and soil samples may affect the conclusions of a risk assessment. One common shortcoming is that many risk assessors have failed to recognize that the concentrations of contaminants in soil, water, and air are usually lognormally rather than normally (Gaussian) distributed.[16,17] Accordingly, it is important that appropriate statistical procedures which account for lognormal distribution of data be followed. In cases where data are lognormally distributed, the best estimate of central tendency and variability are the geometric mean and standard deviation of the geometric mean.[18-20] Should the arithmetic mean and standard deviation for a normal distribution be used to describe the environmental distribution of a contaminant when lognormal statistics are more appropriate, the potential exposure and health risk for the average person will often be significantly overestimated.[7]

Similarly, scientifically defensible approaches to the handling of spatial distributions of contaminants over a site, such as kriging, a weighted moving-average estimation technique based on geostatistics, are rapidly gaining acceptance in predicting environmental distribution and estimating the mean contamination present at a site.[9] Weighted moving-average methods use the notion that data values closest to the point or block where the mean contaminant concentration is being estimated contain more information than data farther away and, hence, have more weight. This method can be used to determine the proper weighting for each data point. Kriging is especially useful in assessing whether additional data are needed in any portion of a contaminated site.[9,16] Such a method clearly has applications in assessing petroleum-contaminated soil.

Estimation of Environmental Fate of Compounds in Soil

During the past decade, it has become clear that the physical and chemical properties of a compound determine its ability to be transported through ecosystems and its fate in the environment.[7] Accordingly, these properties have a direct bearing on the risk to the exposed human population. For example, the potential for environmental contaminants to accumulate in an organism (biological magnification) may be roughly approximated by the ability of a chemical to partition between water and nonpolar solvents similar to body lipids (octanol/water partition coefficient).[21] Plants also assimilate numerous inorganic and organic contaminants. However, research has been limited in the area of uptake of the specific constituents of petroleum.[2]

To date, only a few risk assessments have attempted to account mathematically for the various chemical degradation and bioaccumulation processes that may occur during the 70 year period usually considered in most human lifetime exposure analyses. For example, the significant effect that the environmental half-life of a compound has on the anticipated lifetime dose may be illustrated using exposure to 100 ppb 2,3,7,8-tetrachlorodibenzo-p-dioxin (TCDD) in surface soil.[14]

Using a given set of exposure assumptions for oral, dermal, and inhalation pathways over a 70 year lifetime and a TCDD half-life of 1 year, the maximum anticipated lifetime average daily dose (LADD) would be 0.14 pg/kg/day.[14] In contrast, if the half-life of TCDD was estimated to be 12 years, that LADD would increase tenfold to 1.6 pg/kg/day. It is clear that environmental fate can have a substantial effect on human exposure estimates and should be incorporated into risk assessments whenever possible.

Identification of All Exposure Routes and Potentially Exposed Populations

In order to accurately assess the impact of chemicals in soil on human health, the potentially exposed subpopulations must be identified. Included in this step is the characterization of human exposure conditions at industrial, residential, and recreational sites. One may also need to consider a number of different populations such as workers, adults, and children when evaluating exposure to petroleum contaminated soils.

Exposure to soil contaminants may occur via soil ingestion, dermal contact, inhalation, and ingestion of contaminated plants and animal products (i.e., dairy products). Factors to consider in selecting plausible exposure pathways include the accessibility of the site and the physical characteristics of the contaminants. In some cases, where access to the site is restricted, exposure through one route may be more likely than another. For example, inhalation of volatiles may constitute the primary exposure route when access to the site is restricted and direct contact with the soil is minimized. In an uncontrolled site, there may be multiple important exposure routes. One should consider among these the inhalation of gaseous emissions in residential or occupational indoor air due to contaminated soil or groundwater.

Selection of Exposure Parameters

In conducting risk assessments, selection of exposure parameter values that represent the conditions of human exposure is critical. In assessing exposure to petroleum contaminated soils, the key exposure parameters include soil ingestion, dermal absorption, and inhalation of vapors, dusts, and soil. Indirect routes of exposure to chemicals in soil should also be considered, e.g., ingestion of plants or animal products that have assimilated petroleum products from soil. Rather than simply using a "cookbook" approach, the risk assessor should determine suitable exposure parameter values for each exposure scenario. Wherever possible, site-specific exposure values should be used. In many cases, defensible exposure assumptions may be developed for parameters such as ingestion rates for soil, vegetables, and dairy products that differ significantly from those recommended by the federal and state agencies.

For example, the use of a soil ingestion rate of 200 mg/day[9] may not be appropriate for children between the ages of 0 to 1 year who are under the close supervision of adults. Alternative values may be based on studies that have estimated the soil ingestion rates for children and adults.[14] Calabrese et al.[22] estimated values ranging below the EPA's 200 mg/day value.

Inhalation rates should represent the activity levels and exposure conditions assumed for individuals in each exposure scenario. Although workers may spend 8 hours at a work site, an inhalation rate of 20 m^3/day, which is based on a 24-hour exposure period, would be inappropriate for workers who are only exposed to contaminated fugitive dusts for several hours during the work day. Inhalation rates can be estimated based on information provided in several sources.[23,24]

Likewise, care must be taken when estimating dermal contact rates, or the amount of soil that is in contact with exposed skin. For example, a dermal contact rate (expressed as mg soil per cm^2 of exposed skin surface) for potting soil[9] may not be appropriate in occupational, residential, or recreational scenarios where exposure occurs only to contaminated fugitive dusts deposited indoors or to sandy soils. Again, various estimates of dermal contact rates can be obtained from the scientific literature.[14,23,25]

Recently, there has been a trend in exposure assessments to consider the range of values that could apply to the exposure of a population. This range of values considers both the uncertainty in the estimate, as well as inter-individual variability for a particular factor. The variables potentially influencing exposure and uptake by individuals include age, eating patterns, lifestyle, personal physiological parameters such as respiration rate, skin surface area, body mass, cardiac output, water consumption, body fat content, skin absorption, and consumption of breast milk. Although such rigorous analyses are not routinely needed for most assessments, such factors should be considered for the most important routes of exposure for a given analysis.

A good deal of effort has been expended in recent years to identify the range and best estimate of values to be used in exposure assessments (Table 17.1).[7,23] In light of the fairly broad range for some of the values, exposure assessments should be based on the best estimate of a specific exposure parameter rather than repeatedly adopting the highest one. It is possible to characterize exposure for a population by determining the probability distribution of reasonable values for each of the various parameters. Such analyses make a critical contribution to the accurate evaluation of the risks and to a more thorough understanding of uncertainties which are crucial to the risk characterization step of the risk assessment. The evaluation of both uncertainty and parameter sensitivity for multipathway exposure assessment permits one to generate a probability distribution for the various analytical results. Such an approach offers the hope that, rather than having to state that it is plausible that someone might absorb as much as 100 μg/kg/day of a given chemical in the soil, the risk assessor might be able to state what fraction of that population might reasonably be exposed at that level. In other words, the risk assessor could predict that perhaps 1% of the exposed population might absorb as much as 100 μg/kg/day, 75% might absorb as much as 5 μg/kg/day,

Table 17.1. The Best Estimate (Weight of Evidence) and Frequently Cited Ranges for Exposure Parameters Used in Risk Assessment of Contaminated Soils

Parameter	Best Estimate[a]	Range of Values[a,b]
Uptake of air by typical adult	12 m^3/day	10 to 30 m^3/day
Ingestion of dust and soil by typical toddler (ages 2–6)	25–50 mg/day	25 to 10,000 mg/day
Ingestion of dust and soil by typical adult (ages 8–80)	5–10 mg/day	10 to 500 mg/day
Airborne dust in a neighborhood with a few unpaved roads	200 $\mu g/m^3$	50 to 20,000 $\mu g/m^3$
Ingestion of tomatoes by the average American	30 g/day	10 to 200 g/day

[a]Based on currently available information cited in Paustenbach.[7]
[b]The U.S. EPA *Exposure Factors Handbook*[23] and South Coast Air Quality Management District *Multi-Pathway Health Risk Assessment Input Parameters Guidance Document*[25] are useful resources.

and 24% absorb less than 1 μg/kg/day of the contaminant. Such an approach would allow risk managers to make more rational and informed decisions.

Risk assessments typically include an evaluation of exposure to the maximally exposed individual (MEI). The MEI is evaluated in an effort to ensure that the potential human exposure to an environmental contaminant will not be underestimated. In these cases, extreme situations are hypothesized wherein individuals are continuously exposed throughout their lifetimes to the maximum plausible amounts of a substance via all routes of exposure. Although such an approach may be considered to be reasonable and appropriate for a "screening assessment" in which insignificant exposure pathways may be identified and dismissed, these analyses should not be depicted as an estimation of the likely actual human uptake.[26-29] Rather, risk assessors should use their best professional judgment and and site-specific information in developing reasonable exposure estimates.

Predictably, worst-case assessments do not describe the bulk of the exposed population and often overestimate even the most exposed person. For example, many attempts to describe the MEI have been based on an assumption that the same person who is a 95th percentile consumer of fish was also prone to ingest dirt as a child (95th percentile), reside in the most highly exposed residential area (99th percentile), and eat only vegetables grown in his contaminated private garden (99th percentile). A more accurate and reasonable approach to estimating the dose for the MEI is to select the single 95th percentile parameter responsible for giving the highest risk rather assume that the same person is at the 95th or 99th percentile for three or more parameters,[30] since the result may represent an extreme exposure scenario which may apply to fewer than 1 person in 10,000.

In general, a broad range of values can be selected for any exposure parameter. However, the risk assessor should determine the appropriateness of each parameter to the exposure scenario, based on the conditions of the site, thereby reducing the level of uncertainty that is inherent in any risk calculation.

Bioavailability of Chemicals Following Exposure

A relatively new factor which should be considered in exposure assessments is the bioavailability of a substance in various environmental media. Many past risk assessments erroneously assumed that all substances were 100% absorbed in the gastrointestinal tract, and between 10 to 100% absorbed through the skin. Research conducted over the past five years has shown that the systemic absorption efficiency of chemicals is chemical-specific and varies depending on the matrix (i.e., soil, oil) and route of exposure, i.e., inhalation, ingestion, and dermal contact. For example, the oral bioavailability of dioxin has been shown to be affected by the medium, concentration in the media, particle size, and presence of co-contaminants.

In the case of contaminants in soil, there can be a pronounced effect of the soil matrix. Soil characteristics, e.g., silt, clay, or sand, can significantly influence the degree to which a contaminant is available for uptake. In addition, bioavailability of a chemical on soil is affected by its chemical and physical properties. Large molecular weight chemicals often bind to the soil and are less water soluble, while smaller molecules are frequently water soluble, volatile, and highly bioavailable.

CONCLUSIONS

The risk assessment process has become an integral element in the evaluation and management of contaminated sites. While the four steps of risk assessment have reached varying levels of maturity as a result of the past decade of development, the exposure assessment step remains one area where the individual risk assessment practitioners must exercise the greatest judgment and technical skill. This discussion has identified the various aspects of exposure assessment that contribute to the clear and accurate characterization of potential human health risks associated with contaminated soils.

REFERENCES

1. Lord, H. J. and J. Perwak. ''An Approach for Assessing Risks to Health and Environment at Petroleum-Contaminated Sites'' In: *Soils Contaminated by Petroleum: Environmental & Public Health Effects.* E. J. Calabrese and P. T. Kostecki, eds. (New York: John Wiley & Sons, Inc. 1988), pp. 173–189.
2. Calabrese, E. J., P. T. Kostecki, and D. A. Leonard. ''Public Health Implications of Soils Contaminated with Petroleum Products'' In: *Soils Contaminated by Petroleum: Environmental & Public Health Effects.* E. J. Calabrese and P. T. Kostecki, eds. (New York: John Wiley & Sons, Inc., 1988), pp. 191–229.
3. Ruckelshaus, W. D. ''Science, Risk and Public Policy,'' *Science* 221:1026–1028 (1984).
4. Department of Health and Human Services. ''Determining Risks to Health: Federal Policy and Practice,'' Auburn House, Dover, MA, 1986.

5. Yosie, T. "How Risk Assessments are Used in the Federal Government," *Environ. Sci. Tech.* 21: 526–531 (1987).
6. Young, Frank A. "Risk Assessment: The Convergence of Science and the Law," *Regul. Toxicol. Pharmacol.* 7:179–184 (1987).
7. Paustenbach, D. J. "Important Recent Advances in the Practice of Health Risk Assessment: Implications for the 1990's," *Reg. Toxicol. Pharm.* 10:204–243 (1989).
8. National Academy of Sciences. "Risk Assessment in the Federal Government: Managing the Process," National Academy Press, Washington, DC, 1983.
9. "Risk Assessment Guidance for Superfund: Human Health Evaluation Manual Part A," Office of Emergency and Remedial Response, U.S. Environmental Protection Agency, Washington, DC. EPA/540/1-89/002. (July 1989).
10. Paustenbach, D. J. *The Risk Assessment of Environmental and Human Health Hazards: A Textbook of Case Studies* (New York: John Wiley & Sons, Inc., 1989).
11. "Guidelines For Exposure Assessment," U.S. Code of Federal Regulations, 2984, No. 185. U.S. Environmental Protection Agency 1986, pp. 34042–34050.
12. Haque, R. *Dynamics, Exposure, and Hazard Assessment of Toxic Chemicals* (Ann Arbor, MI: Ann Arbor Science, 1980).
13. Conway, R. A. *Environmental Risk Analyses for Chemicals* (New York: Van Nostrand Publishers, 1982).
14. Paustenbach, D. J. "Assessing the Potential Environmental and Human Health Risks of Contaminated Soil," *Comments Tox.* 1(3–4):185–220 (1987).
15. Peterec, L. J. "A Case Study in Petroleum Contamination: The North Babylon, Long Island Experience" In: *Soils Contaminated by Petroleum: Environmental & Public Health Effects.* E. J. Calabrese and P. T. Kostecki, eds. (New York: John Wiley & Sons, Inc., 1988) pp. 231–255.
16. Gilbert, R. O. *Statistical Methods for Environmental Pollution Monitoring* (New York: Van Nostrand-Reinhold, 1987).
17. Sielken, R. L. "Statistical Evaluations Reflecting the Skewness in the Distribution of TCDD Levels in Human Adipose Tissue," *Chemosphere* 16(8–9):2135–2140 (1987b).
18. "Proposed Guidelines on Exposure Related Measurements for Risk Assessments," U.S. Code of Federal Regulations No. 232, U.S. Environmental Protection Agency, 1988, pp. 48,83048,853.
19. Leidel, N. and K. A. Busch, "Statistical Design and Data Analysis Requirements" In: *Patty's Industrial Hygiene and Toxicology,* Vol. IIIa, 2nd ed. (New York: John Wiley & Sons, 1985).
20. Rappaport, S. M. and J. Selvin, "A Method for Evaluating the Mean Exposure From a LogNormal Distribution," *J. Amer. Ind. Hyg. Assoc.* 48:374–379 (1987).
21. Spacie, A. and L. Hamelink. Bioaccumulation. In: *Fundamentals of Aquatic Toxicology,* (New York: Hemisphere Pub., 1985), Ch. 17.
22. Calabrese, E. J., R. Barnes, E. J. Stanek, H. Pastides, C. E. Gilbert, P. Veneman, X. Wang, A. Lasztity, and P. T. Kostecki. "How Much Soil do Young Children Ingest: An Epidemiologic Study," *Reg. Toxicol. Pharm.* 10:123–137 (1989).
23. "Exposure Factors Handbook," Exposure Assessment Group, Office of Health and Environmental Assessment, U.S. Environmental Protection Agency, Washington, DC. EPA/600/8-89/043, May 1989.
24. "Report of the Task Group on Reference Man," International Commission on Radiological Protection, Publication No. 23 (Pergamon Press, 1974).

25. ''Multi-Pathway Health Risk Assessment Input Parameters Guidance Document,'' South Coast Air Quality Management District, Prepared by Clement Associates, Inc. Contract #8798 (June 1988).
26. Finkel, A. M., and J. S. Evans. ''Evaluating the Benefits of Uncertainty Reduction in Environmental Health Risk Management,'' *J. Air Poll. Cont. Assoc.* 37:1164–1171 (1987).
27. Nichols, A. L. and R. J. Zeckhauser. ''The Perils of Prudence: How Conventional Risk Assessments Distort Regulation,'' *Regul. Toxicol. Pharmacol.* 8:61–75 (1988).
28. Maxim, L. D. ''Problems Associated with the Use of Conservative Assumptions in Exposure and Risk Analysis,'' In: *The Risk Assessment of Environmental and Human Health Hazards: A Textbook of Case Studies,* D. J. Paustenbach, ed. (New York: John Wiley & Sons, 1989).
29. Maxim, L. D. and L. Harrington. ''A Review of the FDA Risk Analysis for PCB in Fish,'' *Regul. Toxicol. Pharmacol.* 4:192–219 (1984).
30. Tollefson, L. ''Methyl Mercury in Fish: Assessment of Risk for United States Consumers'' In: *Risk Assessment of Environmental and Human Health Hazards: A Textbook of Case Studies.* D. J. Paustenbach, ed. (New York: John Wiley & Sons, 1989), Ch. 25.

CHAPTER 18

Health Risks Associated with the Remediation of Contaminated Soils

Gayle Edmisten-Watkin, Harding Lawson Associates
Edward J. Calabrese, University of Massachusetts
Robert H. Harris, Environ Corporation

In recent years, hazardous chemicals have been detected at many sites throughout the United States, often as a result of past waste handling activities. Because of the presence of these chemicals, sites may be listed on the National Priorities List (NPL; Superfund). Under the Superfund process, typically, public health risk assessments are conducted to evaluate the risks associated with "baseline" conditions at a site, in the absence of remediation, and are used to evaluate whether remediation of the site is needed. If it is determined in the baseline assessment that remediation of the wastes at the site is necessary, it is important that the selection process for remedial alternatives consider the potential health risks associated with the proposed remediation process, as these risks may be significant compared with baseline conditions.

To date, selection of the appropriate remedial option for a waste site has typically been driven by cost and/or technical feasibility.[1,2] Evaluation of health risks associated with proposed remedial alternatives has generally been qualitative in nature, although the Environmental Protection Agency's (U.S. EPA's) guidance document for conducting Remedial Investigations and Feasibility Studies (RI/FS)[3] indicates that both short- and long-term risks resulting from the proposed remediation need to be assessed. Short-term risks refer to "any risk that results from excavation . . . that may affect human health (and) threats that may be posed to workers.''[4] Long-term risks refer to the ''magnitude of residual risk remaining from untreated waste or treatment residuals remaining at the conclusion of the remedial activities.[4] In addition, the revised National Oil and Hazardous Substances Pollution Contingency Plan (NCP)[5] requires that the short- and long-term effectiveness of each potential remedy be evaluated. This includes evaluation of potential impacts on the neighboring community, workers, or surrounding environment, including the potential threats to human health and the environment associated with excavation, treatment, and transportation of hazardous substances.

The significance of considering potential health risks in the selection of remedial options for contaminated soils has been observed in a growing number of cases in which remediation of waste sites has created public health concerns distinctly different than the risks associated with the baseline (''no action'') condition. For example, emissions from the remediation process (excavation) at the Brooklawn (Petro Processors) Superfund site in Baton Rouge, Louisiana, produced vapor concentrations at the site boundary that exceeded occupational exposure standards on several occasions. This type of unanticipated exposure may have implications for numerous workers at such sites, such as those in the contractor staging and remediation support areas, and/or at adjacent facilities, since they are not likely to be wearing protective gear while the source area excavation is implemented. In the case of the Rocky Mountain Arsenal Superfund site, excessive emissions during excavation forced workers to evacuate the site and caused nearby residents to complain of astringent fumes (*Washington Post,* December 24, 1988). While these two excavation-based remedies have been delayed or reconsidered based on the unacceptable emissions, the U.S. EPA has utilized public health risk assessment procedures for the rejection of remedies involving excavation at a number of other Superfund sites, including the Tyson's Lagoon Superfund site in Upper Merion Township, Pennsylvania,[6] the Seymour Indiana/Recycling Superfund site,[7] and the Lone Pine Landfill Superfund site in Freehold Township, New Jersey.[8] The risk assessment and subsequent risk management decisions for these sites considered risks during remediation to onsite workers, impacts on offsite air quality, and potential adverse effects from transportation of the wastes.

There are several advantages to considering public health risks prior to adopting a particular remedial option. As discussed above, adverse health impacts may occur as a result of exposure to the chemicals during the remediation process. Human or environmental exposure to chemicals increases the liability of the potentially responsible party (PRP), overall costs may increase, and regulatory concern may be heightened. Public acceptance of a remedy may be minimized if chemical exposures have the potential to occur, and these concerns may activate media interest and possibly generate negative media coverage for the PRP.

The use of risk assessment techniques in the selection of remedial options can assist in predicting those remedial alternatives that are protective of public health and the environment, and/or in predicting if risks would be expected from the proposed activities. If risks are predicted for a specific remediation alternative, it does not necessarily follow that this remedial option should not be considered further, as risk management measures implemented during the remediation operations can be used to reduce the anticipated adverse health effects to acceptable levels.[9]

Assessment of the public health issues associated with remediation of a waste site can be varied and complex. Many decision points must be addressed in the evaluation, each of which must be dealt with separately, with consideration of the final outcome. These decision points may require specific expertise (e.g., toxicology, chemical exposure modeling) to define methods used to evaluate the specific remediation alternatives, and to select models, input variables, and justify assumptions used in the assessment. Information necessary to evaluate the potential health risks from proposed remedial action(s) is based on data used for the baseline assessment and includes previous waste handling and disposal activities at the site, and the age and source of contamination (e.g., pits, underground storage tanks). The results of site investigations (remedial investigations) can provide valuable information on the chemicals present in the various media, chemical concentrations, and the lateral and vertical extent of contamination. Site conditions, such as the geology and soil type(s), site drainage, hydrogeology, and local meteorology are also used to predict the potential release of chemicals during remediation. Evaluation of the characteristics of the contaminated media and the physical and chemical properties of the chemicals permits insight into the potential migration of the chemicals in the future, and is used to predict how chemical exposures may occur during implementation of the remedial measures.

For quantification of potential health risks, the chemical concentrations are estimated at each receptor location, using information on the mechanics of the remedial process to predict potential chemical emissions. Populations that may be exposed during remediation (receptor populations) may include residential or worker populations, children, and/or adults. Estimated chemical intakes (doses to the receptor populations) are calculated using the chemical concentrations in each medium, as well as site-specific and activity-specific intake assumptions, to estimate potential health risks to a specific receptor. If chemical exposures are predicted, the toxicity of the chemical(s) is also evaluated, including the acute and chronic toxicity, or potential carcinogenic, mutagenic and/or teratogenic effects which may result from chemical exposure. A variety of methods may be used to calculate the potential health risks, including EPA and/or state methods (e.g., New Jersey, California). If worker populations are considered in the assessment, Occupational Safety and Health Administration (OSHA), National Institute of Occupational Safety and Health (NIOSH), or American Council of Governmental Industrial Hygienists (ACGIH) methods may also be used in the evaluation.

Decisions regarding the remediation methods to be used will affect both the direction and the final outcome of the risk assessment. For example, if air stripping is considered as the preferred remedial option, volatile emissions may be released during the cleanup activities, and potential health risks may occur from inhalation of these volatiles. If, on the other hand, excavation is chosen as the remedial alternative, both volatile and dust emissions from materials handling activities may result in exposures during remediation.

To illustrate the use of quantitative risk assessment as a predictive tool, a case history of a waste site is presented in the remainder of this chapter. At this site, risk assessment methods were used to guide decisions for the preferred remediation plan. The methods used, the major decision points in the assessment, and the use of risk management measures during remediation field activities are discussed.

Case History

The property discussed in this case history is located in a heavily populated area of a northern California city, with residential and commercial properties in close proximity to the site. Between 1875 and 1930, a gas plant was operated on an adjacent property which generated wastes commonly referred to as "lampblack." The lampblack, a black porous material of low density, was disposed in pits on the property in question. During a preliminary site assessment to evaluate the potential for development of the property, several carcinogenic and noncarcinogenic chemicals were detected in soil and groundwater at the site, including polynuclear aromatic hydrocarbons (PNAs), volatile aromatic compounds (VOCs), and metals. Because the site was proposed for development, two risk assessments were conducted; one to evaluate risks associated with the baseline conditions at the site (to evaluate if remediation of the site was necessary), and one to evaluate potential exposures during and after the proposed site development.

The baseline risk assessment evaluated potential health risks from exposure to chemicals at the site in the absence of remediation. The results of the assessment indicated that public health risks from exposure to groundwater and soil could potentially occur if the chemicals were left in place. For this reason, various alternatives were evaluated for the remediation of soils and groundwater. To remediate the groundwater and prevent further plume migration, a horseshoe-shaped slurry wall was constructed around the site, and a groundwater treatment system was installed to reduce VOC and PNA concentrations in the groundwater to acceptable levels. No risk assessment was conducted for these activities, as all air emissions during remediation were permitted by the state at levels not expected to impact public health or the environment. A separate risk assessment was prepared to evaluate potential health effects from exposure to the contaminated soil during and after the proposed development and remediation activities. Because commercial development was planned for the property, the preferred remediation option for soils was to couple the site development and remediation plans to produce a cost-effective, efficient remediation. The proposed development was a large retail center, planned to encompass approximately 6 acres of the site.

The proposed remediation plan for the site soils included excavation and redeposition of the contaminated soil into an onsite disposal cell in one portion of the property, and capping of all site chemicals within the confines of the slurry wall. To increase the efficiency of these operations, the remediation and development plans were integrated, and potential emissions from all materials handling activities were evaluated in the remediation risk assessment. Where excavation of the contaminated soil would be necessary for construction of buildings at the site (e.g., footings, utilities), these areas would be remediated by overexcavation and backfilling with clean soil before construction began, so that workers would not come into contact with the contaminated soil during construction.[10]

The remediation risk assessment included an evaluation of potential exposure pathways both during and after remediation, the receptor populations that may be exposed during remediation, an estimation of volatile and dust emissions from the site, and quantification of health risks from the predicted exposures. To select the exposure pathways most appropriate for the evaluation of risks during remediation, it was necessary to assess both the contamination at the site and the details of the proposed remediation. Specific locations of the construction activities for site development were identified, including the locations of utility lines, trenches, footings, and buildings. Maps were compiled outlining areas where construction activities would contact contaminated soil, and detailing specific chemical concentrations in each of these areas.

To estimate potential exposures during the soil handling processes, predictive modeling of volatile and dust emissions and the subsequent deposition of dusts onto adjacent properties was conducted. Particulate and volatile emissions were estimated both during and following remediation.[11] Emission sources were determined on the basis of which areas of the site both contained contaminated soil and would be disturbed by construction activities. For the evaluation, the site was divided into a total of 81 grids, each $25 \times 25\ m^2$. It was found that proposed construction activities would disturb contaminated soil in 46 of the 81 grids; therefore, these 46 grids were considered to be potential emission sources from the site during remediation. Details of the evaluation are outlined in other publications.[9,11]

On the basis of the proposed remediation/development plans and the fate of the chemicals at the site, the primary pathways of exposure from the proposed remediation were found to be:

- inhalation of dusts and volatiles
- ingestion of soils, and
- direct dermal (skin) contact with soils

Three receptor populations were evaluated on the basis of the primary wind conditions and direction, proximity to the site, and sensitivity of the receptors, including onsite workers, offsite residential populations, and students and staff at a nearby middle school. The chemical concentrations of dusts and volatiles at each receptor location were modeled using site-specific parameters (e.g., concentration variances at specific depths, soil particle size, etc.) and the results of detailed analyses of the site activities (e.g., bulldozer movements on the site; number of wheels on the heavy equipment; dusts expected to be generated during batch loading, truck movements, etc).[11]

The primary exposure pathway during construction/remediation was considered to be the inhalation of contaminated dusts or volatile compounds. As indicated above, incidental ingestion and dermal contact with onsite soil and incidental ingestion and dermal contact with windblown dusts deposited on adjacent properties (i.e., residential areas and the school yard) during construction were also evaluated. Because the contaminated soil from the trenches and footings areas would be disposed of in an onsite deposition area, and other contaminated soils would be left in place, the remediation risk assessment also evaluated the potential for volatile compounds entering the buildings (e.g., through the concrete foundations) or being emitted at the ground surface outside the buildings (through the asphalt parking lot and landscaped areas) following construction. Modeling of the chemical volatilization from the subsurface soils was performed, and inhalation risks from the predicted airborne concentrations were evaluated.[9]

The results of the risk assessment indicated that incidental ingestion of soil and dermal contact with both onsite and offsite soil during remediation/construction would result in excess carcinogenic risks above the generally accepted risk range of 10^{-4} to 10^{-6}.[11] These results were used to make critical decisions concerning the proposed remediation for the site. Because the results indicated that risks above "acceptable" risk levels were expected, risk management measures were developed for implementation during field activities, to reduce the predicted risks to acceptable levels. These risk management measures included watering of the soil to reduce dusts, utilization of vapor-suppressing foam to reduce volatile emissions, minimizing the area of excavation exposed at any one time to reduce emissions from the site, and the use of personal protective equipment (PPE) to mitigate worker exposures from incidental ingestion and/or dermal contact with the contaminated soils.

The results of the evaluation of risks following construction of the retail center indicated that risks above acceptable levels were not expected to occur. However, to ensure the protection of future public and worker health, additional risk management measures were implemented at the site. Following construction of the retail center, the volatile emissions from the property would be controlled by capping the site with asphalt, concrete, buildings, or clean soil, and sealing the holes in floor slabs where utilities penetrated to prevent migration of vapors into the buildings.[9,11]

The reduction of risks achieved with the use of risk management measures was quantified and compared to the predicted risks without the use of the risk management measures.[9] It was found that, with the use of risk management measures, all risks above acceptable levels were reduced to well below the acceptable levels. This reduction in risks minimized the client costs and future liabilities associated with management of the waste site.

As illustrated in the case history, the use of risk assessment methods during the remedial process allowed the prediction of health risks before they occurred, and assisted in the development of site-specific risk management measures. The risk management measures reduced the predicted health risks to acceptable levels, and ultimately resulted in the use of an acceptable, effective, and permanent remediation for the site.

REFERENCES

1. United States Environmental Protection Agency. ROD (Record of Decision) Annual Report, FY 1987. Hazardous Site Control Division, U.S. EPA, Washington, DC. July, 1988a.
2. United States Environmental Protection Agency. ROD Annual Report, FY 1988. Office of Emergency and Remedial Response. EPA/540/8-89/006. OERR No. 9200.5-210. July, 1989.
3. United States Environmental Protection Agency. Guidance for Conducting Remedial Investigations and Feasibility Studies under CERCLA. Office of Emergency and Remedial Response, Office of Solid Waste and Emergency Response. Washington, DC, Interim Final. (OSWER Directive 9355.3-01). October, 1988b.
4. United States Environmental Protection Agency. National Oil and Hazardous Substances Pollution Contingency Plan; Proposed Rule. Federal Register 53:51394-51520. December 21, 1988c.
5. United States Environmental Protection Agency. National Oil and Hazardous Substances Pollution Contingency Plan; Final Rule. Federal Register. 55:8666-8865. March 8, 1990.
6. United States Environmental Protection Agency. Superfund Record of Decision: Tyson's Lagoon Site, Upper Merion Township, Pennsylvania, 1988d.
7. Environ Corporation. The Risks Associated with Two Alternative Remediation Plans for the Hardage Superfund Site. Washington, DC, June 7, 1989.
8. United States Environmental Protection Agency. Superfund Record of Decision: Lone Pine Landfill, New Jersey. Office of Emergency and Remedial Response. Washington, DC, September, 1984.
9. Edmisten, G. E., J. C. Blasco, T. L. Foster, S. Ardalan, F. I. Cooper, K. A. Macoskey, and S. C. Charjee. From Risk Assessment to Risk Management: Case History of a Hazardous Waste Site. Proceedings of HAZTECH INTERNATIONAL '89, Cincinnati, OH, September, 1989.
10. Edmisten, G. E. It Can Be Done: Managing the Development of Chemically Contaminated Properties. In: *Site Assessment Management Sourcebook,* Environmental Publications Inc., 1990, pp. 86–91.
11. Harding Lawson Associates. Risk Appraisal, San Rafael Retail Project, San Rafael, California. June 21, 1989.

CHAPTER 19

A Preliminary Decision Framework for Deriving Soil Ingestion Rates

Edward J. Calabrese, Environmental Health Sciences Program,
Edward S. Stanek, Biostatistics and Epidemiology Program,
Charles E. Gilbert, Environmental Health Sciences Program,
School of Public Health, University of Massachusetts, Amherst, Massachusetts

INTRODUCTION

The issue of selecting an appropriate soil ingestion value for children and more recently for adults[1] has become a significant challenge to public health and regulatory agencies as well as consultants performing site-specific risk assessments. The public health implications, along with enormous cost considerations, have played a major role in shaping the debate over what the level(s) should be. This chapter will offer guidance on how to approach the problem of selecting an appropriate daily soil ingestion rate in light of regulatory/public health needs within the context of the quality of the present soil ingestion database.

QUALITY OF THE PRESENT SOIL INGESTION DATABASE

It was shown in Calabrese and Stanek[2] (i.e., Part II of the series of articles) that four studies (i.e., Binder et al.[3]; Calabrese et al.[4]; Van Wijnen et al.[5]; Davis et al.[6]) have been published which were designed to provide quantitative evidence for soil ingestion rates among children. Of those four studies, only the Calabrese et al.[4] and Davis et al.[6] reports provided convincing qualitative evidence that the children in their studies actually eat soil. However, only the Calabrese et al.[4] report was able to provide quantitative estimates of soil ingestion based on acceptable precision of the recovery of tracers ingested with soil. Of the eight tracers employed in the Calabrese et al.[4] study, only two (i.e., Ti and Zr) were recovered with acceptable limits of precision (i.e., 100% ± 20% for 2 SD). The current database, therefore, on how much soil is ingested by children, is limited to one study. Prior to the development of the model[7] to estimate soil ingestion detection capacities of soil ingestion studies, it was believed that the soil ingestion studies offered quantitative evidence of soil ingestion and that their general similarity in estimates provided a complementary and stable database upon which reliable estimates of soil ingestion could be made. However, the quantitative foundations for soil ingestion values were challenged with the recent findings reported in this series of papers[2,7] leaving a seriously compromised soil ingestion database for children, with only one study providing quantitative estimates and even for that study,[4] six of the eight tracers employed could not be determined with adequate precision of recovery. Given this situation, how do regulatory/public health officials estimate soil ingestion among children and possibly adults?

The current limited database places great emphasis on the Calabrese et al.[4] study, since it offers the only quantitative estimates of soil ingestion. Within this context one must recognize the limitations that this study offers. While based on observations of a mass-balance study of 64 children over a two-week period, this study offered several methodological advantages including:

- the use of multiple (i.e., 8) soil tracers
- following the subjects over two weeks
- obtaining daily tracer ingestion/excretory values
- validating tracer recovery efficiency with adults.

While this study contributed to helping assure a high reliability in soil ingestion estimates of the study participants, it is important to emphasize that the study has significant limitations with respect to its generalizability to other populations of children, especially those residing in urban areas. In addition, the nonrandom nature of the selected population affects its capacity to be generalized to the community in which it was conducted (i.e., academic community in western Massachusetts). In addition, the study population was observed only over two consecutive weeks. While suggesting a possible magnitude of intrasubject soil ingestion variation, the study provides no insight for variation with respect to seasonality, which may be a significant variable, especially in colder climates. Another factor inadequately considered in the study was the relationship of the extent of grass cover and how that was quantitatively related to soil ingestion.

These collective limitations of the Calabrese et al.[4] study to offer generalizations to other populations of children present a serious challenge to regulatory/public health officials performing soil-based risk assessments. In the absence of an adequate database, it becomes even more imperative to establish rational procedures for deciding what is (are) the most appropriate soil ingestion value(s). Ideally, given the limitations of the current database, a national numerical value for children and adults should not be derived until additional appropriate studies are completed. However, given that soil cleanup activities are rapidly proceeding, and that risk-based criteria are needed to assist in the determination of cleanup levels, it appears that soil ingestion values will be selected regardless of whether there is an adequate database or not. In light of this reality and given the current database, how should a daily soil ingestion rate be selected? Five critical issues need to be addressed in the analysis.

1. Selection of tracer. Which tracer should be selected? Since the two quantitatively acceptable tracer elements in the Calabrese et al.[4] study differ in their median values by a factor of 3.4 (16 for Zr vs 55 for Ti), should they be averaged, or should the highest (i.e., Ti) be selected?
2. Statistic selection. Should the population-based median, geometric mean, or arithmetic mean be selected? Or should a specific percentage such as the value for the upper 75%, 90%, 95%, or 99% be selected?
3. How should soil ingestion rates change with age? Since soil ingestion is likely to be a function of age, how should these values be estimated and interage variations be determined for public health/regulatory purposes?
4. Is it possible to differentiate the urban child from the suburban and rural child with respect to age-dependent and seasonal soil ingestion rates?
5. How should the so-called soil pica child be handled?

SELECTION OF TRACER

As indicated in Calabrese and Stanek,[2] those tracer elements with the smallest confidence interval estimates (expressed as a percent of the estimated median) are considered superior to those with larger confidence intervals. However, it is important that the distribution of respective tracers display a high degree of overlap, since this supports the assumption of intertracer reliability. Both of the above factors must then be seen within the context of the precision of tracer recovery efficiency.

At the present time, estimations of soil ingestion have been tracer-element specific. However, this leads to proliferation of soil ingestion estimates, and possible confusion in interpretation. It is possible to fit novel statistical models (generally seemingly unrelated regression models, using mixed model variance structures) that simultaneously account for prediction of soil ingestion, based on several elements. Such models have a clear appeal, since they may produce a single soil ingestion estimate for a study, and thus produce a better estimate. This is partly because they take advantage of the co-variance of soil ingestion estimates from different elements, and in part because they weight different elements proportionally to their recovery precision. Such models provide perhaps the most economical way of gaining additional insight into soil ingestion. This approach has yet to be implemented, but should be explored with the eight tracers of the Calabrese et al.[4] study as a feasible technique to selecting soil ingestion values.

Since the above analysis has yet to be completed and evaluated, which tracer(s) should be selected for use? The present evidence indicates that only Ti and Zr in the Calabrese et al.[4] study were seen with a reliable degree of precision. For these two tracers, acceptable estimates of soil ingestion exist. For the remaining six tracers in the Calabrese et al.[4] and Davis et al.[6] studies, upper bound values of soil ingestion per tracer element are provided. Consequently, the estimates offered for Ti and Zr are the most appropriate for further consideration. Statistical analyses[2] revealed that the distribution of median values for Zr were tighter for Zr than Ti. In addition, Zr displayed a more favorable (i.e., lower) food to soil ratio than Ti and a more sensitive (i.e., lower) level of soil ingestion detection capacity in the Calabrese et al.[4] study. These factors collectively suggest that Zr is more likely to be a better choice than Ti. We therefore recommend that Zr be viewed as the preferred tracer upon which soil ingestion estimations for the average child in the Calabrese et al.[4] study be based. We caution that selection of the preferred tracer will be study-specific and may vary according to the population statistic selected.

STATISTIC SELECTION

For regulatory/risk assessment purposes, some have recommended using a mean estimate rather than median. The mean has the advantage of being simpler and more easily understood. We feel that this is a weak argument for using the mean, since the mean will be strongly influenced by extreme values. In contrast, the median has been shown to be much more robust than the mean,[8] in the analysis of the Calabrese et al.[4] data with a variety of various assumptions in the formulation of different population estimates. However, the use percentiles (such as the 75%) would be a sound measure if some estimate larger than the median were desired. If some estimate of the mean is insisted upon, a better estimate will be the geometric mean, since this estimate will account for the skewness of daily soil ingestion values observed among children. In fact, in forming confidence intervals for soil ingestion, better coverage can be obtained by forming confidence intervals based on the geometric mean (using log base e), rather than the simple arithmetic mean. However, the use of the geometric mean reduces soil ingestion estimate simplicity.

If it were necessary to select a single estimate for soil ingestion we would recommend an estimate on a combined soil and dust element concentration, and consider the best estimate to be based on the median. The rejection of the arithmetic mean is based not only on its instability as a measure of central tendency of the population, but also that it does not have any precise meaning in the assessed population. For example, with variables that have skewed distributions, such as the log normal distribution of the soil ingestion estimates, the mean does not represent any benchmark in terms of a percentile.

AGE-RELATED CHANGES

It has been generally assumed by various state/federal regulatory and public health agencies that all human age groups ingest soil. It has been concluded, based principally on professional judgment, that children ingest more soil than adults and that children with high hand-to-mouth activity (i.e., ages 1–4) ingest more soil than children of other ages.

Analysis of the Calabrese et al.[4] data revealed that soil ingestion increased linearly with age for all tracers. This was particularly evident for Ti, while much less for Zr.[9] The fact that the slopes of these two tracers, (i.e., the most reliable tracers) differ to such an extent precludes being able to make a confident decision for the Calabrese et al.[4] study population that soil ingestion increases as children increase in age from 1 to 4. An attempt to estimate soil ingestion in the six adults used in the three-week tracer recovery validation study indicated an average soil ingestion rate of approximately 40 mg/day.[10] However, the estimated rate of soil ingestion in this pilot study was considerably below the actual soil ingestion detection limits of that particular study because of the small sample size and high food tracer to soil tracer ratio (see Stanek and Calabrese[7]). Thus, the reported soil ingestion estimates for adults in Calabrese et al.[4] could not be seen with sufficient recovery precision to derive accurate judgments about soil ingestion rates in adults. The database, therefore, does not include confident values for soil ingestion for adults, while no clear evidence exists that age-related changes in soil ingestion occurred in the Calabrese et al.[4] study. The age analysis for children is also further compromised because the sample size for each age group was small (i.e., ~10 for six-month age intervals). Because of the small sample size for age-specific soil ingestion values, the study's soil ingestion detection limits are far higher than values that were reported.

In light of the inadequacies of the soil ingestion database, how are age adjustments in soil ingestion to be made? It would appear logical that adults should ingest significantly less soil than young children. It would seem reasonable, in the absence of reliable quantitative data, to assume that an "average" adult ingests from 25% to 10% of the "average" child (1–6 years old), based on diminished hand-to-mouth activity and other maturational and social factors.

Based on this rationale, it is recommended that children 6–12 years of age be assumed to ingest 25% of the soil ingestion value of a 1–6 year old child, while those >12 years of age be assumed to ingest 10% of the 1–6 year old child.

RURAL VS URBAN/SUBURBAN CHILDREN

No quantitative data exist on the comparative soil ingestion rates of children from rural, urban, and suburban areas. This remains an important data gap to be filled. At present, any attempt to make a distinction in soil ingestion rates would be speculative. There may be a number of potential factors affecting the differential rate of soil ingestion among rural, urban, and suburban children, such as time spent outdoors, degree of grass cover of outdoor play areas, quantity of dust in home, and others. However, in the absence of adequate information on these variables, the present emphasis will focus on the extent of grass cover, because of the obvious direct access to contact with soil. Under such circumstances it is not unreasonable to suggest the incorporation of an uncertainty factor analogous to those used in risk assessment activities for noncarcinogens. Since this represents concern with interindividual variation, an uncertainty factor of approximately 1–3 is selected, depending on the degree of grass cover in areas where children play. If grass cover is extensive (>90%) then a UF of 1 would be appropriate. However, if grass cover is more limited (<90%) in areas of access, then a threefold factor would be recommended.

SEASONALITY

The Calabrese et al.[4] study was conducted in the fall (Sept./Oct.) in Massachusetts. It may be speculated that ingestion of soil may be highest in the summertime and lowest in the winter, based on the premise that children play longer outdoors in the summer, with greater direct contact with soil. However, it may be argued that soil contact may actually be greater in the spring before the growth of grass becomes significantly thickened, or during seasons with more rain, such as spring and winter, depending on geographical locations. Thus, it is possible that seasonal effects may markedly vary according to a variety of factors, and that ingestion may not be highest in the summer months in all locations. In addition, there may be seasonal variation in the tracking in of dust within the home, with perhaps more mud being tracked into the home during the seasons with more rain. In the absence of information to clarify these uncertainties in the database, no seasonal effect is recommended at this time.

IDENTIFICATION OF PICA CHILDREN

The consumption of nonfood items, especially by young children, is a very common activity; when this activity is excessively performed it becomes characterized as pica. The range of nonfood items that such children may ingest is extremely variable, including: clothing, books/paper, crayons, soil, cigarettes, household furnishings, and other items.

The prevalence of pica behavior appears to be highly variable, being contingent on the definition of pica and the population assessed, among other factors. Table 19.1 reveals that the prevalence of pica behavior can range from 10% in Caucasian children to 66% in institutionalized psychotic children. It appears, therefore, that children 1 to 6 years old display a pica prevalence that is between 15% and 30%, with no obvious significant variation between males and females.

Table 19.1. Range of Pica Behavior Prevalence

Group Description	No. of Subjects	% of Pica	Reference No.
Retarded children	30	50	11
Black children > 6 mo.	386	27	12
White children > 6 mo.	398	17	
Black, 1–6 years	486	32	13
White, 1–6 years	294	10	
Children, low income	859	55	14
Children, high income		30	
Children, 1–6 (interview)	439	15	15
Children, 1–6 (mail survey)	227	50	
Institutionalized, psychotic 3–13 years	40	66	16
Spanish American children (California)	21	32	17
Children (Mississippi)	115	16	18

Source: Charles Gilbert, School of Public Health, University of Massachusetts, Amherst, MA, 1989.

The identification of pica children presents a major initial stumbling block, since there are no definitive criteria for this behavior. The present literature often represents subjective judgments based on individual perceptions of what comprises pica behavior, with limited standardized behavioral norms concerning whether children display pica behavior.

Some evidence exists suggesting that there is considerable variation among pica behaviors, especially concerning what items are preferentially ingested.[15,19] More specifically, a child with pica behavior may ingest only selective items, to the exclusion of others. For example, a pica child ingesting books or paper may not ingest other items such as cigarettes or soil. In contrast, a child with a preference for soil may not ingest other nonfood items. In fact, Harvey et al.[19] has observed that soil pica behavior comprises but a small subset of childhood pica behaviors. Their data suggest that there is an age-dependence in object selectivity and that such selectivity increases with age. The potential implications of the Harvey et al.[19] data are that not all children with pica ingest soil and, in fact, only a subset of children with pica ingest greater than average amounts of soil. If one quarter (25%) of children display pica behavior, it may be reasonable to assume that about 25% of those children are soil pica children.[19] This would result in about 6.25% of the population aged 1–6 years displaying soil pica behavior. Given this theoretical estimate of 6.25% for an estimated soil pica prevalence, it would be important to compare this value with estimates based on soil tracer ingestion studies for children in the 1–6 year age range. The four available soil ingestion studies[3-6] were examined for evidence of pica soil evidence. These collective studies have provided daily soil ingestion on 517 children. If soil pica were subjectively defined in quantitative terms as consumption of greater than 1 gm of soil per day, then 10 individuals would be identified as having displayed this behavior. This would amount to a prevalence of 1.9%. This would be considerably lower than the earlier noted theoretical estimate of 6.25% for soil pica in 1–6 year olds.

This soil tracer estimation of the prevalence of soil pica children of course rests on very limited data. The nine individuals in the Van Wijnen et al. study[5] displayed the pica-like behavior (≥1000 mg/day) on only a single observation day over a 2–5 day period. The soil ingestion values of these subjects were not adjusted downward for food ingestion of the tracer elements, thus leading to variable overestimates of soil ingestion. The one child pica subject in the Calabrese et al. study was observed over an 8 day period and displayed soil-pica behavior only in the second of the two week period of observation. These data suggest that those displaying soil pica behavior do so irregularly and thus would not be predicted to consistently ingest ≥1 gram of soil per day. If one accepted that about 2% of children 1–6 years of age occasionally ingest ≥1.0 gram of soil per day, what percentage would ingestion ≥5 grams per day? The Calabrese et al.[4] data suggest that only 1 child of the 517 (0.2%) displayed this level of soil ingestion if based on Al, Si, Ti, V, and Y as tracers (note that approximately 1300 mg/day was estimated for Zr). It should be noted that soil ingestion estimates for this soil pica child were seen with a precision of 100% ± 20% or less for Al, Si, Ti, and Zr. At present, no adequate explanation accounts for the lack of intertracer reliability, with Zr displaying about 20% of the soil ingestion value estimated for Al, Si, and Ti. The low values for Zr are expected since the adult percent recovery studies suggest that some Zr was lost during sample preparation and chemical analysis. Possible toothpaste consumption could contribute to the elevated values for Al and Si.[2] As for Ti, the reliability of the precision of recovery is markedly variable, depending on whether one uses the median versus the mean in the food to soil ratio, i.e., food soil ratio for median (0.00118) and for mean (0.2112). So variable is this estimate that with the median value, the precision of recovery is 100% ± 5% while 100% ± >50% if using the mean! In contrast, using either the mean or median value for Zr results in a recovery of 100% ± <20%. Based on these collective findings it would appear that the Zr value is the most reliable of the four available. If this were the case, then the value of 1 gm/day would be clearly preferable to the 5–6 gm/day value.

Over what duration of normal life span would one display soil pica? While there are no adequate data to resolve this issue, it is generally believed that pica behavior is of limited duration, with the prevalence in the population being highest over 1–3 years of age, but declining to 1–5% by the age of 6.[15] If soil pica declined accordingly, then the soil pica prevalence at age 6 might be predicted to be .02–1.0%.

Despite the inadequate database concerning soil ingestion of soil-pica children, it is becoming necessary to offer tentative guidance in this area. Soil pica is a subset of pica behavior, and has a prevalence in the 1–6 age population of under 8%, based on survey methods.[19] The quantitative tracer methodology for soil ingestion is believed superior to the qualitative survey information of soil pica prevalence, since it offers definably precise estimates of exposure. It is, therefore, more likely that the actual prevalence of soil-pica children is far below the 8% figure given above. It should be noted that the soil tracer methodology estimates that 2% will occasionally ingest at least 1 gram of soil on a given day. The 2% figure is likely to have been considerably lower if Van Wijnen et al.[5] had adjusted for food intake. The soil tracer methodology estimates that 0.19% ingest up to 5 grams of soil per day if the less reliable tracers (i.e., Al, Si, and Ti) of the Calabrese et al. study[4] were used) and 0.0% if based on Zr. The duration of exposure for these estimates is most likely restricted to ages 1–6 (i.e., 5 years). While it is possible that soil pica may be observed in children beyond age 6 for some individuals, the prevalence of this behavior is expected to rapidly decrease as the individual ages. Note that the prevalence calculations employed above (i.e., arriving at the 0.19% value) used the ages of greatest prevalence and would be markedly lower for ages 5 and 6.

It is, therefore, generally recognized that an inadequate database exists with respect to the prevalence of soil pica, the amount of soil such children ingest, and over what duration this is observed. However, what limited data do exist suggest that soil pica, as defined by approximately 1000 mg/day, may have an upper bound of 2% of children aged 1–6. A small subset of this population (~ 0.2%) is speculated to ingest up to 5 gm/day.

With the notable exception of the one child in the Calabrese et al.[4] study, no conclusive data exist that any other pica children (~1000 mg/day) were observed, since food ingestion was not adjusted by Van Wijnen et al.,[5] nor were these behaviors seen on repeated days.

If one were to err on the side of safety with a speculative upperbound daily soil ingestion rate, one possible course of action might be the following:

1. Assume that 2% of children aged 1–6 years exhibit soil pica of 1 gram of soil per day.
2. Assume that 0.2% of children ingest 5 grams of soil from 1–6 years of age.

If, however, a more realistic estimate of soil ingestion in soil pica children were desired, then the following suggestions may be followed:

1. Assume that 1% of children exhibit soil pica of 1 gram of soil per day for 4 days per week and 500 mg/day for 3 days per week for 4 years.
2. Assume that 0.2% of children ingest 5 grams of soil for 4 days per week and 500 mg of soil for 3 days per week for 4 years.

DISCUSSION

Selection of the "correct" soil ingestion number is not a very wise goal, as much as simple solutions are desired. Soil ingestion is likely to be influenced by a variety of factors that need to be assessed and then quantitatively incorporated within a soil ingestion derivation procedure. This chapter provides a guide for how to proceed along such a soil ingestion derivation process. It attempts to identify the critical issues and to show how such factors may affect soil ingestion values and how they may be quantitatively dealt with in the context of available data or in default values. Table 19.2 presents the structure of the decision framework recommended to lead to a rational and defensible soil ingestion rate. This approach is designed to assist, but not replace, professional judgment by public health/regulatory risk assessment specialists in this site-specific approach to assessing soil ingestion by children.

Table 19.2. Decision Framework for Deriving Soil Ingestion Rate

Tracer	1. Needs to have acceptable precision in recovery. 2. Narrow C. I. for distribution of median. 3. High inter-tracer reliability.
Statistic	1. Stable measure of central tendency (median, geometric mean). 2. Select percentile of choice.
Age	1. Data are inadequate to differentiate age. 2. Age related behavioral changes suggest that older children and adults ingest from 1/4 to 1/10 that of children.
Urban/suburban/rural	1. Use UF approach—use UF of up to 3 if grass cover is limited (i.e., < 90% covered).
Seasonality	1. This is also unknown; no adjustment is recommended at this time since reasonable cases can be made for different seasons providing greater risk of soil ingestion.
Dust/soil	1. Recommend using a combined soil/dust measurement.
Pica	1. Assume 1/200 children ingest about 1 gm soil 4 days/wk for 4 years during the 1–6 age span.

REFERENCES

1. Porter, J. W. U.S. Environmental Protection Agency Office of Solid Waste and Emergency Response. Memorandum to regional administrator, Region I–X, regarding interim final guidance on soil ingestion rates, January 27, 1989.
2. Calabrese, E. J., and E. J. Stanek III. "A Guide to Interpreting Soil Ingestion Studies. II. Qualitative and Quantitative Evidence of Soil Ingestion," *Reg. Toxicol. Pharm.* (Submitted, 1990.)
3. Binder, S., D. Sokal, and D. Maughan. "Estimating the Amount of Soil Ingested by Young Children Through Tracer Elements," *Arch. Environ. Health* 41:341–345 (1986).
4. Calabrese, E. J., R. Barnes, E. J. Stanek, H. Pastides, C. E. Gilbert, P. Veneman, X. Wang, A. Lasztity, and P. T. Kostecki. "How Much Soil Do Young Children Ingest: An Epidemiologic Study," *Reg. Toxic. Pharm.* 10:123–131 (1989).

5. Van Wijnen, J. H., P. Clausing, and B. Brunekreef. "Estimated Soil Ingestion by Children," *Env. Res.* 51:147–162 (1990).
6. Davis, S., P. Waller, R. Buschbom, J. Ballou, and P. White. "Quantitative Estimates of Soil Ingestion in Normal Children Between the Ages of 2 and 7 Years: Population-Based Estimates Using Aluminum, Silicon, and Titanium as Soil Tracer Elements," *Arch. Env. Hlth.* 45:112–122 (1990).
7. Stanek, E. J., III, and E. J. Calabrese. "A Guide to Interpreting Soil Ingestion Studies. I. Development of a Model to Estimate the Soil Ingestion Detection Level of Soil Ingestion Studies," *Reg. Toxicol. Pharm.* (Submitted, 1990.)
8. Stanek, E. J. III, E. J. Calabrese, and C. E. Gilbert. "Choosing a Best Estimate of Children's Daily Soil Ingestion," in *Petroleum Contaminated Soil.* Vol. 3. P. T. Kostecki and E. J. Calabrese, Eds. (Chelsea, MI: Lewis Publishers, 1990), pp. 341–348.
9. Stanek, E. J. III, E. J. Calabrese, and L. Zheng. "Soil Ingestion Estimates in Children: Influence of Age and Sex," *J. Soil Geochem. Hlth.* (Submitted, 1990.)
10. Calabrese, E. J., E. J. Stanek, C. E. Gilbert, and R. M. Barnes. "Preliminary Adult Soil Ingestion Estimates: Results of as Pilot Study," *Reg. Tox. Pharm.* 12:88–95 (1990).
11. Kanner, L. *Child Psychiatry,* (Springfield, IL: Charles C. Thomas, 1937) , pp. 340–353.
12. Cooper, M. *Pica* (Springfield, IL: Charles C. Thomas, 1957).
13. Millican, F. K., E. M. Layman, R. S. Lourie, L. Y. Rakahashi, and C. C. Dublin. "The Prevalence of Ingestion and Mouthing of Non-Edible Substances by Children," *Clin. Proc. Child. Hosp.* (Wash.), 18:207–214 (1962).
14. Lourie, R. S., E. M. Layman, and F. K. Millican. "The Epidemiology of Lead Poisoning and Children," *Arch. Pediat.*, 79:72–76 (1963).
15. Barltrop, D. "The Prevalence of Pica," *Am. J. Dis. Child.*, 112:116–123 (1966).
16. Oliver, B. E., and G. O'Gorman. *Develop. Med. Child. Neurol.*, 8:704–706 (1966).
17. Bruhn, C. M., and R. M. Pangborn. *J. Am. Diet. Assoc.* 58:417–420 (1971).
18. Vermeer, D. E., and D. A. Frate. "Geophagia on Rural Mississippi: Environmental and Cultural Contexts and Nutritional Implications," *Am. J. Clin. Nutr.*, 32:2129–2135 (1979).
19. Harvey, P. G., A. Spurgeon, G. Morgan, J. Chance, and E. Moss. "A Method for Quantifying Hand-to-Mouth Activity in Young Children," *J. Child. Psychol.* (1986).

CHAPTER 20

Development and Application of a Decision Tool for Managing Petroleum Contaminated Soils

Katherine K. Connor, Jennie S. Rice, and
Justin L. Welsh, Decision Focus Incorporated, Los Altos, California

INTRODUCTION

Decision Focus Incorporated (DFI) is working with the Electric Power Research Institute (EPRI) to develop tools to assist the utility industry in managing petroleum contaminated soils, one of the largest volume waste streams for many utilities. Major sources of petroleum contamination include underground and aboveground storage tanks, pipelines, maintenance operations, and fleet vehicles. In the past, utilities have typically managed this waste stream by excavation and disposal in Class III landfills. However, recent classification of contaminated soils as a hazardous waste in some states, together with the 1990 land ban, has made landfilling this waste considerably more difficult and more expensive. As a result, identifying and evaluating new options for managing petroleum contaminated soils has become a priority for utility managers.

PREVIOUS WORK ADDRESSING SOILS RISK MANAGEMENT

Experience in Risk Management of Contaminated Soils

DFI, a management consulting firm with broad experience in the areas of risk assessment and risk management of contaminated soils and other waste streams, has developed several easy-to-use tools for personal computers that have been applied to risk management decisions for the utility and other industries. Current software tools address various sources of contamination including underground storage tanks, polychlorinated biphenyl (PCB) equipment and spills, chemical spills, and manufactured gas plant sites. These tools have helped utilities identify and evaluate alternative risk management strategies by analyzing the benefits and costs of each option. These tools are described below.

Contaminated Site Risk Management System

The Contaminated Site Risk Management System (SITES) analyzes the health and environmental risks at hazardous waste sites and evaluates alternative remediation strategies. SITES is an interactive software tool for use on personal computers. Pull-down menus guide users through data entry, analysis, and report generation. Once a user develops the initial site risk analysis, SITES enables alternative assumptions to be tested quickly and easily. The SITES model reports detailed and aggregate doses and risks in both tables and graphs that may be easily transformed into visual aids for presentation. SITES has been used to study contaminated soil problems at manufactured gas plant waste sites and sites contaminated by petroleum leaks from fuel tanks.

QuickTANKS and QuickPIPES

In September 1988, the Environmental Protection Agency (EPA) released new regulations for underground tank systems, including pipes, that provide schedules for replacing or upgrading existing underground tanks and for beginning leak detection programs. DFI's QuickTANKS and QuickPIPES decision models help environmental managers respond to the considerable freedom to select the most appropriate action. The models, using risk management methodology, balance known capital and operating costs against uncertain leak and future replacement costs. Depending on the specific underground storage tank system and site characteristics, replacing tanks and piping before being required to by EPA can save thousands of dollars. Specifically, the models identify:

- the best time to replace or upgrade existing tanks and piping
- the lowest cost replacement tank or piping system
- the best leak-detection strategy

Screening and Priority-Setting Systems

A formal priority-setting program using an analytical ranking system can help a company manage both the internal and external pressures of site management. Under the circumstances of limited information about a large number of sites, computer-based models can be particularly useful. DFI has developed two models to help prioritize contaminated sites: the Site Screening and Priority-Setting System (SSPS) and the PCB Spill Priority-Setting Model (PSPM).

The SSPS is a scoring system modeled after EPA's Hazard Ranking System (HRS). The SSPS improves on the HRS by incorporating more of the information available about a site and by correcting many of the HRS's functional shortcomings. Like the HRS, the SSPS assigns scores to site attributes and then combines the scores to produce ranking values. For each site the SSPS calculates a value for the level of human health risk, the level of perceived risk, and the cost of remediation. Companies can efficiently focus their management efforts on the sites with the highest risk scores. Although originally applied at manufactured gas plant sites, the SSPS can be used to analyze almost any type of site.

The PSPM uses site-specific data, information about environmental transport and fate of PCBs, and risk assessment techniques to produce three scores for comparing sites contaminated with PCBs.

THE NONCOMBUSTION WASTE RISK MANAGEMENT PROJECT

What are Noncombustion Wastes?

DFI is currently working for EPRI on a project called Risk Management for Noncombustion Wastes. The goal of this three-year project is to develop materials and software tools to help utilities manage noncombustion wastes. Noncombustion wastes, wastes that do not result from combustion processes, have varied sources. This project examines the following waste streams:

- petroleum contaminated soils
- spent solvents
- boiler chemical cleaning wastes
- wood poles
- PCB-contaminated oil and equipment
- ethylene glycol
- used oil
- batteries
- asbestos
- paint, paint sludges, and containers
- surface impoundments
- oily wastewater
- oily debris

Why Are Noncombustion Wastes Important?

In the past, management of noncombustion wastes has been relatively straightforward, with many of them being sent to landfills for ultimate disposal. However, evolving solid waste regulations and large cleanup and liability costs incurred for past practices are rapidly changing the options and costs for managing these wastes.

The Federal Resource Conservation and Recovery Act (RCRA)

Recent solid and hazardous waste legislation at both the federal and state levels has provided a strong impetus for utilities to reexamine their noncombustion waste management practices. Enacted in 1976, the RCRA regulates the generation, transportation, storage, treatment, and disposal of hazardous and nonhazardous wastes. Statistics show that 133 million metric tons per year of hazardous waste were being disposed of in surface impoundments, 30 million metric tons per year injected into underground wells, and 5 million metric tons per year in landfills or waste piles.[1] Regulatory activity under RCRA has focused primarily on the management practices of these wastes.

The Hazardous Waste and Solid Waste Amendments of 1984

Congress has amended RCRA six times. The most important amendments, the Hazardous and Solid Waste Amendments of 1984 (HSWA), contain specific provisions designed to reduce land disposal of hazardous waste. The ultimate goal of these regulations is to have no untreated hazardous wastes in landfills, hence HSWA is often referred to as the "land ban." Other provisions of the amendments include the following:

- stringent technical design standards for landfills and surface impoundments
- regulation of small waste generators (100 to 1000 kg/month) and owners, operators, and distributors of underground storage tanks
- proof of the existence of waste reduction plans

Impact on Electric Utility Industry

The phased ban on land disposal of hazardous wastes, which EPA is expected to complete in 1990, may be the most significant of the new hazardous waste regulations from the electric utility industry perspective. Its impact may be enormous, leading utilities to seek alternative, probably more costly, methods of disposal. In the meantime, enhanced design requirements for land disposal facilities have increased costs and caused a shortage of available waste treatment capacity. Recent reports indicate that the costs of land disposal have increased two to six times since HSWA was enacted.[2] By November 1985, 1,000 of the approximately 4,000 land disposal facilities in the United States had lost their interim status and were forced to file closure plans.[3] Further contributing to the shortage of waste management alternatives, 20 states have no commercial hazardous waste treatment, storage, or disposal facilities.[4]

State Regulations

Evolving regulations at the state level also may have significant impacts on the cost and options available for noncombustion waste management. Several states have enacted solid and hazardous waste management regulations that are more stringent than RCRA. In addition, many states have imposed severe restrictions on the construction and use of hazardous waste treatment and disposal facilities within their borders. Another emerging trend at both state and federal levels is emphasis on waste minimization and source reduction. Many states have initiated waste reduction programs and the federal EPA has established a new Office of Pollution Prevention and added waste minimization staff within its Office of Solid Waste.

Summary of Regulatory Impact

In summary, state and federal regulations have resulted in escalating costs and an increasingly risky environment for utility management of noncombustion wastes, including petroleum contaminated soils. Regulatory pressures, combined with growing public concerns, provide strong motivation for utility managers to understand and incorporate into their decisionmaking, the costs and risk associated with current and future management options for noncombustion wastes.

The Need for Noncombustion Waste Management Decision Tools

Utility noncombustion waste management decisions are characterized by

- numerous interrelated noncombustion waste-generating processes
- a large number of different waste types requiring many different management approaches
- complex and evolving regulations that affect or may soon affect these wastes
- costs and other considerations in waste generation and management that are multidimensional, and are incurred at a variety of different levels within a utility
- limited data on utility noncombustion wastes and the waste-generating processes
- many uncertainties in all facets of the management problem
- potentially very high cost impacts of poor management

As shown in Figure 20.1, the noncombustion waste management problem is characterized by a number of interrelated decisions. Further, each of these decisions must be evaluated with respect not only to its direct cost implications, but also with respect to potential impacts on human health and the environment, workers, public perceptions, and long-term liabilities. In order to minimize costs and

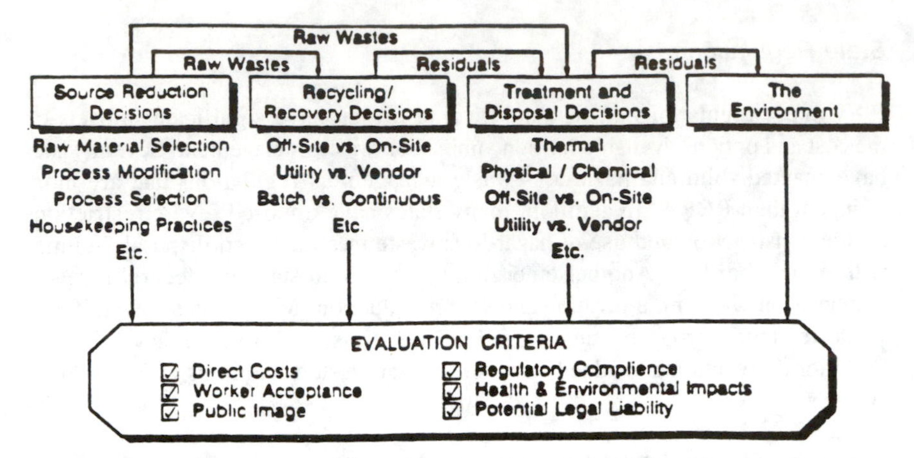

Figure 20.1. The noncombustion waste management problem.

risks, utilities must address the potential impacts of multiple management decisions regarding source reduction, recycling/recovery, and treatment and disposal of noncombustion wastes in an integrated and consistent manner.

In order to address complex noncombustion waste management problems, utility managers will need access to comprehensive databases of waste characteristics relating chemical composition to impacts on human health and the environment, databases of process-specific waste reduction techniques, and databases of waste-specific management alternatives (for both current and emerging technologies). For each noncombustion waste, it is important to understand its source (the details of the waste-generation process), frequency of generation, chemical composition, and physical characteristics. Onsite waste handling practices may also be an important component, since mixing streams can change the overall waste characteristics and, subsequently, change the regulatory and risk management requirements. For example, the use of appropriate segregation practices may be an important consideration in developing a waste management strategy. Uncertainties in cost, performance, availability, effects on utility operations, and applicable regulations also may impact a utility's waste reduction and management decisions.

What is Risk Management?

Risk management is the process of weighing alternatives and selecting the most appropriate management action, integrating the results of risk assessment with engineering data and with social, economic, and political concerns to reach a decision. An important concept in risk management is that of a risk portfolio—the idea that at any point in time a group (or individual) is "carrying" a portfolio of risks to which the group is exposed. Risk assessment is the use of the factual base to define the health effects of exposure of individuals or populations to hazardous materials and situations. Hence, risk assessment is concerned with the art of characterizing what a risk portfolio looks like while risk management is concerned with characterizing the relationship between decisions that can be made and the effect on a risk portfolio of those decisions.

Waste management is concerned with the amount of waste present and alternative steps that can be taken to reduce the quantity or environmental impacts of the waste. Waste management usually does not explicitly consider the risks associated with alternative waste management options.

Waste management alternatives have been divided into three broad categories:

- source reduction
- recycle/reuse/recovery
- treatment and disposal

Traditionally, utility waste management decisions have emphasized treatment and disposal. In contrast, the primary emphasis of the Risk Management for Noncombustion Wastes project is on the use of source reduction and recycle/reuse/recovery options when they can cost-effectively reduce risks.

Timeline of the Risk Management for Noncombustion Waste Project

Figure 20.2 illustrates the primary components of this three-year project. Each of the primary steps are summarized below.

Synthesize Information

In 1988 the project team, using utility visits and surveys, gathered information about noncombustion wastes and prepared succinct decision-oriented formats called

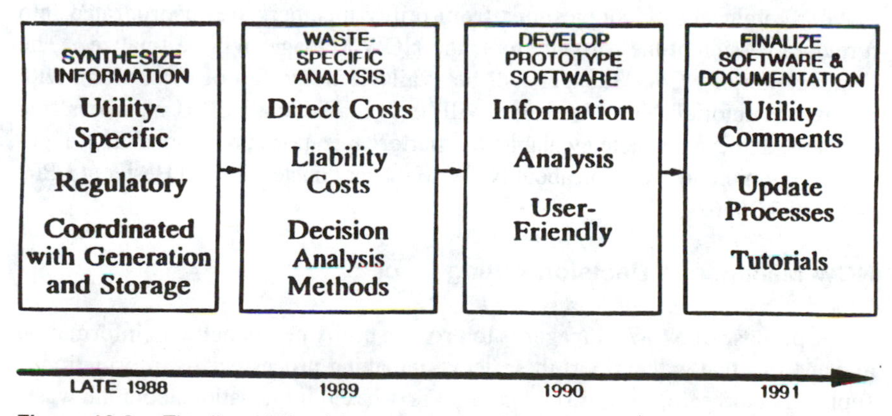

Figure 16.2. Timeline of the risk management for noncombustion waste project.

Decision Summaries that characterized the wastes. In a survey designed to obtain information about petroleum-contaminated soils, soil remediation vendors were asked to characterize their technology and describe the factors that influence the cost of the technology.

Waste-Specific Analyses

In 1989 and 1990, the project team undertook several waste-specific analyses with utility companies. The purpose of these analyses is to apply risk management modeling techniques to actual utility situations and determine the most appropriate waste management strategies for these utilities. These analyses focused on one waste stream at a time and gave the project team a better understanding of company goals, important uncertainties, and the costs associated with alternative waste management strategies.

Software Development

The waste modeling performed for the individual utilities during each case study was expanded to be applicable to the utility industry as a whole. Next, a prototype software tool was developed. Based on the industry-wide waste management modeling, this software tool allows utilities to perform risk management analyses on their own. This software tool, called NCW Manager, has three primary features:

- text-based information
- decision analysis and reporting
- communication of corporate policy

Finalizing Software and Documentation

After comments and suggestions from utility members are incorporated into a revised version of the software tool, the NCW Manager will be finalized. The final version of NCW Manager will be available in the fall of 1991 along with a new-user tutorial. NCW Manager will be available to all EPRI members free of charge and can be made available to other organizations under special arrangements. For more information about NCW Manager contact Gordon Hester at EPRI (415) 855-2696.

NCW Manager: A Decision-Aiding Tool

The purpose of NCW Manager is to provide utility personnel with information and analysis that will support their decisionmaking process. This information is supplied in three different forms: general text-based information about the waste stream, information about waste management policy (set at the corporate level and communicated to decisionmakers at the field level), and information based on the results of analyses performed comparing alternative waste management strategies for a particular situation at a particular site. These features are shown in Figure 20.3.

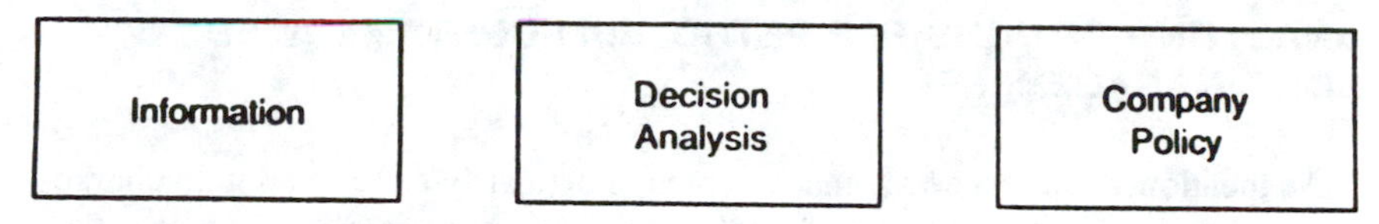

Figure 20.3. NCW manager features.

Text-Based Information

For each of the 13 waste streams contained in this tool, a 20 to 30 page Decision Summary has been prepared that contains critical information for making a decision about the most appropriate waste management strategy. A user can view this text simply by selecting the Information option from the Main Menu. Once the text is displayed on the screen, a user can use the cursor keys to scroll through the text or use the cursor keys to choose certain "Hot Keys" which, when selected, will display text on a particular topic. This feature can be used to gain access to relevant information more efficiently.

Company Policy

The Company Policy feature helps disseminate corporate-level information to company field personnel. While using NCW Manager, an individual at the corporate level can create a text file that describes policy for a particular waste stream. This text is then saved as a semipermanent file and becomes that company's version of NCW Manager. The modified version of NCW Manager can then be distributed to company field personnel giving them the convenience of online company policy text.

Decision Analysis

This feature of NCW Manager can be used to perform an analysis of alternative waste management strategies. The analysis uses a decision analysis approach in which all technically feasible waste management alternatives are first identified. Cost distributions for the various cost components (i.e., fixed cost, variable cost, permitting cost, and liability cost) are then supplied by a user if such information is available. If a user does not have this information, he or she responds to a series of qualitative questions. NCW Manager uses the answers to these questions to define cost distributions for each cost component, performing standard decision analysis computations that determine the waste management strategy with the lowest expected total cost. A report, which can be displayed on the screen or printed out to a printer, summarizes the results of this analysis. Appendix A describes the topic of decision analysis in more detail.

ANALYZING OPTIONS FOR PETROLEUM-CONTAMINATED SOILS MANAGEMENT

As mentioned earlier, waste management is primarily concerned with characterizing alternative steps that can be taken to reduce the amount of waste or environmental impact of a waste. DFI has extensive experience performing waste management analyses to help manage contaminated soils. These analyses frequently involve computer modeling and decision analysis.

Steps in a Waste Management Analysis Process

There are five key steps in the process of a waste management analysis:

- determining technically feasible technologies
- characterizing cost components for feasible technologies
- performing decision analysis computations
- performing question sensitivity computations
- communicating results

Each of the steps of a waste management analysis can be performed by a competent risk management team.

The Spreadsheet Prototype

DFI has developed a prototype in a LOTUS 1-2-3 spreadsheet of an automated petroleum-contaminated soils management tool that performs all of the steps of a petroleum-contaminated soils management analysis which will be incorporated into the existing NCW Manager. A description of the capabilities of the spreadsheet prototype accompanies the discussion of each step in the waste management process below.

Step 1: Determining Technically Feasible Technologies

The first step of the waste management process for the remediation of petroleum-contaminated soils is to identify all technically feasible alternatives. A number of general alternatives exist from which to choose.

- In Situ Technologies
  - Volatilization
  - Biodegradation
  - Leaching and Chemical Reaction
  - Vitrification
  - Passive Remediation
  - Isolation and Containment

- Non-In Situ Technologies
  - Land Treatment
  - Thermal Treatment
  - Solidification and Stabilization
  - Asphalt Incorporation
  - Chemical Extraction
  - Excavation and Landfill

Some alternatives may not be technically feasible for a particular company or situation. Factors that influence feasibility include site characteristics, waste characteristics, current and future regulations, and corporate policy. A set of questions that address these factors can be used to determine which of the alternatives are indeed technically feasible for a given situation.

The Screening Analysis and the Detailed Analysis

The spreadsheet prototype has two analysis procedures: a screening analysis procedure and a detailed analysis procedure. The screening analysis procedure includes asking feasibility questions. The detailed analysis procedure incudes both feasibility questions and cost questions. Logic computations are performed on the user's responses to feasibility questions to determine which technologies are technically feasible. Sample feasibility questions for petroleum-contaminated soil are:

- What is the permeability of the soil (in cm/sec): __________
- What is the maximum amount of time that can be taken for this remediation (months): __________
- What is the current concentration of VOCs in the soil (in ppm): __________

The spreadsheet prototype user may elect to run either the screening analysis procedure or the detailed analysis procedure. When the user selects the screening analysis procedure the model asks each of the feasibility questions and, based on the answers to these feasibility questions, gives a list of feasible technologies. Additionally, the spreadsheet gives a rough "order of magnitude" cost estimate for each technology. Finally, a list of qualitative characteristics is presented for each feasible technology.

The purpose of the screening analysis is to give the user a quick summary of the feasible technologies and an approximate idea of each technology's cost. The user can improve upon the approximation by executing the detailed analysis procedure. Figure 20.4 shows a sample of the screening analysis output.

```
SCREENING ANALYSIS/Report/1
==========================================================================
Below is a matrix displaying the results of the screening analysis.
The cost estimates are very rough "order of magnitude" estimates and
SHOULD NOT BE USED FOR DECISION MAKING.  PLEASE REVIEW THE QUALITATIVE
CRITERIA TABLE AS A PART OF THE TECHNOLOGY COMPARISON.  The costs
listed include potential liability costs.  Costs are         1989 net
present value dollars for a     5000 ton site.

                           FEASIBILITY     SCREENING COST ESTIMATES
TECHNOLOGY                   STATUS        Total Cost      Cost per Ton
                                       Pres. Val. 1989 $        1989 $
-----------------------  ----------  -----------------  -------------
In Situ Volatilization   Feasible          $1,000,000            $200
Mobile Incineration      Feasible          $3,500,000            $700
Excavation and Landfill  Feasible          $1,784,240            $357
```

Figure 20.4. Sample screening analysis output.

Step 2: Characterizing Cost Components for Feasible Technologies

The second step of the waste management analysis is to determine values for all components of the total cost. The cost components that have been identified for petroleum contaminated soils include capital costs, operation and management (O&M) costs, indirect costs, and liability costs. Capital costs consist of equipment, testing, setup, and effluent treatment costs. Indirect costs are company reporting costs and regulatory costs.

Sometimes a cost component may be known for certain; for example, a vendor may give a company a specific bid to remediate a site. At other times some uncertainty may exist as to the value of a cost component. Uncertain cost component values can be characterized by probability distributions. A simple distribution consists of a high, a medium, and a low value with a probability assigned to each value. The probabilities must add up to one.

In the detailed analysis procedure of the prototype, the user is asked a series of cost questions. The prototype uses the answers to these questions to define the values of the cost components. Depending on the answers a user gives, the cost components are represented either by a single number or a cost distribution. Sample cost questions for petroleum contaminated soil are:

- What is the amount of contaminated soil (in tons): ________
- What is the moisture concentration in the unsaturated zone (fraction): ________
- Indicate the contaminated soil classification (i.e., either Hazardous, Not Hazardous, or Unknown): ________

The prototype determines cost components, either single value or probability distributions, based on the user's answers to these questions. Published literature and a survey of soil remediation vendors provide the basis for the cost figures. Single value and probability distributions on cost are generated automatically in the spreadsheet prototype.

Step 3: Performing Decision Analysis Computations

After all of the cost components are defined, as a single value or a probability distribution, for each of the feasible technologies, these costs are incorporated into a decision analysis representation called a "decision tree." A decision tree is used to represent alternative courses of action, uncertain variables, and outcome values. The alternative courses of action in the petroleum contaminated soils analysis are the alternative feasible treatment and disposal technologies. The uncertain variables are the four cost components. The outcome values are the total costs associated with using a technology to effectively remediate a site. Figure 20.5 shows a sample of the decision tree used in this analysis.

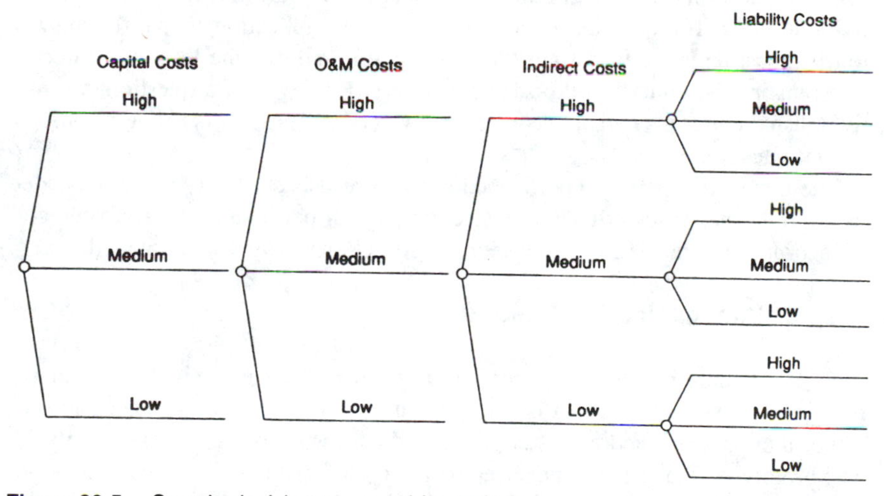

Figure 20.5. Sample decision tree used in analysis.

The decision tree is structured to facilitate the "rolling back" of all possible outcome values to determine the *expected value* for each feasible technology. In the tree used in this analysis the expected value is the probability-weighted sum of all of the possible outcome values. This expected value is a single number approximation of a technology's cost—a number that is useful when comparing technologies. The topic of decision analysis is described in more detail in Appendix A.

All of the decision analysis calculations are performed automatically in the spreadsheet prototype. The expected values are printed in a summary report.

Step 4: Performing Sensitivity Calculations

Step 4 of the waste management process entails performing a *question sensitivity analysis* to determine how alternative answers to the feasibility and cost questions would impact the results of the analysis. The primary result of the decision analysis is the identification of the least-cost treatment and disposal technology. This least-cost technology is based on the set of answers to the detailed analysis questions. Thus, a different answer could lead to another technology being identified as the least-cost technology.

In this question sensitivity step, all of the questions are assigned the user's answer except for one question where an *alternative* answer is substituted. The decision analysis calculations are performed again to see if a different technology becomes the least-cost technology. If a different technology is identified, then this particular question is said to be "sensitive." This procedure is performed for each question and each alternative response, thus identifying all of the sensitive questions. If a question is not sensitive, the implication is that the answer to that question is not important to the selection, and that the least-cost alternative remains the same for all possible answers. However, if a question is sensitive, then the user may wish to consider how certain he or she is about the answer to that question.

Question sensitivity is not performed in the spreadsheet prototype, but it is done in NCW Manager for other waste streams. When petroleum-contaminated soil is incorporated into NCW Manager, question sensitivity will be added.

Step 5: Communicating Results

This fifth and final step, communicating results of the analysis, is particularly important. A summary report is generated automatically in the spreadsheet prototype that presents the key assumptions and the results of the decision analysis. The prototype creates a one-page report for each feasible technology and a one-page comparison summary of all the feasible technologies. These reports present *both* the assumptions and results in a form that is easy to understand. These reports can help facilitate discussions about the appropriateness of assumptions and the tradeoffs associated with using resources to reduce contaminated soil and the risks associated with contaminated soil. The report includes

- the name of the technology
- the volume of contaminated soil
- the total cost for treatment
- the cost for each component of the total cost
- the cost per ton for treatment
- the cost per ton excluding liability costs
- a list of assumptions

Figure 20.6 shows an example of the first page of a single technology report.

```
COST ANALYSIS/MOBILE INCINERATION/Report 1
=====================================================================
    For the 5000 ton site that you have described, the costs are:

  Capital= $1,546,528      (These costs represent
      O&M= $1,339,333      total present value in 1989 $)
 Indirect=   $156,667
Liability=   $468,607
           ___________
    Total= $3,511,134

Cost/ton=          $702 per ton in 1989 $.

Cost/ton excluding liability costs=          $609 per ton in 1989 $.

To see a graph of the cash flow, select GRAPH.  Press RETURN after
viewing graph.

To see the assumptions used to determine this cost estimate select
CONTINUE.
```

Figure 20.6. Summary report for one technology.

The spreadsheet prototype user is able to view a graph of the cash flow that results from a remediation project. This cash flow often includes potential liability costs that occur several years in the future.

The NCW Manager report consists of three sections: a comparative summary of all the feasible technologies, a breakdown of cost components, and a list of all the questions and answers from the detailed or screening analysis phase. The first section is a comparative summary of all of the feasible technologies. In this section, the total cost and the cost for each component for all feasible technologies are shown in one table. The second section shows, for each feasible technology, which cost components are modeled nominally (by a single value) and which components are modeled probabilistically (using probability-weighted high, medium and low values). A *deterministic sensitivity analysis* step is included in the decision analysis computations in NCW Manager to determine which costs components can be modeled nominally and which components must be modeled probabilistically. The results of this deterministic sensitivity analysis are reported in this section. The third section lists all of the questions that were asked and the answers that were given. This aids in case documentation. In addition, this section shows the results of the question sensitivity. For each response that was not given, a description is given of all implications in the event that response had been given. For example, if another technology would become the least-cost technology, then that implication is described.

CONCLUSION

Evolving solid waste regulations and large cleanup and liability costs incurred for past practices are rapidly changing the options utilities and others have for managing petroleum contaminated soils and other wastes. DFI has responded to these challenges by developing easy-to-use personal computer based decision tools that provide text-based information, decision analysis and reporting capabilities, and corporate policy communication. These tools take advantage of powerful decision analysis techniques, such as sensitivity analysis, and provide results in a simple summary format designed to facilitate waste management decisionmaking.

APPENDIX A

Review of Decision Analysis

Decision Analysis is a comparatively new technique of formal analysis—less than 40 years old—and the concepts and methods of decision analysis are still relatively unfamiliar to most decisionmakers and many analysts. Some of the primary features of decision analysis are that it

- represents a formalization of common sense for decisions that are too complex for the informal use of common sense
- seeks a rational, logical, orderly, systematic framework for choosing among alternative actions or policies when the consequences of these alternatives are uncertain
- attempts to describe, quantify, and clarify tradeoffs among the relative advantages and disadvantages of alternative actions and policies
- decomposes complex decision problems into their component, and more manageable, parts
- synthesizes information produced by the analysis into a single number reflecting the overall value of a proposed action or policy

Steps in Decision Analysis

Vincent Covello has broken the decision analysis process into eight basic steps:[5]

1. Identifying decision alternatives and structuring the decision problem.
2. Defining decision objectives.

3. Defining performance measures or variables for quantifying decision objectives.
4. Identifying critical uncertain variables.
5. Assessing probabilities.
6. Specifying value judgments, preferences, and tradeoffs.
7. Evaluating alternative actions or policies.
8. Conducting sensitivity analyses and value of information analyses.

Each step is briefly described below.

Identifying Decision Alternatives. This process involves identifying existing decision alternatives or creating new decision alternatives. Alternatives may be broad or specific. The status quo alternative is frequently included in the analysis.

Defining Decision Objectives. Decision objectives are the criteria or attributes on which the decision will be made. Decision objectives can include any consequences of the decision judged to be important to decisionmakers or to persons or groups identified by decisionmakers as being part of the decision process.

Defining Performance Measures or Variables. This step involves defining performance measures (or variables) that quantify the decision objective identified above. Adverse health impacts, for example, might be measured by a range of values reflecting work days lost, expected annual fatalities, or loss of life expectancy.

Identifying Critical Uncertain Variables. Critical uncertain variables are those variables whose variations across their range of uncertainty produce the greatest change in consequences and whose uncertainty is most critical to the decision (e.g., whose variations can change the decision). Critical uncertain variables are distinct from variables that are sensitive but whose variations across their range of uncertainty are not critical to the decision.

Assessing Probabilities. In this step, the decision analyst works closely with experts in quantifying uncertainty over each critical variable. Uncertainties are typically quantified in a decision analysis through probabilities. Probabilities are obtained through a formal process of quantifying the judgments of experts—called probability encoding—and through the analysis of statistical data.

Specifying Value Judgments, Preferences, and Tradeoffs. In this step, the decision analyst assists decisionmakers in making subjective value judgments about the relative ''utility''—that is, the value, worth, or desirability—of all relevant consequences. The concept of utility and the conduct of utility analyses are basic components in a decision analysis. One objective of utility analysis is to construct a ''utility function'' that represents a scaling of the subjective values assigned by the decision maker to the consequences of a decision alternative.

Evaluating Alternative Actions or Policies. In this step the analyst integrates the various components of the analysis to evaluate each decision alternative. This process involves two procedures. First, each consequence of a decision alternative is assigned a utility (worth or desirability) value weighted by an estimate of its probability of occurrence. Second, these combined values are added across all consequences to provide a single number representing the overall expected utility (worth or desirability) of the alternative. By applying this process to every alternative, a comparison can be made of the expected utility of all decision alternatives. As Fischhoff et al.[6] note, the basic decision rule in decision analysis is: "The alternative with the greatest expected utility is the indicated choice." This does not mean, however, that the indicated choice must be adopted. Because of simplifying assumptions and limitations of the analysis, the results of a decision analysis are best viewed as an aid to decisionmaking rather than as an absolute prescription for action.

Conducting Sensitivity and Value-of-Information Analyses. In this step the analyst systematically varies the value of critical uncertain variables across their range of uncertainty and then notes the effects on the overall expected utility and the recommended decision. A negligible effect indicates low sensitivity, while a substantial effect indicates high sensitivity. The basic strategy in sensitivity analysis is to reexamine the results of the analysis by repeating the calculations using alternative values for uncertain variables. The results of this reexamination indicate the robustness of the overall results of the analysis. The objective of value-of-information analysis is to determine the monetary value of additional information that would reduce or eliminate uncertainty in those variables for which better information is important.

REFERENCES

1. Skinner, J. "Banning Waste from Land Disposal," *First Public Brief on the* 1984 *Amendments to the RCRA,* U.S. EPA Office of Solid Waste and Emergency Response.
2. Fromm, C. H., M. S. Callahan, H. M. Freeman, and M. Drabkin, "Succeeding at Waste Minimization," *Chem. Eng.* September 14, 1987.
3. "Less Than One-Third Get EPA Permit of On-Site RCRA Facilities," *Hazardous Material Control Research Institute Focus,* Hazardous Material Control Research Institute, Silver Springs, MD, July 1987.
4. U.S. Environmental Protection Agency, *Waste from the Combustion of Coal by Electric Utility Power Plants,* March 10, 1988 Report to Congress.
5. Covello, V. T., "Decision Analysis and Risk Management Decision Making: Issues and Methods," *Risk Analysis,* 17:2 (1987).
6. Fischhoff, B., S. Lichtenstein, P. Slovic, S. L. Derby, and R. L. Keeny, *Acceptable Risk* (New York: Cambridge University Press, 1981).

CHAPTER 21

A Progress Report on the Council for Health and Environmental Safety of Soils—CHESS

Paul T. Kostecki and **Edward J. Calabrese,**
Environmental Health and Sciences, University of Massachusetts, Amherst

BACKGROUND

The Council for the Health and Environmental Safety of Soils (CHESS) is a national coalition which was created to provide leadership in soil contamination issues by:

- providing consensus guidelines on analytical techniques, risk assessment methodologies, and remediation of contaminated soils,
- conducting scientific evaluation, making analyses, and providing recommendations for courses of action,
- exchanging technical information,
- providing education and training functions,
- encouraging dialogue among affected groups.

CHESS is making significant progress toward achieving these goals. With CHESS' initial area of concentration set on petroleum contamination, work is well under way with expert committees to provide consensus guidelines in the areas of analysis, environmental fate, risk assessment, and remediation.

The committees are reviewing an array of methodologies and will be developing their scientific evaluations and recommendations over the next few months.

They hope to have the results of their efforts available to the scientific and regulatory communities by the winter of 1990.

AGENDA

The Council met in early 1989 to assess the most effective approach for developing practical solutions concerning contaminated soils. To this end, the Council determined that the first initiative should focus on a specific soil contamination issue, and that the approach should be a product that is constructed within an expert decision framework. The rationale for focusing on a single type of soil contamination problem was that, although it may be possible to develop a generic methodology applicable to all soil contamination instances, it is not feasible within an acceptable time frame, and would have a high chance for failure. Therefore, the Council decided that focusing on a specific area would markedly increase the chance for success and facilitate the development of a model approach that could be applied to other pollutant classes.

By utilizing an expert system, the Council believes that the product will be scientifically and technically sound and easy to use. Moreover, this type of system could provide the user with conceptual and theoretical information basic to the technical input without sacrificing its overall utility.

Discussions concerning the first area of concentration revolved around the collective experiences of Council and Governing Board members. In addition, input was solicited from various state agencies regarding the significance of soil contamination problems in terms of dollars, manpower, and time. Specific information was provided via videotape from representatives from the Orange County Health Care Agency, Santa Ana, California; Arizona Department of Environmental Quality, Phoenix, Arizona; and Massachusetts Department of Environmental Quality Engineering, Springfield, Massachusetts. In addition, the results from an ongoing national survey being conducted at the University of Massachusetts assessing the significance of petroleum soil contamination were provided to the Council by Dr. Paul Kostecki.

States were in unanimous agreement that the most pressing soil contamination problem facing them is petroleum contamination. Agencies indicated that anywhere from 60% to 90% of their resources were being directed to petroleum contaminated soils. The Council also indicated that the ubiquitous nature of petroleum contamination has direct and immediate public health implications when the health hazard of the components of petroleum products and the exposure potential for large segments of the population are considered. The Council therefore overwhelmingly voted to have CHESS' first efforts directed to this important area.

The orientation toward petroleum contaminated soils will enhance the likelihood that environmental and public health groups and agencies will adopt the CHESS product.

EXPERT COMMITTEE'S MAKEUP

With CHESS' first mission clearly outlined, the Council set out to create the four Expert Committees:

- Analysis and Environmental Fate
- Environment and Health
- Remediation
- User Application

CHESS policy dictated that Council members serve as chairman for each expert committee to provide direct oversight and guidance as well as continuity. Committee nominations were approved by the Council in the summer and the full committees were established. They are:

Analysis and Environmental Fate

Chairman: James Dragun, The Dragun Corporation
Members: Bruce Bauman, American Petroleum Institute
Marc Bonazountas, Technical University of Athens, Greece
Dwayne Conrad, Texaco Research
Donald Mackay, University of Toronto
Thomas Potter, University of Massachusetts

Environment and Health

Chairman: Dennis Paustenbach, MacLaren/Hart Environmental Associates
Co-Chairmen: Renate Kimbrough, U.S. Environmental Protection Agency
Barbara Beck, Gradient Corporation
Members: Cynthia Harris, Agency for Toxic Substances and Disease Registry
David Layton, Lawrence Livermore National Laboratory
Richard McKee, Exxon Environmental Science Corporation
David Rosenblatt, Argonne National Laboratory
Randall Roth, ARCO
John Schaum, U.S. Environmental Protection Agency
Jeffrey Wong, California Department of Health Services

Remediation

Chairman: Allen Hatheway, University of Missouri
Co-Chairman: William Kucharski, Woodward-Clyde Consultants

Members: James Franz, ARCO
David Leu, Mittelhauser Corporation
John Matthews, U.S. Environmental Protection Agency

User Applications

Chairman: Vinay Kumar, U.S. Environmental Protection Agency
Co-Chairman: John Hills, City of Anaheim, California
Members: David Chen, American Petroleum Institute
Audrey Eldridge, Massachusetts Department of Environmental Protection
Jeanmarie Haney, Arizona Department of Environmental Quality

EXPERT COMMITTEE ACTIVITIES

The Council has charged the committees with evaluating relevant methodologies and approaches, and providing recommendations as to their use in the development of a comprehensive expert decision methodology. These methodologies and approaches were provided to the committees by the CHESS Chairman and Managing Director, and represent state-of-the-art approaches in a particular area. They are:

Analysis and Environmental Fate

SGS Compositional Multiphase Model[1-3]
PCB Onsite Spill Model (POSSM)[4,5]
Pesticide Root Zone Model (PRZM)[6-8]
GEOTOX multimedia compartment model[9-12]
Seasonal Soil Compartment Model (SESOIL)[13]
Personal Computer-Graphical Exposure Modeling System (PCGEMS)[14-16]
Preliminary Pollutant Limit Value (PPLV) Approach[17]
AERIS model[18-20]
Leaking Underground Fuel Tank (LUFT) Manual[21]
Risk Assistant/Fate and Transport (RAFT) Manual[22]
Risk Assistant[23]
Various EPA analytical methodologies[24]

Environment and Health

AERIS model[18-20]
Preliminary Pollutant Limit Value (PPLV) Approach[17,25,26]
Risk Assistant/Fate and Transport (RAFT) Manual[22]

GEOTOX multimedia compartment model[9-12]
Superfund Human Health Assessment Manual[27]
Massachusetts Contingency Plan (MCP)[28]
Leaking Underground Fuel Tank (LUFT) Manual[21]
California Site Mitigation Decision Tree Manual[29]
New Jersey's Soil Cleanup Criteria[30,31]

Remediation

Lewis Publishers' book, *Remedial Technologies for Underground Storage Tanks*[32]
API's *Guide to the Assessment and Remediation of Underground Petroleum Releases*[33]
EPA's *Soil Advisor*[34]
EPA's *Assessing UST Corrective Action Technologies*[35]
EPA's *Cleanup of Releases from Petroleum USTs*[36]

Since the ultimate goal is to review the best approaches in each discipline, the committees welcome independent parties or individuals to make them aware, through the Managing Director, of additional approaches or methodologies. CHESS would be glad to provide expert independent evaluation of your approaches.

MILESTONES

The Expert Committees will be finishing their reviews over the summer of 1990. This effort will result in over 700 reviews, 35 methodologies reviewed by 22 members; Chairmen will then sort through their committee's reviews and synthesize member comments into an executive summary which will also include recommendations for the committee's next course of action. Those activities may include modifying and/or consolidating some methodologies or parts of methodologies. It is also possible that committees may have to create new methodologies. It is anticipated that the Chairmen reports will be ready by the fall of 1990.

INTERNATIONAL ACTIVITIES

Although conceived in this country, CHESS was created as an international council whose mission is global. To date, international outreach has taken the form of special meetings with Environment Canada personnel and Canadian scientists and discussions with World Health Organization officials. The AERIS (Aid for Evaluating the Development of Industrial Sites) program being reviewed by our Analysis and Environmental Fate and Environment and Health Committees is a Canadian approach that was identified through our meetings.

Plans are under way to meet with members of the European community's environmental and related directorates, including West German environmental protection officials, to establish formal ties in the spring of 1990. CHESS already has direct European involvement at the expert committee level with the efforts of William Kucharski who works for Woodward-Clyde Consultants in Frankfurt, Germany, and Marc Bonazountas of Epsilon International located in Marousi, Greece.

SUPPORT

CHESS would not be possible without the generous support of a variety of federal agencies and private companies. We wish to thank the following sponsors for their continued financial support:

Agency of Toxic Substances and Disease Registry, Atlanta GA
Ashland Oil Inc., Ashland KY
Chevron Inc., Richmond CA
Eastman Kodak Company, Kingsport TN
Electric Power Research Institute, Palo Alto CA
Ford Motor Company, Dearborn MI
General Electric Company, Fairfield CT
Gillette Company, Gaithersburg MD
Goodyear Tire and Rubber Company, Akron OH
Hercules Inc., Wilmington DE
Hoechst Celanese Corporation, Somerville NJ
Morton Thiokol Inc., Chicago IL
Public Service Electric & Gas, Newark NJ
Shell Oil Company, Houston TX
Texaco Inc., Beacon NY
Union Carbide Corporation, Danbury CT
The Environmental Protection Agency, Washington DC

CHESS continually seeks additional support to expedite and broaden its activities.

REFERENCES

1. Baehr, A., and M. Corapcioglu. "A Compositional Multiphase Model for Groundwater Contamination by Petroleum Products," *Water Resour. Res.* 23(1):191–213 (1987).
2. Baehr, A. "Selective Transport of Hydrocarbons in the Unsaturated Zone due to Aqueous and Vapor Phase Partitioning," *Water Resour. Res.* 23(10):1926–1938 (1987).

3. Baehr, A., G. Hoag, and M. Marley. ''Removing Volatile Contaminants from the Unsaturated Zone by Inducing Advective Air-Phase Transport,'' *J. Contam. Hydrol.* (4):1–26 (1989).
4. Brown, S., and S. Boutwell. ''Chemical Spill Exposure Assessment Methodology'' Prepared by CH2M HILL for Electric Power Research Institute, January 1988.
5. Brown, S., and A. Silvers. ''Chemical Spill Exposure Assessment,'' *Risk Analysis,* 6(3):(1986).
6. Carsel, R., C. Smith, L. Mulkey, J. Dean, and P. Jowise. ''User's Manual for the Pesticide Root Zone Model (PRZM),'' USEPA Environmental Research Lab, Athens, GA. December 1984.
7. Dean, J., and R. Carsel. ''Agricultural Chemical Use,'' Prepared by Woodward-Clyde Consultants for USEPA, Athens, GA, 1988.
8. Enfield, C., R. Carsel, S. Cohen, T. Phan, and D. Walters. ''Approximating Pollutant Transport to Groundwater,'' *Ground Water,* 20:(6) (1982).
9. McKone, T. E., L. B. Gratt, M. J. Lyon, and B. W. Perry. ''Geotox—User's Guide and Supplement,'' Lawrence Livermore National Laboratory/U.S. Army Medical Research and Development Command. Project Order 83PP3818. May 1987.
10. McKone, T. E., and D. W. Layton. ''Screening the Potential Risks of Toxic Substances Using a Multimedia Compartment Model: Estimation of Human Exposures,'' *Regulatory Toxicology and Pharmacology,* 6:359–380 (1986).
11. McKone, T. E., and D. W. Layton. ''Exposure and Risk Assessment of Toxic Waste in a Multimedia Context,'' Lawrence Livermore National Laboratory. Presented at Air Pollution Control Association. May 1986.
12. McKone, T. E. ''Geotox—Simulating Contaminant Behavior and Human Exposure,'' E&TR. 14–20. May 1987.
13. Bonazountas, M., and J. Wagner. '' 'Sesoil,' A Seasonal Compartment Model,'' ADL and DIS/ADLPIPE. For U.S. Environmental Protection Agency, Office of Toxic Substances. May 1984.
14. Wagner, J., and M. Bonazountas. ''Potential Fate of Buried Halogenated Solvents via SESOIL,'' ADL for U.S. Environmental Protection Agency, Office of Toxic Substances. January 1983.
15. Watson, D., and S. Brown. ''Testing and Evaluation of the SESOIL Model,'' Anderson-Nichols & Co. for U.S. Environmental Protection Agency, Environmental Research Lab, Athens, GA. August 1985.
16. ''Personal Computer Version of the Graphical Exposure Modeling System.'' User's Guide. Prepared by General Sciences Corp. for USEPA/OTS Contract #68024281. September 1989.
17. Small, Mitchell. ''The Preliminary Pollutant Limit Value User's Manual,'' U.S. Army Biomedical Research and Development Laboratory, Ft. Detrick, Frederick, MD. Technical report 8918. July 1988.
18. Senes Consultants (Prepared for the Decommissioning Steering Committee). ''The Development of Soils Cleanup Criteria in Canada,'' Volume 2—Interim Report on the ''Demonstration'' Version of the AERIS Model. 1988-12-15.
19. Senes Consultants (Prepared for the Decommissioning Steering Committee). ''Contaminated Soil Cleanup in Canada,'' Volume 5—Development of the AERIS Model, Final Report. September 1989.
20. Senes Consultants (Prepared for the Decommissioning Steering Committee). ''Contaminated Soil Cleanup in Canada,'' Volume 6—User's Guide for the AERIS Model. September 1989.

21. "Leaking Underground Storage Tank Manual: Guidelines for Site Assessment, Cleanup, and Underground Storage Tank Closure." Prepared by LUFT Task Force, State of California, Sacramento, CA, May 1988.
22. RAFT—User's Manual for Risk Assessment/Fate and Transport (RAFT) Modeling System. Prepared for the Pennsylvania Bureau of Waste Management by the Scientific Services Section. 1989.
23. RISK ASSISTANT—Overview of Microcomputer Software to Facilitate Assessments of Hazardous Waste Sites. Prepared for USEPA and NJDEP by Hampshire Research Institute. May 1989.
24. Potter, T. "Analysis of Petroleum Contaminated Soil and Water: An Overview," in *Petroleum Contaminated Soils,* Volume 2, P. T. Kostecki and E. J. Calabrese, Eds. (Chelsea, MI: Lewis Publishers, Inc., 1989).
25. Kostecki, P. T., E. J. Calabrese, and H. Horton. "Review of Present Risk Assessment Models for Petroleum Contaminated Soils," In *Petroleum Contaminated Soils,* P. T. Kostecki and E. J. Calabrese, Eds. (Chelsea, MI: Lewis Publishers, Inc., 1988).
26. Rosenblatt, D. H., Dacre, J. C., and Cogley, D. R. "An Environmental Fate Model Leading to Preliminary Pollutant Limit Values for Human Health Effects," *Environmental Risk Analysis for Chemicals,* R. A. Conway, Ed., (New York: Van Nostrand Reinhold Co., 1981).
27. Environmental Protection Agency, "Risk Assessment Guidance for Superfund: Volume I—Human Health Evaluation Manual (Part A)," Interim Final, December 1989.
28. "Draft Interim Guidance for Disposal Site Risk Characterization—In Support of the Massachusetts Contingency Plan," prepared by Massachusetts Department of Environmental Quality Engineering, Office of Research and Standards. October 3, 1988.
29. Leu, D. "California Site Mitigation Decision Tree Manual," prepared by the Dept. of Health Services, Toxic Substances Control Division. May 1986.
30. Hawley, J. K. "Assessment of Health Risk from Exposure to Contaminated Soil," *Risk Analysis,* 5(4) (1985).
31. Stokman, S., and R. Dime. "Soil Cleanup Criteria for Selected Petroleum Products," *Risk Assessment,* 342–345 (1986).
32. Preslo, L., J. Robertson, D. Dworkin, E. Fleischer, P. T. Kostecki, and E. J. Calabrese. *Remedial Technologies for Underground Storage Tanks.* EPRI and EEI. (Chelsea, MI: Lewis Publishers, Inc., 1988).
33. "Guide to the Assessment and Remediation of Underground Petroleum Releases," American Petroleum Institute Publication No. 1628. Washington, D.C.
34. Foskett, W. "Soil Treatment Advisor," Part of EPA/OUST's Corrective Action Triage Software (CATS).
35. Rosenberg, M. et al. "Assessing UST Corrective Action Technologies. Site Investigation and Selection of Soil Treatment Technologies," EPA/OUST. Contract No. 68-03-3409. November 1988.
36. Environmental Protection Agency, Office of Underground Storage Tanks. "Cleanup of Releases from Petroleum USTS: Selected Technologies," EPA/530/-UST-88/001. April 1988.

Glossary of Acronyms

AAL	applied action level
ABTA	Applied BioTreatment Association
ACGIH	American Council of Governmental Industrial Hygienists
AEHS	Association for the Environmental Health of Soils
API	American Petroleum Institute
ARAR	applicable or relevant and appropriate
BACT	best available control technology
BAT	best available technology
BOP	blowout prevention
BTXE	benzene, toluene, xylene, and ethylbenzene
BWT	Ballast Water Treatment (facility, Valdez, Alaska)
CAL	corrective action level
CDC	Centers for Disease Control
CDOG	California State Division of Oil and Gas
CERCLA	Comprehensive Environmental Response, Compensation, and Liability Act
CF	calibration factor
CHESS	Council for the Health and Environmental Safety of Soils
CS	cleanup standards
CSDWTE	California Safe Drinking Water and Toxic Enforcement (Act)
DFI	Decision Focus Incorporated
DHS	Department of Health Services
EPA	Environmental Protection Agency
EPRI	Electric Power Research Institute
FID	flame ionization detector
GEMs	genetically engineered microorganisms
GEMS	Graphic Exposure Modeling System
GL	guidance level
GRO	gasoline range organics
HCS	hydrocarbon contaminated soils
HSWA	Hazardous and Solid Waste Amendments (of 1984)
HRS	Hazard Ranking System
LADD	lifetime average daily dose
LCS	laboratory control sample
LTU	land treatment unit
LUFT	leaking underground fuel tank
MCL	maximum contaminant level
MCP	Massachusetts contingency plan

MEI	maximally exposed individual
NAPL	nonaqueous phase liquid
NAS	National Academy of Sciences
NCP	National Oil and Hazardous Substances Pollution Contingency Plan
NIOSH	National Institute of Occupational Safety and Health
NPL	National Priorities List
OSHA	Occupational Safety and Health Administration
PAH	polynuclear aromatic hydrocarbon
PCB	polychlorinated biphenyl
PCE	tetrachloroethylene
PCGEMS	personal computer-graphical exposure modeling system
PCP	pentachlorophenol
PID	photoionization detector
P-PK	physiologically based pharmacokinetic
PNA	polynuclear aromatic hydrocarbon
POSSM	PCB onsite spill model
PPE	personal protective equipment
PPLV	preliminary pollutant limit value
PQL	practical quantitation limit
PRP	potentially responsible party
PRZM	pesticide root zone model
PSPM	PCB Spill Priority-Setting Model
QA/QC	quality assurance/quality control
RAFT	risk assistant/fate and transport
RCRA	Resource Conservation and Recovery Act
RfD	reference dose
RI/FS	remedial investigations and feasibility studies
ROD	Record of Decision
RSD	relative standard deviation
RWQCB	regional water quality control board
SCS	surrogate control standard
SFS	Santa Fe Springs
SITES	Contaminated Site Risk Management System
SSPS	Site Screening and Priority-Setting System
STLC	soluble threshold limit concentration
SVOC	semivolatile organic chemical
TCDD	2,3,7,8-tetrachlorodibenzo-p-dioxin
TCE	trichloroethylene
TPH	total petroleum hydrocarbons
TSCA	Toxic Substances Control Act
TTLC	total threshold limit concentration
UST	underground storage tank
VOC	volatile organic compounds

List of Contributors

Michael A. Acton, Applied Geosciences, Inc., 17321 Irvine Boulevard, Tustin, CA 92680

John H. Barkach, The Dragun Corporation, 3240 Coolidge Highway, Berkley, MI 48072-1634

Richard A. Becker, California Department of Health Services, Toxic Substance Control Program, 400 ''P'' Street, Sacramento, CA 94234-7320

Charles E. Bell, Environmental Health Sciences Program, School of Public Health, University of Massachusetts, Amherst, MA 01003

Angelo J. Bellomo, McLaren/Hart, 100 N. First Street, Burbank, CA 91502

David A. Benson, Applied Geosciences, Inc., 17321 Irvine Boulevard, Tustin, CA 92680

Marc Bonazountas, Epsilon International, Kifissias Avenue 16, 15125 Athens, Greece

Edward G. Butler, California Department of Health Services, Toxic Substance Control Program, 400 ''P'' Street, Sacramento, CA 94234-7320

Dallas L. Byers, Shell Development, Westhollow Research Center, P.O. Box 1380, Houston, TX 77251-1380

Edward J. Calabrese, Environmental Health Sciences Program, School of Public Health, University of Massachusetts, Amherst, MA 01003

Bruce L. Cliff, VAPEX Environmental Technologies, Inc., 480 Neponset Street, Canton, MA 02021

Katherine K. Connor, Decision Focus, Inc., 4984 El Camino Real, Los Altos, CA 94022

Seth J. Daugherty, Orange County Health Care Agency, Environmental Health, 2009 E. Edinger St., P.O. Box 355, Santa Ana, CA 92702

James Dragun, The Dragun Corporation, 3240 Coolidge Highway, Berkley, MI 48072-1634

Gayle Edmisten-Watkin, Harding Lawson Associates, 200 Rush Landing Road, P.O. Box 578, Novato, CA 94948

Charles E. Gilbert, Environmental Health Sciences Program, School of Public Health, University of Massachusetts, Amherst, MA 01003

John D. Hanby, Hanby Analytical Laboratories, Inc., 4400 South Wayside, Suite 107, Houston, TX 77087

Robert H. Harris, Environ Corporation, 210 Carnegie Center, Suite 201, Princeton, NJ 08540

John J. Hills, Public Utilities Department, Environmental Service Division, 909 E. Vermont Avenue, Anaheim, CA 92803

George E. Hoag, University of Connecticut, The School of Engineering, 191 Auditorium Road, RM-376, Storrs, CT 06269-3210

Mike Hoffman, ENSECO, Inc., Rocky Mountain Analytical Laboratory, 4955 Yarrow Street, Arvada, CO 80002

Paul C. Johnson, Shell Development, Westhollow Research Center, P.O. Box 1380, Houston, TX 77251-1380

Paul T. Kostecki, Environmental Health Sciences Program, University of Massachusetts, School of Public Health, Amherst, MA 01003

Jean B. Kulla, McLaren/Hart, 16755 Von Karman Avenue, Irvine, CA 92714

A. C. Lazzaretto, Santa Fe Springs, Department of Environmental Management, 11710 Telegraph Road, Santa Fe Springs, CA 90670

Michael C. Marley, VAPEX Environmental Technologies, Inc., 480 Neponset Street, Canton, MA 02021

Sharon A. Mason, The Dragun Corporation, 3240 Coolidge Highway, Berkley, MI 48072-1634

Jayne M. Michaud, ChemRisk—A Division of McLaren/Hart, Stroudwater Crossing, 1685 Congress Street, Portland, ME 04102

Peter Nangeroni, VAPEX Environmental Technologies, Inc., 480 Neponset Street, Canton, MA 02021

Jerry L. Parr, ENSECO, Inc., Rocky Mountain Analytical Laboratory, 4955 Yarrow Street, Arvada, CO 80002

Alan H. Parsons, ChemRisk—A Division of McLaren/Hart, Stroudwater Crossing, 1685 Congress Street, Portland, ME 04102

Dennis J. Paustenbach, ChemRisk—A Division of McLaren/Hart, 1135 Atlantic Avenue, Alameda, CA 94501

Michael R. Piotrowski, Woodward-Clyde Consultants, 1550 Hotel Circle North, San Diego, CA 92108

Donald W. Rice, Los Angeles Harbor Department, P.O. Box 151, San Pedro, CA 90733

Jennie S. Rice, Decision Focus, Inc., 4984 El Camino Real, Los Altos, CA 94022

Stephen R. Ripple, ChemRisk—A Division of McLaren/Hart, 1135 Atlantic Avenue, Alameda, CA 94501G

G. Michael Schum, California Department of Health Services, Toxic Substance Control Program, 400 ''P'' Street, Sacramento, CA 94234-7320

Terry Sciarrotta, Southern California Edison Company, P.O.Box 800, 2244 Walnut Grove Avenue, Rosemead, CA 91770

Douglas A. Selby, Las Vegas Valley Water District, 3700 West Charleston Boulevard, Las Vegas, NV 89153

Edward S. Stanek, Biostatistics and Epidemiology Program, School of Public Health, University of Massachusetts, Amherst, MA 01003

Curtis C. Stanley, Shell Oil Company, Westhollow Research Center, P.O. Box 1380, Houston, TX 77251-1380

Gary Walters, ENSECO, Inc., Rocky Mountain Analytical Laboratory, 4955 Yarrow Street, Arvada, CO 80002

Justin L. Welsh, Decision Focus, Inc., 4984 El Camino Real, Los Altos, CA 94022

Jeffrey J. Wong, California Department of Health Services, Toxic Substance Control Program, 400 ''P'' Street, Sacramento, CA 94234-7320

Index

AALs, *See* Applied Action Levels
adsorbed phase hydrocarbons, 28
adsorption, 183
advection, 171, 172
agricultural amendments, 162
air flow modeling, 188–191
air permeability test, 264–269
air phase modeling, 192
aliphatic hydrocarbons, 171, 240
alkanes, 163
alkanoic acids, 163
alkanoids, 163
alkanols, 16
alkyl sulfides, 173
Alyeska Ballast Water Treatment Facility, 213–216
amines, 173
applicable or relevant and appropriate requirements (ARARs), 13–14, 40
Applied Action Levels (AALs), 11, 26
aquatic bioassays, 248–249
aquatic equilibrium modeling, 183
aqueous phase hydrocarbons, 28
aquifer physical properties, 181
aquifer treatment, 217–225
ARARs, *See* applicable or relevant and appropriate requirements
aromatic amines, 173
aromatic hydrocarbons, 133–147, 171
    colorimetry for, 135, 140, 145–146
    extraction methods for, 135, 140, 145–146
    instrumental parameters for detection of, 141
    mononuclear, 240
    Ohio River study of, 137–138
    polynuclear (PNAH), 163, 239, 240, 241, 246, 247
    purge and trap methods for, 136, 145
    Valdez oil spill and, 138–140
asphalt, 239–251
    air emissions and, 251
    chemistry of, 240–242
    environmental testing of, 248–249
    feasibility study of, 243–246
    future of, 250–251
    integrity of, 242–243
    liquid, 240–241
    mixtures of, 242–243
    regulation and, 249
    road building and, 250
    as surface seal, 253
asphalt cement, 240
assays, 248–249, *See also* specific types
atmospheric deposition, 162

background concentrations, 161
BACT, *See* best available control technology
bacterial biodegradation, 29
BAT, *See* best available technology
benzene, 48, 140, *See also* benzene, toluene, xylene, and ethylbenzene (BTXE)
    mobility of in soil, 29
    persistence of, 165
    solubility of, 28
benzene, toluene, xylene, and ethylbenzene (BTXE), 37–38, 39, 40, 49, 161
    aromatics and, 136
    GROs and, 111, 113, 114
    naturally occurring, 163
    in petroleum distillates, 162
    state levels of, 78
best available control technology (BACT), 42
best available technology (BAT), 42
bioaccumulation, 173, 175
bioassays, 248–249, *See also* specific types

bioavailability of chemicals, 290
biodegradation, 29, 32, 173, 253
biological activity, 183
biological transformation, 171
bioremediation, 203–236, *See also* remediation approaches to, 203, *See also* specific types
assessment of, 207–213
aquifer treatment and, 217–225
at Ayeska Ballast Water Treatment Facility, 213–216
conceptual framework for decisionmaking in, 230–234
land treatment studies and, 226–227
long-term implications of, 234–236
microbial ecology approach to, *See* microbial ecology
microbiological approach to, *See* microbiological remediation
wood preservatives and, 216–227
biota, 163–165
biotransformation, 29, 33, *See also* transformation processes
Bloomfield, 103
blowout prevention (BOP) devices, 100
BOP, *See* blowout prevention
BTXE, *See* benzene, toluene, xylene, and ethylbenzene
bulk phase hydrocarbons (nonaqueous phase liquid (NAPL)), 28, 34, 35, 43, 171, 175, 182, 183
bulk properties, 172

capacitance models, 183
capillary chromatography, 134, 151, 247
capillary zone immiscible contaminants, 189, 195
capillary zone/unsaturated zone interaction, 191–192
carcinogenicity, 26
cation exchange, 172, 183
cement, 240
CERCLA, *See* Comprehensive Environmental Response, Compensation, and Liability Act
chelation (complexation), 172, 173, 183
chemical oxidation, 173
chemical speciation, 183
chemical transformation, 171
CHESS, *See* Council for Health and Environmental Safety of Soils
chlorine, 171
chlorophenols, 171
chromatography, 134, *See also* specific types
capillary, 134, 151, 247
gas, *See* gas chromatography (GC)
cleanup numbers, *See* soil cleanup numbers
coagulation, 183
cold mix asphalt paving materials, *See* asphalt
colorimetry, 135, 140, 145–146
community needs, 5
compartmental modeling, 176, 178
complexation, 172, 173, 183
Comprehensive Environmental Response, Compensation, and Liability Act (CERCLA), 7, 18
constant attenuation/availability factor multiplier, 39
constant capacitance models, 183
cooperation, 91–98
cost-effectiveness, 5
costs, 7, 8, 14, 17
loss-of-income, 66–67
remediation, 66, 93
Council for Health and Environmental Safety of Soils (CHESS), 331–336
cresols, 171
cumulative effects, 26
cyanides, 163
cyclic alkanes, 163

decay coefficient, 181
decision analysis, 321–322, 325–326, 328–330
de minimis risk, 4–5
desorption, 253
deterministic models, 176
diesel, 140
diffusion, 33, 181
diffusive flux, 44
dilution, 33
dioxins, 9, 78, 286, 287, *See also* specific types
direct purge and trap method, 110, *See also* purge and trap method

dispersion, 33, 171, 172
dissolved pollutant modeling, 175–181
distribution of contaminants, 171
documentation, 27
dose-response relationships, 2, 3, 5, 284
droplets, 28

electrical resistivity surveys, 152
electric utility industry, 316
electron transfer reactions, 174
empirical studies, 37
engineering feasibility, 5
environmental assessment, 161
equilibrium, 188
equilibrium distribution coefficients, 35
equilibrium modeling, 183
ethylbenzene, *See* benzene, toluene, xylene, and ethylbenzene (BTXE)
expert-based systems, 18
exposure
  assessment of, *See* exposure assessment
  bioavailability of chemicals after, 290
  pathways of, *See* exposure pathways
  site-specific conditions of, 2
  USTs and, 25–26
exposure assessment, 3, 43–47
  components of, 285–290
  critical factors in, 283–290
  objectives of, 284–285
exposure pathways, 15, 30–31, 51, 58, *See also* specific types
  identification of, 287
  multiple, 25–26
extent of contamination, 2
extraction, 135, 140, 145–146
  vacuum, *See* soil venting
  vapor, *See* vapor extraction
Exxon Valdez oil spill, 138–140

fate of hydrocarbons, 28–34, 167–183
  chemistry and, 172–175
  CHESS and, 333, 334
  distribution and, 171
  environmental factors in, 172–175
  estimation of, 286–287
  migration parameters and, 183
  modeling of, 28–34, 167–183
    aquatic equilibrium, 183
    dissolved pollutant, 176–181
    immiscible contaminant, 181–183
    issues in, 180–181
    mathematical, 167, 175–176
    TDE, 176–178, 179, 180
  nature of contamination and, 171
  sources and, 171
  transport and, 171
fauna, 163
feasibility, 5, 7
fertilizers, 162
FID, *See* flame ionization detection
Field Test Kit, 136, 140
fill material, 162
first-order kinetics, 172
fission, 165
flame ionization detection (FID), 106, 107, 112, 113, 145, 151
  soil venting and, 257, 273
floaters, 28, 171
flora, 163
free-liquid gasoline, 259
fresh gasoline, 259, 260–261
Friedel-Crafts chemistry, 133, 135
funnel effect, 53

ganglia, 28
gas chromatography (GC)
  in aromatic detection and measurement, 133, 134, 136, 145
  capillary, 134, 151, 247
  in GRO detection, 106, 107, 112, 113, 115–121, 126–129
  packed column, 134
  in PNAH analysis, 247
  purge and trap, 136
  soil venting and, 257, 273
gaseous phase hydrocarbons, 28
gasoline control standards, 114, 121
gasoline range organics (GROs), 105–132
  defined, 113
  direct purge and trap method and, 110
  headspace method and, 110

methanol extraction and, 110
method for determination of, 112–132
apparatus for, 115–121
application of, 112–113
interferences in, 114–115
materials for, 115–121
performance of, 131–132
procedures for, 123–130
quality control in, 130–131
reagents for, 121–122
safety issues in, 115
sampling in, 123
scope of, 112–113
Phase I study of, 106
Phase II study of, 106–110
Phase III study of, 110–112
purge and trap method and, 110, 113, 128
GC, *See* gas chromatography
GEMs, *See* genetically engineered microorganisms
generic soil cleanup numbers, 1–19
applications for, 8–9
benefits of, 8–9
future of, 14–19
guidelines for levels of, 49
history of, 10–17
origins of, 10–13
problems with, 8–9
genetically engineered microorganisms (GEMs), 206, 211–212
geophysical methods, 152
globules, 28
glycolysis, 164
GROs, *See* gasoline range organics
ground penetrating radar, 152
groundwater modeling, 167, 175, 179
groundwater monitoring, 33, 37
groundwater sampling, 151–152, 155–159

Hanby Field Test Method, 135, 136, 140
hazard identification, 3, 284
Hazardous Waste and Solid Waste Amendments of 1984, 316
Hazard Ranking System (HRS), 315
headspace method, 110
health based risk assessment, 36, 50
health risks, 283–290
quantification of, 295
of remediation, 293–298
heavy metals, 247–248
Henry's law, 43, 151
HRS, *See* Hazard Ranking System
human activities, 162–163, 170, *See also* specific types
hydrolysis, 171, 173
hydroxylation, 165

immiscible contaminants
capillary zone, 189, 195
density of, 196, 198, 200
distribution of, 193
modeling of, 181–183
subsurface condition of, 197, 199
immiscible fluids, 173
immiscible phase boundary conditions, 196
incineration, 253
indoor air, 25–26, 43–47
induced polarization, 152
industrial point sources, 161
inert fill material, 162
inert metals, 174
ingestion of soil, *See* soil ingestion rates
input exposure parameters, 15
ion-cation exchange, 172
ion exchange, 172
ionization, 172
irrigation, 170

kinetics, 172, 173, 174

laboratory analyses, 78
laboratory control samples, 114
LADD, *See* lifetime average daily dose
landfills, 170
land surface models, 167
land treatment studies, 226–227
lead, 78
Leaking Underground Fuel Tank (LUFT) Field Manual, 24, 37–38, 41, 49, 53, 54, 111, *See also* underground storage tanks (USTs)
leaks, 170
liability, 69–70, 93, *See also* potentially responsible parties (PRPs)
lifetime average daily dose (LADD), 287

ligand substitution reactions, 174
liquid asphalt, 240–241
liquid phase hydrocarbons (nonaqueous phase liquid, NAPL), 28, 34, 35, 43, 171, 175, 182, 183
local equilibrium, 188
LUFT, *See* Leaking Underground Fuel Tank

macronutrients, 175, *See also* specific types
manganese, 175
man-made sources, 162–163, 170, *See also* specific types
mass phase hydrocarbons (nonaqueous phase liquid, NAPL), 28, 34, 35, 43, 171, 175, 182, 183
mass spectroscopy, 247
mathematical modeling, 32–33, 167, 175–176
maximally exposed individual (MEI), 289
maximum contaminant levels (MCLs), 10, 11, 19, 26, 36, 52, 53
MCLs, *See* maximum contaminant levels
medium curing liquid asphalts, 240
MEI, *See* maximally exposed individual
metals, 165, 171, 174, 247–248, 253, 296, *See also* specific types
methanol extraction, 110
methyl alkanones, 163
microbial ecology, 206, 207
  assessment of, 212–213
  case studies of, 213–230
    Alyeska Ballast Water Treatment Facility, 213–216
    aquifer treatment and, 217–225
    wood preservatives in, 216–227
  indicators for, 232–234
microbiological bioremediation, 204–206, *See also* superbugs
  assessment of, 207–212
  indicators for, 230–232
micronutrients, 175, *See also* specific types
migration parameters, 183
Millington-Quirk equation, 44
miscible inorganic compounds, 173
miscible organic compounds, 172–173
modeling, 167–183, *See also* specific types
  adsorption, 183
  air flow, 188–191
  air phase, 192
  applications of, 180, 181, 182–183
  aquatic equilibrium, 183
  bioaccumulation, 175
  biological process, 178
  cation-exchange, 183
  chemical process, 178
  compartmental, 176, 178
  complexation, 183
  concepts in, 182
  constant capacitance, 183
  deterministic, 176
  dissolved pollutant, 176–181
  equilibrium, 183
  fate, *See* fate of hydrocarbons, modeling of
  features in, 180
  groundwater, 167, 175, 179
  hydrocarbon fate, *See* fate of hydrocarbons, modeling of
  immiscible contaminant, 181–183
  immiscible fluid, 173
  input data for, 180
  issues in, 180–181
  land surface, 167
  mathematical, 32–33, 167, 175–176
  miscible inorganic compound, 173
  miscible organic compound, 172–173
  numerical soil, 180
  output validation in, 180–181
  physical process, 178
  porous media air flow, 188–191
  probabilistic, 178
  ranking, 180
  saturated zone, 167, 179
  selection in, 180
  slower reaction, 174
  soil ingestion rate, 302
  soil quality, 175
  soil-sediment interaction, 174
  spatial resolution in, 180
  speciation, 174
  stochastic, 176, 178
  surface complexation, 183

TDE, 176–178
temporal resolution in, 180
unsaturated soil zone, 176–178
validation in, 180
mononuclear aromatic hydrocarbons, 240
multiphase partitioning, 28–29
multiple exposure pathways, 25–26
multiple potentially responsible parties (PRPs), 91–98

naphthalene, 140
NAPL, *See* nonaqueous phase liquid
''nasal appraisals,'' 150
National Environmental Policy Act of 1969 (NEPA), 167
National Oil and Hazardous Substance Pollution Contingency Plan (NCP), 294
National Primary Drinking Water Regulations, 19
National Priorities List (NPL), 293
naturally occurring sources, 163–165, *See also* specific types
nature of contamination, 2, 171, 262
NCP, *See* National Oil and Hazardous Substances Pollution Contingency Plan
NCW Manager, 320–322, 326, 327
NEPA, *See* National Environmental Policy Act of 1969
nitrates, 175
nitration, 165
nonaqueous phase liquid (NAPL), 28, 34, 35, 43, 171, 175, 182, 183
noncombustion waste risk management project, 315–322
non-site-specific soil cleanup numbers, *See* generic soil cleanup numbers
NPL, *See* National Priorities List
numerical soil models, 180

odors, 171
Ohio River study, 137–138
oil, 249
oil field brines, 162
oil spills, 138–140
organic cyanides, 163
oxidation, 173, 174, 243
oxidation-reduction reactions, 172, 183

pattern recognition standards, 114, 125–126
paving materials, *See* asphalt
PCBs, *See* polychlorinated biphenyls
pedestrian surveys, 150, 153
pentachlorophenols, 10, 11
pesticides, 162, 170
petroleum distillates, 162, *See also* specific types
petroleum products, 171, *See also* specific types
phases of hydrocarbons, 28
phenols, 171, 173
phosphates, 175
photoionization detection (PID), 106, 113, 114, 151
pica children, 306–309, *See also* soil ingestion rates
PID, *See* photoionization detection
PNAH, *See* polynuclear aromatic hydrocarbon
point sources, 161, 170
polarization, 152
polychlorinated biphenyls (PCBs), 78, 162, 253, 314, 315
polymer liners, 253
polynuclear aromatic hydrocarbons (PNAH), 163, 239, 240, 241, 246, 247, 296
porous media air flow modeling, 188–191
ports, 71–75
potentially responsible parties (PRPs), 91–98, 294
PQL, *See* practical quantitation limit
practical quantitation limit (PQL), 110
precipitation of solids, 174
pre-packaged superbugs, 205, 209
Primary Drinking Water Regulations, 19
priority-setting systems, 314–315
private sector perspectives, 65–70
proactive cooperation, 92–93
probabilistic modeling, 178
property values, 67–68
PRPs, *See* potentially responsible parties
pseudo-first-order kinetics, 173
P&T, *See* purge and trap
public expectations, 27
public health, 9, 101–102
purge and trap method, 110, 113, 128, 136, 145

quality control, 130–131
QuickPIPES, 314
QuickTANKS, 314

radar, 152
ranking modeling, 180
rapid curing liquid asphalt, 240
RCRA, *See* Resource Conservation and Recovery Act
receptor types, 15
recyclable oil, 249
reduction reactions, 172, 183
reference dose (RfD), 11
Remedial Investigation and Feasibility Studies (RI/FS), 14, 293
remediation
  bio-, *See* bioremediation
  CHESS and, 333–334, 335
  costs of, 66, 93
  criteria for, 50–51
  goals for, 78
  health risks of, 293–298
  options in, 262–263
  PRPs and, 91–98
  soil venting and, 262–263
  total exposure approach to, 15
residential development, 48
resource availability, 68–69
Resource Conservation and Recovery Act (RCRA), 35, 246, 247, 316
resource limitation, 5
resource protection, 36
retardation, 171
RfD, *See* reference dose
RI/FS, *See* Remedial Investigation and Feasibility Studies
ring fission, 165
risk assessment, 284, 289, *See also* specific steps
  application of, 283
  boundary between risk management and, 7
  complexity of, 18
  components of, 284
  flowchart for, 6
  goals of, 284
  health based, 36, 50
  history of, 2
  results of, 2–4
  role of, 9
  site-specific, 301
  steps in, 3
  for Superfund, 24
  uniform guidelines for, 2
  for USTs, 40–43
risk characterization, 284
risk estimation, 5
risk management, 1, 298
  boundary between risk assessment and, 7
  complexity of, 18
  decisionmaking in, 8
  defined, 318–319
  experience in, 313–315
  expert-based system of, 18
  flexibility of, 17
  flowchart for, 6
  goals of, 4
  history of, 2
  need for, 5
  for noncombustion waste, 315–322
  previous work in, 313–315
  priority-setting systems in, 314–315
  problems in, 4–8
  screening in, 314–315
  timeline of, 319–320
risk reduction, 4, 5
road building, 250
run-offs, 170

sampling, 114, 123, 134, *See also* specific types
  groundwater, 151–152, 155–159
  soil, 151, 154–155, 161
  soil gas, 150–151, 153–154
Sante Fe Springs, 89–104
saturated zones, 167, 179, 261
scheduling, 8, 17
screening, 314–315
sediment-soil interactions, 174
semiarbitrary numerical standards, 38–39
semivolatile organic chemicals (SVOCs), 162
sensitivity calculations, 326, 330
separation, 243
septic systems, 162
SESOIL Model, 33, 178

sewage, 162, 170
shared responsibility, 91-98
significant risk, 5
site assessment methods, 149-160, *See also* specific types
  applications of, 159-160
  categories of, 150
  evaluation of, 152-159
  geophysical, 152
  groundwater sampling type, 151-152, 155-159
  investigation in, 150-152
  pedestrian survey type, 150, 153
  soil gas survey type, 150-151, 153-154
  soil sampling type, 151, 154-155
  standard, 149
Site Risk Management System (SITES), 314
SITES, *See* Site Risk Management System
Site Screening and Priority-Setting System (SSPS), 315
site-specific exposure conditions, 2
site-specific superbugs, 205, 211
slow curing liquid asphalts, 240-241, 250
sludge, 162
soil biota, 163-165
soil cleanup numbers
  generic, *See* generic soil cleanup numbers
  non-site-specific, *See* generic soil cleanup numbers
  state, *See* state programs
soil compartment, 12
soil fauna, 163
soil flora, 163
soil gas concentrations, 263-264
soil gas surveys, 150-151, 153-154
soil ingestion rates, 301-310
  age-related changes in, 305
  database on, 301-303
  models of, 302
  rural vs. urban, 305-306
  seasonality of, 306
  tracer selection in study of, 303-304
soil quality modeling, 175
soil sampling, 151, 154-155, 161
soil-sediment interactions, 174
soil vapor extraction, *See* vapor extraction
soil venting, 253-280
  air permeability test and, 264-269
  contaminant delineation and, 257-259
  days 1-83 in, 277-278
  days 83-121 in, 278
  days 121-135 in, 278
  days 149-266, in 278
  days 266-370 in, 279-280
  future of, 180
  monitoring in, 272-273
  operation in, 273-280
  remediation options and, 262-263
  review of, 255-259
  saturated zone instrumentation and, 261
  site characterization for, 255-259
  soil gas concentrations and, 263-264
  subsurface characteristics and, 257
  system design in, 269-272
  system monitoring in, 272-273
  system operation in, 273-280
  vadose zone monitoring and, 261
soil washing, 253
solidification, 253
solids precipitation, 174
soluble threshold limit concentration (STLC), 9, 247
sorption, 134, 171, 172
sources of contaminants, 161-165, 170, *See also* specific types
  industrial point, 161
  man-made, 162-163, 170
  naturally occurring, 163-165
  point, 170
speciation, 174, 183
specific gravities, 171
spectrometry, 136, 140, 141, 247, *See also* specific types
spills, 138-140, 170, 314
spray irrigation, 170
spreadsheet prototype, 322-328
SSPS, *See* Site Screening and Priority-Setting System

stabilization, 253
state programs, 5, 77–89, 317
    categories of cleanup levels in, 79–88
    history of, 77–78
    interpretation of cleanup levels in, 78, 89
    standards for cleanup levels in, 78
STLC, *See* soluble threshold limit concentration
stochastic modeling, 176, 178
subsoil, 25–26
substitution reactions, 174
subsurface heterogeneity, 26
subsurface hydrocarbons, 30–31
subsurface transport, 171
sulfides, 173
superbugs, 204, *See also* microbiological bioremediation
    genetically engineered, 206, 211–212
    pre-packaged, 205, 209
    site-specific, 205, 209–211
Superfund, 5, 24, 293, 294
surface complexation models, 183
surface run-offs, 170
surface seals, 253
surrogate control samples, 114
SVOCs, *See* semivolatile organic chemicals
syneresis, 243

tastes, 171
TCDD, *See* 2,3,7,8-tetrachlorodibenzo-p-dioxin
TCLP methods, 146
TDE, *See* Traditional Differential Equation
terrestrial plants, 175
2,3,7,8-tetrachlorodibenzo-p-dioxin (TCDD), 286, 287
thermal desorption, 253
toluene, 29, 140, *See also* benzene, toluene, xylene, and ethylbenzene (BTXE)
total petroleum hydrocarbons (TPHs), 9, 37–38, 39, 40, 43, 49
    GROs and, 111
    public health concerns and, 102
    soil venting and, 253, 257
    state levels of, 78
total threshold limit concentration (TTLC), 9, 247
toxicological data, 2, 15
TPHs, *See* total petroleum hydrocarbons
tracers in soil ingestion rate studies, 303–304
Traditional Differential Equation (TDE) modeling, 176–178, 179, 180
transformation processes, 29–32, 33, 171, *See also* biotransformation
transport, 29, 32, 33, 170, 171
TTLC, *See* total threshold limit concentration

ultraviolet spectrophotometry, 136, 140, 141
underground storage tanks (USTs), 23–58
    acceptable cleanup levels for, 27
    cleanup criteria guidelines for, 48–51
    documentation and, 27
    exposure assessment for, 43–47
    hydrocarbon fate and, 28–34
    leaking, 24, 37–38, 41, 49, 53, 54, 111
    monitoring of, 40–43
    nature of contaminants in, 26
    number of sites for, 27
    numerical standards for, 40–43
    public expectations of protection from, 27
    regulatory approaches to, 34–39
        illustration of, 51–55
        monitoring in, 40–43
        numerical standards in, 40–43
        remediation technology in, 40–43
        risk assessment in, 40–43
    regulatory concerns about, 25–27
    remediation criteria for, 50–51
    remediation technology for, 40–43
    residential development and, 48
    risk assessment of, 40–43
unleaded gasoline, 140
unsaturated zone-capillary zone interaction, 191–192

unsaturated zones, 167, 176–178
used oil, 249
USTs, *See* underground storage tanks

vacuum extraction, *See* soil venting
vadose zone monitoring, 261
Valdez oil spill, 138–140
vapor extraction, 187–201
    research milestones in, 188–189
    research needs in, 191–198
vegetation, 163
venting, *See* soil venting
VIS spectrophotometry, 136, 140, 141
VOCs, *See* volatile organic chemicals
volatile organic chemicals (VOCs), 162, 296
volatilization, 172, 243

waste extraction test (WET), 247
watershed models, *See* land surface models
WET, *See* waste extraction test
wood preservatives, 216–227
WORLDPORT L.A., 71–75

xylenes, 29, 140, *See also* benzene, toluene, xylene, and ethylbenzene (BTXE)

zero risk, 5, 38